Barron's Review Course Series

Let's Review:
Physics

Barron's Review Course Series

Let's Review:

Physics

Miriam A. Lazar
Albert S. Tarendash
Stuyvesant High School
New York, New York

BARRON'S

Illustration Acknowledgments

page 264	From *Physics,* by Halliday and Resnick. Copyright © 1962 by John Wiley & Sons, Inc. Reprinted by permission of John Wiley & Sons, Inc.
page 268	Adapted from Giancoli, PHYSICS, 3e, 1992, p. 516. Reprinted by permission of Prentice Hall, Upper Saddle River, New Jersey.
pages 269, 272, 275	Adapted from *Physics: Principles and Problems,* Transparency Masters component, by Paul Zitzewitz (© 1992, Glencoe Division, Macmillan/McGraw-Hill). Reprinted by permission of Glencoe/McGraw-Hill.
pages 449, 450, 452, 453, 454	Adapted from Lightman, A., GREAT IDEAS IN PHYSICS (1992), McGraw Hill. Reproduced with permission of The McGraw-Hill Companies.

All inquiries should be addressed to:
Barron's Educational Series, Inc.
250 Wireless Boulevard
Hauppauge, NY 11788
http://www.barronseduc.com

Library of Congress Catalog Card No. 96-20740

International Standard Book No. 0-8120-9606-1

Library of Congress Cataloging-in-Publication Data
Lazar, Miriam A.
    Let's review. Physics / Miriam A. Lazar, Albert S. Tarendash.
      p.   cm. — (Barron's review course series)
    Includes index.
    ISBN 0-8120-9606-1
    1. Physics—Outlines, syllabi, etc. 2. Physics—Examinations—
Study guides. 3. High schools—New York (State)—Examinations—
Study guides. I. Tarendash, Albert S. II. Title. III. Series.
QC31.L39   1996
530′.76—dc20   96-20740
                                          CIP

PRINTED IN THE UNITED STATES OF AMERICA

9 8 7 6 5 4 3 2 1

TABLE OF CONTENTS

PREFACE

To the Student:

This book has been written to help you understand and review high school physics. Physics is not a particularly easy subject, and no book—no matter how well written—can give you *instant* insight into the subject. Nevertheless, if you read this book carefully and do all of the problems and review questions, you will have a pretty good understanding of what physics is all about.

We have designed this book to be your "physics companion." It is somewhat more detailed than some review books, but is probably much less detailed than your hardcover text. The book is divided into 16 chapters. Each chapter beings with *Key Ideas* and *Key Objectives* and ends with multiple-choice questions drawn from past New York State Regents physics examinations. The answers to these questions appear after the glossary. The appendices contain various items of information that are useful in the study of physics and in answering free-response questions.

While the book covers all of the topics listed in the New York State Regents physics syllabus, there is also a significant amount of other material. Since many of you will be taking the New York State Regents Examination in Physics at the end of your course, we have included general information about this examination and a topic outline for each unit. These items can be found in Appendix 4.

If you have any comments about this book, we would appreciate hearing from you. Please write to us in care of the publisher, whose name and address are given on the copyright page.

To the Teacher:

Another review book in physics? Every author feels that he or she has a unique contribution to make, and we're certainly no exception to this rule. It is our belief that a physics review book should be more than an embellished outline of a particular syllabus with questions and problems added. Introductory physics is a difficult subject that requires diligent effort on the part of the student, and any book worth its salt must provide careful and detailed explanations of the material. We wrote the book with these thoughts in mind, and we hope that it will be successful for both you and your students.

We have divided the book into 16 chapters because we believe that the material is best presented in this fashion. Solid-state physics is included, not only because it is part of the New York State Regents physics syllabus, but because we believe it to be essential to understanding some of the basics of modern technology. Imagine a physics student at the end of the twentieth century not knowing what a transistor is! We have also included sections on special relativity (Chapter 15) and fundamental particles (Chapter 16) because we believe that all introductory physics students should be exposed to these topics.

You may not feel that our division of the material is appropriate for your teaching style or course, but there is no one correct way to order the material in a physics text. If you are more comfortable with another approach, then use it, by all means. Nevertheless, the correlation of chapter material with the New York State Regents syllabus is clearly indicated in the Appendix.

We spent considerable time in working out free-response problems because we believe that this is the area where students experience their greatest difficulties with physics. We chose *not* to pad each section with multiple-choice questions but reserved them for the end of each chapter. These questions, unusually large in number, were drawn from past New York State Regents examinations spanning 25 years. We also included multiple-choice questions that were rejected for the Regents examinations because of their difficulty. ("[One's] reach should exceed [one's] grasp, Or what's a heaven for?"*)

We are aware that some review books include SAT II-type questions. We considered this option but decided to focus on free-response problems and the more standard multiple-choice questions. Students who wish to take the SAT II examination in physics can use this book in conjunction with one of the SAT II preparation books currently available.

We hope that this book will give you some new insights and enable you to plan and execute your physics lessons with a spirit of inquiry. If you have any comments or corrections, please write to us in care of the publisher, whose name and address are given on the copyright page.

We wish to thank our editors at Barron's, Mr. Joseph Lazar for his excellent art work, and our students at Stuyvesant, who provided us with invaluable assistance in the preparation of this manuscript.

This book is lovingly dedicated to the memories of Dr. Murray Kahn and Dr. Matthew Litwin, late of Stuyvesant High School. Both men were outstanding chemistry teachers whose first and last thoughts were always of their students.

Miriam A. Lazar
Albert S. Tarendash

*With apologies to Robert Browning.

Chapter One	# INTRODUCTION TO PHYSICS

KEY IDEAS

Science depends on our ability to measure quantities. The SI metric system is used internationally as the standard for scientific measurement. It consists of seven basic quantities (e.g., length and time) and many derived quantities (e.g., speed and density).

Every measurement has a degree of uncertainty that is related to the limits of the measuring instrument. The concept of significant digits helps us to evaluate the degree of uncertainty in a particular measurement.

A graph of data points can indicate whether there is regularity within a set of data and can be used to predict how variables will behave under given conditions.

KEY OBJECTIVES

At the conclusion of this chapter you will be able to:
- State the fundamental quantities of measurement in the Système International (SI) and the metric units associated with them.
- Perform calculations using scientific notation.
- Determine the number of significant digits in a measurement.
- Incorporate significant digits within calculations.
- Determine the order of magnitude of a measurement.
- Plot a graph from a series of data points.
- Determine proportional relationships within data.
- Calculate the slope of a straight-line graph.
- State the common mathematical relationships in a right triangle.

1.1   PHYSICS IS . . . ?

It's not easy to come up with a precise definition of physics because this subject is so broad. The best one we've heard so far is this: "Physics is what physicists do." But *what* do they do? Physicists study the universe, from the smallest parts of matter (the particles that make up atoms) to the largest (the galaxies and beyond). They study the interactions of matter and energy, using both experimental (laboratory) and theoretical (mathematical) techniques. In this chapter, we will introduce some of these techniques in order to prepare us for the material that is covered in later chapters.

1.2 MEASUREMENT AND THE METRIC SYSTEM

The basis of all science lies in the ability to measure quantities. For example, we can easily measure the length of a table. In principle, we can even design an experiment to measure the distance from Earth to the nearest star. Therefore, these lengths have scientific meaning to us. However, modern physics theorizes that an electron has no measurable diameter, and consequently no experiment or apparatus can ever measure this length. As a result, we say that the diameter of an electron has no scientific meaning.

To be able to measure quantities we must have a *system* of measurement. We use the Système International (*SI*) because scientists all over the world express measurements in these metric units. The SI recognizes seven fundamental quantities upon which all measurement is based: (1) length, (2) mass, (3) time, (4) temperature, (5) electric current, (6) luminous intensity, and (7) number of particles. Each of these quantities has a unit of measure based on a standard that can be duplicated easily and does not vary appreciably.

Length

The unit of length is the *meter* (m), which is approximately 39 inches. The standard is based on the speed of light, which is absolutely constant and has an assigned value of 2.99792458×10^8 meters per second.

Mass

The unit of mass is the *kilogram* (kg), which has an approximate weight (on Earth) of 2.2 pounds. The standard is a platinum-iridium cylinder that is kept at constant temperature and humidity in a dustless vault in Sèvres, France. (Note that mass and weight are *not* the same quantities. Mass is the measure of the matter an object contains, while weight is the force with which gravity attracts matter.)

Time

The unit of time is the *second* (s). The standard is based on the frequency of vibrations of cesium-133 atoms under certain defined conditions.

Temperature

The unit of temperature is the *kelvin* (K). The standard is based on the point at which solid, liquid, and gaseous water coexist simultaneously (the "triple" point, which has an assigned value of 273.16 K).

Electric Current

The unit of electric current is the *ampere* (A). The standard is based on the mutual forces experienced by parallel current-carrying wires.

Luminous Intensity

The unit of luminous intensity is the *candela* (cd). The standard is based on the amount of radiation emitted by a certain object, known as a black-body radiator, at the freezing temperature of platinum (2046 K).

Number of Particles

The unit of number of particles is the *mole* (mol). The standard is based on the number of atoms contained in 0.012 kilogram of carbon-12 (6.02×10^{23} atoms).

1.3 SCIENTIFIC NOTATION

Once we have established a system of measurement, we need to be able to express small and large numbers easily. Scientific notation accomplishes this purpose. In scientific notation, a number is expressed as a power of 10 and takes the form

$$M \times 10^n$$

M is the *mantissa*; it is greater than or equal to 1 and is less than 10 ($1 \leq M < 10$). The *exponent*, *n*, is an integer. For example, the number 2300 is written in scientific notation as 2.3×10^3 (not as 23×10^2 or 0.23×10^4). The number 0.0000578 is written as 5.78×10^{-5}.

To write a number in scientific notation, we move the decimal place until the mantissa is a number between 1 and 10. If we move the decimal place to the left, the exponent is a positive number; if we move it to the right, the exponent is a negative number.

If we wish to *multiply* two numbers expressed in scientific notation, we *multiply* the mantissas and *add* the exponents. The final result must always be expressed in proper scientific notation. Here are two examples:

$$(2.0 \times 10^5)(3.0 \times 10^{-2}) = \underline{6.0 \times 10^3}$$
$$(4.0 \times 10^4)(5.0 \times 10^3) = 20. \times 10^7 = \underline{2.0 \times 10^8}$$

To *add* two numbers expressed in scientific notation, both numbers must have the same exponent. The mantissas are then *added* together. The following example illustrates the application of this rule:

$$(2.0 \times 10^3) + (3.0 \times 10^2) = (2.0 \times 10^3) + (0.30 \times 10^3) = \underline{2.3 \times 10^3}$$

1.4 ACCURACY, PRECISION, AND SIGNIFICANT DIGITS

As stated in Section 1.2, every scientific discipline, including physics, is concerned with making measurements. Since no instrument is perfect, however, every measurement has a degree of uncertainty associated with it. A well-designed experiment will reduce the uncertainty of each measurement to the smallest possible value.

Accuracy refers to how well a measurement agrees with an accepted value. For example, if the accepted density of a material is 1220 kilograms per meter^3 and a student's measurement is 1235 kilograms per meter^3, the difference (15 kg/m^3) is an indication of the accuracy of the measurement. The smaller the difference, the more accurate is the measurement.

Precision describes how well a measuring device can produce a measurement. The limit of precision depends on the design and construction of the measuring device. No matter how carefully we measure, we can *never* obtain a result more precise than the limit of our measuring device. A good general rule is that the limit of precision of a measuring device is equal to plus or minus one-half of its smallest division. For example, in the diagram below:

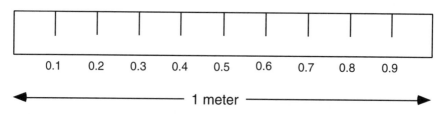

the smallest division of the meter stick is 0.1 meter, and the limit of its precision is ±0.05 meter. When we read any measurement using this meter stick, we must attach this limit to the measurement, for example, 0.27 ± 0.05 meter.

Significant digits are the digits that are part of any valid measurement. The number of significant digits is a direct result of the number of divisions the measuring device contains. The following diagram represents a meter stick with no divisions on it. How should the measurement, indicated by the arrow, be reported?

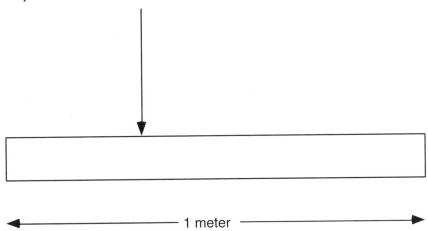

Since there are no divisions, all we know is that the measurement is somewhere between 0 and 1 meter. The best we can do is to make an *educated guess* based on the position of the arrow, and we report our measurement as 0.3 meter. This meter stick allows us to measure length to *one* significant digit.

Suppose we now use a meter stick that has been divided into tenths and repeat the measurement:

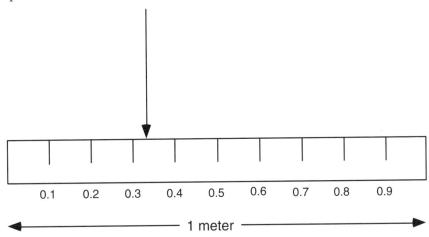

Now we can report the measurement with less uncertainty because we *know* that the indicated length lies between 0.3 and 0.4 meter. If we allow ourselves one guess, we could report the length as 0.33 meter. This measurement has *two* significant digits. The more significant digits a measurement has, the more confidence we have in our ability to reproduce the measurement because only the *last* digit is in doubt.

Measurements that contain zeros can be particularly troublesome. For example, we say that the average distance between Earth and the Moon is 238,000 miles. Do we *really* know this number to *six* significant digits? If so, we would have measured the distance to the nearest *mile*. Actually, this measurement contains only *three* significant figures. The distance is being reported to the nearest *thousand miles*. The zeros simply tell us how *large* the measurement is.

To avoid confusion, a number of rules have been established for determining how many significant digits a measurement has.

Rules for Determining the Number of Significant Digits in a Measurement

1. All nonzero numbers are significant. The measurement 2.735 meters has four significant digits.
2. Zeros located *between* nonzero numbers are also significant. The measurements 1.0285 kilograms and 202.03 seconds each have five significant digits.
3. For numbers greater than or equal to 1, zeros located at the *end* of the measurement are significant only if a decimal point is present. The measurement 60 amperes has one significant digit. In this case, the zero indicates the *size* of the number, not its significance. The measurements 60. amperes and 60.000 amperes, however, have two and five significant digits, respectively.
4. For numbers less than 1, the leading zeros are *not* significant; they indicate the size of the number. Thus, the measurements 0.002 kilogram, 0.0**2**0 kilogram, and 0.000**200** kilogram have one, two, and three significant digits, respectively. (The significant digits are indicated in **bold** type.)

If we use scientific notation, we need not become involved with the preceding rules because the mantissas always contains the proper number of significant digits; the size of the number is absorbed into the exponent. For example, the measurement 3.10×10^{-4} meter has three significant digits.

Using Significant Digits in Calculations

Significant digits are particularly important in calculations involving measured quantities and it is crucial that the *result* of a calculation does not imply a greater precision than any of the individual measurements. Calculators routinely give us answers with ten digits. It is incorrect to believe that the results of most of our calculations have this many significant digits.

When two measurements are multiplied (or divided), the answer should contain as many significant digits as the *least* precise measurement. For example, if the measurement 2.3 meters (two significant digits) is multiplied by 7.45 meters (three significant digits), the answer will contain two significant digits:

$$(2.3 \text{ m})(7.45 \text{ m}) = 17.135 \text{ m}^2 = \underline{17 \text{ m}^2}$$

(Note that the units are also multiplied together.)

When two measurements are added (or subtracted), the answer should contain as many *decimal places* as the measurement with the *smallest* number of decimal places. For example, when 8.11 kilograms and 2.476 kilograms are added, the answer will be taken to the second decimal place:

$$8.11 \text{ kg} + 2.476 \text{ kg} = 10.586 \text{ kg} = \underline{10.59 \text{ kg}}$$

(Note that the answer has been *rounded* to two decimal places.)

If *counted* numbers (such as 6 atoms) or defined numbers (such as 273.16 K) are used in calculations, they are treated as though they had an infinite number of significant digits or decimal places.

1.5 ORDER OF MAGNITUDE

There are times when we are interested in the *size* of a measurement rather than its actual value. The *order of magnitude* of a measurement is the power of 10 closest to its value. For example, the order of magnitude of 1284 kilograms (1.284×10^3) is 10^3, while the order of magnitude of 8756 kilograms (8.756×10^3) is 10^4. Orders of magnitude are very useful for comparing quantities, such as mass or distance, and for estimating the answers to problems involving complex calculations.

1.6 GRAPHING DATA

We have all heard the expression "A picture is worth a thousand words." This adage is particularly true when we wish to present experimental data. The following table of data gives the stretched lengths of a spring (in meters) when different weights (in a unit called newtons) are placed on it.

Weight (N)	Length (m)
0.0	0.0
5.0	0.04
10.	0.11
15	0.13
20.	0.18
25	0.27

Now plot these points and draw a graph.

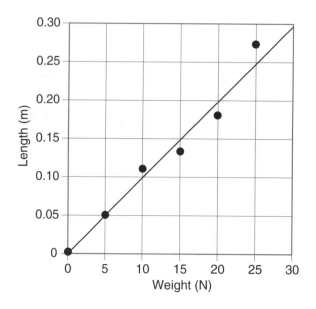

Note how the graph is constructed. First, the axes are drawn so that the data can be displayed over the entire graph. Next, the axes are labeled with the names of the quantities and their units of measure. It is traditional to place the quantity that is varied by the experimenter (the *independent variable*) along the x-axis, and the result of the experiment (the *dependent variable*) along the y-axis. Each item of data is then entered in the correct place on the graph.

Note that we do *not* play "connect the dots." Rather, we look for some regular relationship between the data points. We then draw a smooth line (or curve) that will fit this relationship as closely as possible. The "scatter" of the data in this example implies that we are dealing with a straight-line relationship. It is entirely possible that the graph may not pass through *any* of the data points. We require only that the data points above and below the graph be evenly distributed, as shown in the example. This graph is known as a *best-fit straight line*.

If a graph yields a straight line, we can conclude that the variables change uniformly. In the example given above, the length of the spring increases regularly as heavier weights are applied to the spring, and the steadily rising line of the graph reflects this direct relationship. If a graph is a curve, however, the variables (here, weight and length) change at a rate that is not uniform.

1.7 DIRECT AND INVERSE PROPORTIONS

Two quantities are *directly proportional* to one another if a change in one quantity is accompanied by an identical change in the other. For example, the table given below represents the *ideal* relationship between the mass of a substance and its corresponding volume. If the mass changes by a factor of 2 or 3, the volume changes by the same factor. This means that the *ratio* of the two quantities is a constant.

Mass (kg)	Volume (m³)
0.0	0.0
2.0	0.040
4.0	0.080
6.0	0.12
8.0	0.16
10.	0.20

When we plot this relationship, we obtain a straight-line graph that passes through the origin.

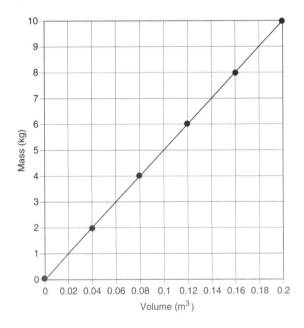

Frequently, the *slope* of a straight line provides us with additional information. Recall that we calculate the slope of a line by selecting two data points $(x_1, y_1$ and $x_2, y_2)$ and finding the ratio $\Delta y / \Delta x$:

9

===================== **PHYSICS CONCEPTS** =====================

$$\text{Slope} = m = \frac{\Delta y}{\Delta x} = \frac{y_2 - y_1}{x_2 - x_1}$$

The slope of the line representing a direct proportion is known as the *constant of proportionality,* and it provides the ratio between the variables. (In the mass-volume graph, the constant is known as the *density* of the substance.)

Two quantities are *inversely proportional* to one another if a change in one quantity is accompanied by a reciprocal change in the other. For example, the table given below represents the ideal relationship between the pressure of a gas and its corresponding volume at constant temperature. If the pressure changes by a factor of 2 or 3, the volume changes by a factor of ½ or ⅓. This means that the *product* of the two quantities is a constant.

Pressure (atm)	Volume (m³)
0.5	0.24
1.0	0.12
2.0	0.060
3.0	0.040
4.0	0.030
5.0	0.024

If we plot this relationship we will obtain a curve known as a *hyperbola.* As shown in the diagram, this curve will approach both the *x-* and the *y*-axis but will not intersect the axes.

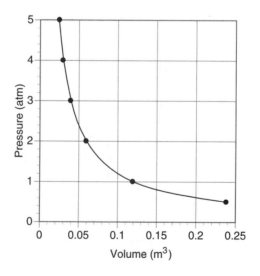

1.8 MATHEMATICAL RELATIONSHIPS WITHIN RIGHT TRIANGLES

Right triangles play an important part in the solution of physics problems. In this section, we summarize some of the more important relationships common to these triangles.

Consider the right triangle drawn below, where h indicates the hypotenuse and x and y are the two shorter sides.

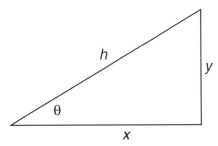

Three well-known trigonometric relationships relate the value of acute angle θ to the lengths of the sides of the triangle:

PHYSICS CONCEPTS

$$\sin \theta = \frac{y}{h} \qquad \cos \theta = \frac{x}{h} \qquad \tan \theta = \frac{y}{x}$$

In addition, the Pythagorean theorem relates the lengths of the sides of the triangle:

PHYSICS CONCEPTS

$$x^2 + y^2 = h^2$$

QUESTIONS

1. The unit of mass in the SI metric system is the
   (1) gram (2) kilogram (3) newton (4) slug

2. A meter stick has millimeter divisions marked on it. The limit of precision of this instrument is
   (1) 0.005 m (2) 0.01 m (3) 0.0005 m (4) 0.001 m

3. An ideal standard of measurement should be
   (1) variable, but not accessible
   (2) accessible, but not variable
   (3) variable and accessible
   (4) neither variable nor accessible

4. What is the order of magnitude for the measurement 72 meters per second?
   (1) 10^{-2} (2) 10^1 (3) 10^2 (4) 10^4

5. How many significant digits are contained in the measurement 500,000 kilometers?
   (1) 1   (2) 2   (3) 3   (4) 6

6. How many significant digits are contained in the measurement 406.200 seconds?
   (1) 6   (2) 5   (3) 3   (4) 4

7. How many significant digits are contained in the measurement 0.000300 volt?
   (1) 1   (2) 2   (3) 3   (4) 6

8. When the measurements 33.972 kilograms and 0.21 kilogram are added, the answer, to the correct number of significant digits, is
   (1) 34 kg   (2) 34.2 kg   (3) 34.18 kg   (4) 34.182 kg

9. When the measurements 8.14 meters and 2.1 meters are multiplied, the answer, to the correct number of significant digits, is
   (1) 17 m^2 (2) 17.0 m^2 (3) 17.09 m^2 (4) 17.094 m^2

10. How can the measurement 0.00567 liter be expressed to one significant digit?
    (1) 0.005 L   (2) 0.0050 L   (3) 0.006 L   (4) 0.0060 L

11. The approximate height of a high school physics student is
    (1) 10^1 m   (2) 10^2 m   (3) 10^0 m   (4) 10^{-2} m

12. The reading of the ammeter in the diagram below should be recorded as

    (1) 1 A   (2) 0.76 A   (3) 0.55 A   (4) 0.5 A

13. Which is the most likely mass of a high school student?
 (1) 10 kg (2) 50 kg (3) 600 kg (4) 2500 kg

14. What is the approximate thickness of this piece of paper?
 (1) 10^1 m (2) 10^0 m (3) 10^{-2} m (4) 10^{-4} m

<table>
<tr><td>

Chapter Two

</td><td>

MOTION IN ONE DIMENSION

</td></tr>
</table>

Motion is the change of position in time. *Displacement* is the directed change of an object's position. *Velocity* is the time rate of change of displacement, and *acceleration* is the time rate of change of velocity.

Under conditions of constant acceleration (also known as uniform acceleration), the motion of an object is governed by a set of interrelated equations. Objects that fall freely near the surface of the Earth are uniformly accelerated by gravity.

The motion of an object can be described by a series of motion graphs. The object's position, velocity, and acceleration can be plotted as functions of time. These graphs can then be used to illustrate various aspects of the object's motion at every point in time.

KEY OBJECTIVES

At the conclusion of this chapter you will be able to:

- Define the terms *motion, distance, displacement, average velocity, speed, instantaneous velocity,* and *acceleration,* and state their SI units.
- Solve problems involving average velocity and constant velocity.
- Distinguish between average velocity and instantaneous velocity, and relate these terms to a position-time graph.
- Solve problems involving the equations of uniformly accelerated motion.
- Solve problems involving freely falling objects.
- Interpret the data provided by motion graphs and solve problems related to them.

2.1 MOTION DEFINED

How do we know when an object is in motion? If we look at the hour hand of a watch, it does not appear to be moving, yet over a period of time we see a change in its position. Therefore, a reasonable definition of **motion** is the change of an object's position in time.

2.2 GRAPHING AN OBJECT'S MOTION

Graphs are especially useful for analyzing an object's motion. The position of the object is plotted along the y-axis, and the elapsed time along the x-axis. Here is a very general motion graph:

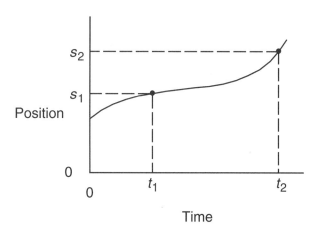

The origin of the graph (0, 0) marks the reference point for both position and time. When we say "zero time" we mean the time when we begin the event—the time when we "start the clock," so to speak. Similarly, "zero position" means the specific place where we begin measuring. It may be the ground or a table top or a spot on the wall. Generally, we use the letters s to represent position and t to represent time. If we know that we are measuring *horizontal* position, we may use the letter x, and we may use y to represent *vertical* position.

The dotted lines on the graph tell us where the object is at a given time. At time t_1, the object is at position s_1; at time t_2, the object is at position s_2.

2.3 DISPLACEMENT

The **displacement** of an object is the change in its position and is measured in units of length (such as meters or inches). In the graph in Section 2.2, the displacement of the object between times t_1 and t_2 is given by the relationship $\Delta \mathbf{s} = \mathbf{s}_2 - \mathbf{s}_1$. Since we are subtracting two coordinates ($\mathbf{s}_1$ from $\mathbf{s}_2$), we are not primarily concerned about the exact path taken by the object between these points. We assume that the magnitude of the displacement is given by the length of a straight line between $\mathbf{s}_1$ and $\mathbf{s}_2$ along the axis representing position.

Displacement is known as a *vector* quantity because, in addition to magnitude, it has *direction.* (Vector quantities, which are discussed in detail in Chapter 4, are set in boldface type.) The magnitude of displacement is known as *distance.*

15

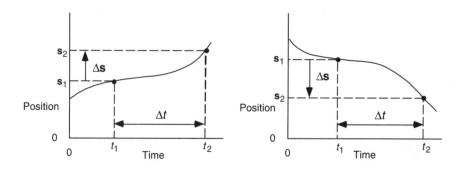

In the first graph above, the displacement (Δs) from t_1 to t_2 is positive because its magnitude (indicated by the arrow) is measured in the positive direction; in the second graph, the displacement is negative. Whenever we are working with one-dimensional motion, we can use positive and negative signs to represent opposite directions.

In each of the graphs, the *elapsed time* is given by the relationship $\Delta t = t_2 - t_1$ and is measured in units of time, such as seconds or hours. We always read the time axis from left to right since normally we do not travel backward in time.

2.4 VELOCITY

In the graph below, the slope of the straight line connecting the two points on the graph is given by the relationship $\Delta s / \Delta t$, and it indicates *how rapidly* the position of the object (Δs) has changed over the time interval (Δt). This quantity is known as the **average velocity** ($\bar{v}$) of the object and is measured in units such as meters per second (m/s) or miles per hour (mph). Mathematically, the average velocity of the object is defined by the equation

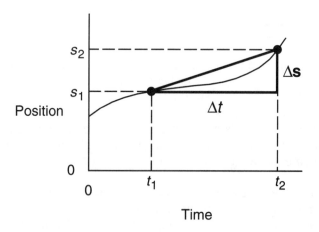

PHYSICS CONCEPTS

$$\bar{\mathbf{v}} = \frac{\Delta \mathbf{s}}{\Delta t} = \frac{\mathbf{s}_2 - \mathbf{s}_1}{t_2 - t_1}$$

Velocity, like displacement, is a vector quantity because it has both magnitude and direction. As with displacement, we can use positive and negative signs to represent motion in opposite directions. The magnitude of the velocity is known as **speed**.

PROBLEM

The position of an object is +35 meters at 2.0 seconds and is +87 meters at 15 seconds. Calculate the average velocity of the object.

SOLUTION

$$\bar{\mathbf{v}} = \frac{\Delta \mathbf{s}}{\Delta t} = \frac{\mathbf{s}_2 - \mathbf{s}_1}{t_2 - t_1}$$

$$= \frac{+87 \text{ m} - (+35 \text{ m})}{15 \text{ s} - 2.0 \text{ s}} = \frac{+52 \text{ m}}{13 \text{ s}} = +4.0 \text{ m/s}$$

This object is traveling in the positive direction with an average speed of 4.0 m/s.

We use the term *average velocity* because we do not know exactly what is happening *between* the two points in question. For example, suppose we traveled by automobile due west 1000 miles and the trip took 20 hours. When we calculate our average velocity for the trip, we obtain

$$\bar{\mathbf{v}} = \frac{1000 \text{ mi [W]}}{20 \text{ h}}$$

$$= 50 \text{ mph [W]}$$

Does this mean that we traveled the entire distance at a constant speed of 50 miles per hour? Not necessarily! We would probably have had to add fuel, eat, pay tolls, or engage in other activities on the trip. There might have been construction delays or reduced speed zones. All we can say with certainty is that our average speed was 50 miles per hour and our direction of travel was west.

How then could we measure our velocity at any *point* on our trip—our **instantaneous velocity**? (This is the value that we read on the speedometer of our car.) One way would be to measure our average velocity over *smaller and smaller* time intervals. To accomplish this, however, we would need to use mathematical techniques that are beyond the scope of this book. The other way is to use a position versus time graph. The instantaneous velocity at any point on the graph is the slope of a line drawn *tangent* to the graph at that point, as shown in the diagram.

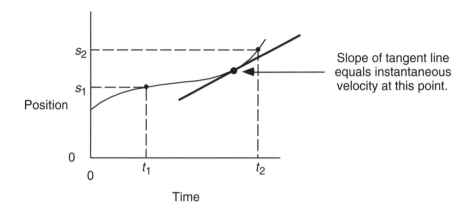

Slope of tangent line equals instantaneous velocity at this point.

2.5 ACCELERATION

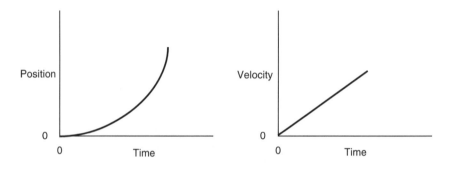

Consider the two graphs shown above. The first graph represents the position of an automobile as a function of time. Note that the graph becomes steeper (curves upward) as time passes. This occurs because the automobile's instantaneous velocity is *increasing.*

The second graph represents the instantaneous velocity of the same automobile as a function of time. Note that the graph is a straight line directed upward. This graph also shows that the instantaneous velocity of the automobile is increasing with time.

Actually, the graphs represent the motion of the automobile from two different viewpoints: that of position (as measured by the automobile's *odometer*) and that of velocity (as measured by the automobile's *speedometer*).

The *slope* of the velocity–time graph is given by the relationship $\Delta v / \Delta t$ and indicates the rate at which the velocity of the object (Δv) has changed over the time interval (Δt). This quantity is known as the **acceleration (a)** of the object and is measured in units such as meters per second^2 (m/s^2). Since the graph is a straight line, the acceleration in this case is constant or *uniform.* Mathematically, the uniform acceleration of the object is defined by the equation

$$\boxed{\text{Physics Concepts}}$$

$$a = \frac{\Delta \mathbf{v}}{\Delta t} = \frac{\mathbf{v}_2 - \mathbf{v}_1}{t_2 - t_1}$$

Acceleration is also a vector quantity because it has both magnitude and direction. A positive acceleration means that the velocity of an object is becoming more positive with time; a negative acceleration, that the velocity of the object is becoming more negative with time.

PROBLEM

The velocity of an object is +47 meters per second at 3.0 seconds and is +65 meters per second at 12.0 seconds. Calculate the acceleration of the object.

SOLUTION

$$a = \frac{\Delta \mathbf{v}}{\Delta t} = \frac{\mathbf{v}_2 - \mathbf{v}_1}{t_2 - t_1}$$

$$= \frac{+65 \text{ m/s} - (+47 \text{ m/s})}{12.0 \text{ s} - 3.0 \text{ s}} = \frac{+18 \text{ m/s}}{9.0 \text{ s}}$$

$$= +2.0 \text{ m/s}^2$$

This object's velocity is becoming more positive by 2.0 m/s each second. Since the magnitude of the velocity is increasing, the object is speeding up. The table below shows how this occurs.

t (s)	0.0	1.0	2.0	3.0	4.0	5.0	6.0	7.0	8.0	9.0
v (m/s)	+47	+49	+51	+53	+55	+57	+59	+61	+63	+65

2.6 THE EQUATIONS OF UNIFORMLY ACCELERATED MOTION

In the problem we solved in Section 2.5, an object accelerates uniformly, at 2.0 meters per second^2, from 47 meters per second to 65 meters per second in 9.0 seconds. There is a great deal of additional information about the object we might wish to learn. For example:

1. What is the *average velocity* of the object over 9.0 seconds?
2. What is the *displacement* of the object at the end of 9.0 seconds?
3. What is the *instantaneous velocity* of the object at any given time (at 6.0 s, for example)?

To solve these problems, we use a set of five equations that describe the motion of an object undergoing uniform acceleration. In each of these equations, we use the subscripts i (for *initial* value), f (for *final* value), and av (for *average* value). Your textbook or teacher may use different subscripts or notation, but they all yield the same results. The equations for uniformly accelerated motion are as follows:

PHYSICS CONCEPTS

1. $\mathbf{v}_{av} = \dfrac{\Delta \mathbf{s}}{\Delta t}$

2. $\mathbf{v}_{av} = \dfrac{\mathbf{v}_i + \mathbf{v}_f}{2}$

3. $\mathbf{v}_f = \mathbf{v}_i + \mathbf{a} \cdot \Delta t$

4. $\Delta \mathbf{s} = \mathbf{v}_i \cdot \Delta t + \dfrac{1}{2}\mathbf{a} \cdot \Delta t^2$

5. $\mathbf{v}_f^2 = \mathbf{v}_i^2 + 2 \cdot \mathbf{a} \cdot \Delta \mathbf{s}$

Equation 1 is the definition of average velocity. Equation 2 tells us that, under uniform acceleration, the average velocity lies midway between the initial and final velocities. Equation 3 is just the definition of acceleration rearranged in a more convenient form for solving problems. Equations 4 and 5 are relationships that have been derived from the first three equations.

Which equation should you use to solve a particular problem? The answer depends on the data you are given. In the problem we have been considering, an object with an initial velocity of 47.0 meters per second accelerates uniformly at 2.0 meters per second² for 9.0 seconds. Suppose we wish to calculate the *displacement* of this object at the end of 9.0 seconds. We list the variables that are part of the problem, along with their values:

$$\begin{aligned}
\mathbf{v}_i &= 47.0 \text{ m/s} \\
\mathbf{a} &= 2.0 \text{ m/s}^2 \\
\Delta t &= 9.0 \text{ s} \\
\Delta \mathbf{s} &= ???
\end{aligned}$$

If we examine the list of equations given above, we see that equation 4 contains the four variables that form the basis of our problem. We solve the problem by substituting the values and calculating the answer:

$$\Delta \mathbf{s} = \mathbf{v}_i \cdot \Delta t + \frac{1}{2}\mathbf{a} \cdot (\Delta t)^2$$

$$= (47.0 \text{ m/s})(9.0 \text{ s}) + \frac{1}{2}(2.0 \text{ m/s}^2)(9.0 \text{ s})^2$$

$$= 504 \text{ m}$$

2.7 FREELY FALLING OBJECTS

The following table represents the motion of an object falling from rest near the surface of the Earth when air resistance is ignored.

Δt (s)	0.00	1.00	2.00	3.00	4.00	5.00	6.00
v (m/s)	0.00	9.80	19.6	29.4	39.2	49.0	58.8
Δs (m)	0.00	4.90	19.6	44.1	78.4	122	176

If we analyze this motion, we see that the speed of the object increases uniformly by 9.8 meters per second for each second of travel. This suggests that the object is subject to a constant acceleration of 9.8 meters per second^2. The distance traveled by the object over time verifies that the object's acceleration is constant.

PROBLEM
How does the distance traveled by the object described above over time verify that the object's acceleration is constant?

SOLUTION
The distance traveled by an object under uniform acceleration is given by the equation:

$$\Delta s = \mathbf{v}_i \cdot \Delta t + \frac{1}{2}\,\mathbf{a} \cdot (\Delta t)^2$$

Since the object starts from rest, this equation reduces to:

$$\Delta s = \frac{1}{2}\,\mathbf{a} \cdot (\Delta t)^2$$

If we substitute each of the corresponding values of Δs and Δt given in the table, we find that the acceleration in each case is 9.8 m/s^2.

If we were to investigate further, we would find that *all* objects falling near the surface of the Earth experience a constant acceleration of 9.8 meters per second^2 if air resistance is ignored. This phenomenon is due to the presence of gravity, which affects each and every object. If we were to travel to the Moon, we would find that all objects also fall to its surface with a uniform acceleration. However, this acceleration is only 1.6 meters per second^2 because of the Moon's weaker gravitational forces.

Since free fall involves uniform acceleration, the five equations we developed in Section 2.6 can be used to solve all free-fall problems. We need only remember that objects can move up as well as down in the presence of gravity. Therefore, we assign the up direction as positive and the down direction as negative. Since gravity *always* points downward (i.e., toward the Earth), its value is taken to be –9.8 meters per second^2. Gravitational acceleration is denoted by the lower-case letter g.

PROBLEM
An object is dropped from rest from a height of 49 meters.
(a) How long does the object take to hit the ground?
(b) What is its speed as it hits the ground?

SOLUTION
(a) Since the initial velocity is zero, we can use the equation

$$\Delta s = \frac{1}{2}\mathbf{a} \cdot (\Delta t)^2$$

The *displacement* is –49 ms (since we measure the distance in a downward direction), and the acceleration due to gravity is –9.8 m/s² (since gravity points downward). Therefore:

$$\Delta s = \frac{1}{2}\mathbf{a} \cdot (\Delta t)^2$$

$$-49 \text{ m} = \frac{1}{2}\left(-9.8\,\frac{\text{m}}{\text{s}^2}\right)(\Delta t)^2$$

$$(\Delta t)^2 = 10\text{s}^2$$

$$\Delta t = 3.2 \text{ s}$$

(b) We can use the relationship $\mathbf{v}_f = \mathbf{v}_i + \mathbf{a} \cdot \Delta t$, which reduces to $\mathbf{v}_f = \mathbf{a} \cdot \Delta t$ since the initial velocity is zero. Then

$$\mathbf{v}_f = (-9.8 \text{ m/s}^2)(3.2 \text{ s}) = -31 \text{ m/s}$$

2.8 MOTION GRAPHS REVISITED

Throughout this chapter we have used motion graphs as aids to understanding the concept of motion. In this section, we take a more detailed look at these graphs and the information they can provide. We shall examine three types of graphs: position–time, velocity–time, and acceleration–time.

Position–Time Graphs

The following graph illustrates the position of an object as a function of time.

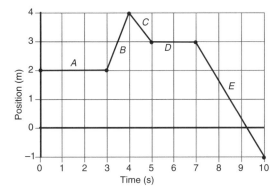

We recall from Section 2.2 that "zero time" represents the start of an event and that "zero position" represents some arbitrary reference point. We have divided the position-time graph into five sections: *A, B, C, D,* and *E.* Since each section is a straight-line segment, the velocity within each section is *constant* and the acceleration over each section is *zero.* We will learn how to interpret this graph by considering the following problem.

PROBLEM

1. What is the displacement over each section of the graph?
2. What is the velocity over each section of the graph?
3. What is the displacement over the entire trip (0–10 seconds)?
4. What is the average velocity over the entire trip (0–10 seconds)?

SOLUTION

1. To calculate the displacement (Δs) we *subtract* the initial position from the final position.
   - The displacement over section *A* is 0 m because the object has not changed its position.
   - The displacement over section *B* is +2 m because the object has changed its position from +2 m to +4 m.
   - The displacement over section *C* is –1 m because the object has changed its position from +4 m to +3 m.
   - The displacement over section *D* is 0 m because the object has not changed its position.
   - The displacement over section *E* is –4 m because the object has changed its position from +3 m to –1 m.
2. The velocity over each section is found by *dividing* the displacement by the elapsed time $\left(\dfrac{\Delta s}{\Delta t}\right)$
   - The velocity over section *A* is 0 m/s (0 m/3 s).
   - The velocity over section *B* is +2 m/s (+2 m/1 s).
   - The velocity over section *C* is –1 m/s (–1 m/1 s).

- The velocity over section D is 0 m/s (0 m/2 s).
- The velocity over section E is –1.3 m/s (–4 m/3 s).
3. The displacement over the entire trip is –3 m because the object changed position from +2 m at $t = 0$ s to –1 m at $t = 10$ s.
4. The average velocity over the entire trip is –0.3 m/s (–3 m/10 s).

Velocity–Time and Acceleration–Time Graphs

The graph below illustrates the velocity of an object as a function of time.

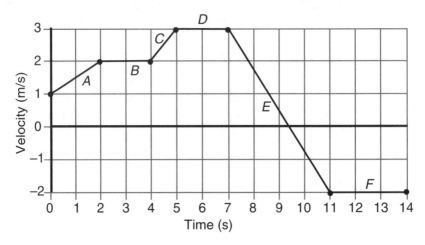

The values on the y-axis represent the *instantaneous* velocities of the object at the times marked on the x-axis. It is as though we were looking at a car's speedometer at various times. We have divided the graph into six sections: A, B, C, D, E and F. Since each section is a straight-line segment, the object's acceleration within each section is *constant*. We will learn how to interpret this graph by considering the following problem.

PROBLEM
1. What is the average velocity within each section of the graph?
2. What is the acceleration within each section of the graph?
3. When does the object come to rest?
4. When does the object reverse the direction of its motion?
5. What is the displacement within each section of the graph?
6. What is the displacement over the entire trip (0–14 seconds)?
7. What is the average velocity over the entire trip (0–14 seconds)?
8. What is the shape of the corresponding *acceleration versus time* graph?

SOLUTION

1. The average velocity for each section is calculated by finding the *mid-point* of each line, that is, by adding the initial and final velocities within each section and dividing this sum by 2:

$$\left(\mathbf{v}_{av} = \frac{\mathbf{v}_i + \mathbf{v}_f}{2} \right)$$

We must remember to take both positive and negative signs into account when we add the velocities. For example, in section *E*, the initial velocity is +3.0 m/s and the final velocity is –2.0 m/s. Therefore, the average velocity is

$$\frac{+3.0 \text{ m/s} + (-2.0 \text{ m/s})}{2} = +0.5 \text{ m/s}$$

The table summarizes the results of the calculations for the six sections:

section	A	B	C	D	E	F
$\mathbf{v}_{av}$ (m/s)	+1.5	+2.0	+2.5	+3.0	+0.5	–2.0

2. The acceleration within each section is found by calculating the *slope* of each of the lines:

$$\left(\frac{\Delta \mathbf{v}}{\Delta t} = \frac{\mathbf{v}_f - \mathbf{v}_i}{\Delta t} \right)$$

For example, in section *A*, $\mathbf{v}_i = 1.0$ m/s, $\mathbf{v}_f = 2.0$ m/s, and $\Delta t = 2.0$ s. The acceleration is calculated to be

$$\left(\frac{+2.0 \text{ m/s} - 1.0 \text{ m/s}}{2.0 \text{ s}} \right) = +0.5 \text{ m/s}^2$$

The table summarizes the results of the calculations for the six sections:

section	A	B	C	D	E	F
a (m/s²)	+0.5	0	+1.0	0	–1.25	0

3. The object comes to rest when its velocity is *zero*. Referring to the graph, we estimate that zero velocity corresponds to an approximate time of 9.5 s. (Actually, the time is 9.4 s; we could have *calculated* this value by using the graph and the equation $\mathbf{v}_f = \mathbf{v}_i + \mathbf{a} \cdot \Delta t$.)
4. Before 9.4 s, the velocity of the object is always positive; after 9.4 s, its velocity is negative. Therefore the object reverses the direction of its motion at 9.4 s.
5. There are two ways to calculate the displacement of the object.

First, we could multiply the average velocity of each section by the time elapsed in that section. For example, in section E the average velocity is +0.5 m/s and the elapsed time is 4.0 s. Therefore, the displacement of the object is +2.0 m (+0.5 m/s · 2.0 s). A *positive* displacement means that the object traveled the distance in the positive direction.

Second, we could calculate the displacement by measuring the *area* between the section line and the *x*-axis. (In mathematics, this is known as calculating the area "under the curve."). We will use this method to calculate the displacement for section F of the graph, as follows.

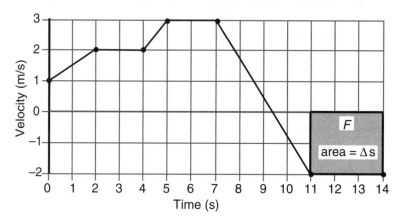

The shaded area is defined by a rectangle whose length is 3.0 s and whose height is –2.0 m/s (this value is negative because section F lies *under* the *x*-axis). The area is the *product* of these two values (3.0 s · –2.0 m/s), that is, –6.0 m.

The table summarizes the results of the calculations for the six sections:

section	A	B	C	D	E	F
Δs (m)	+3.0	+4.0	+2.5	+6.0	+2.0	–6.0

6. The displacement over the entire trip is found by *adding* the displacements for all the sections:

$$\Delta s_{total} = +11.5 \text{ m}$$

(Refer to the table above.)

7. The average velocity for the entire trip is found by *dividing* Δs_{total} by the total time (14 s):

$$v_{av \text{ (total)}} = +0.82 \text{ m/s}$$

8. The acceleration versus time graph for this object is constructed by referring to the accelerations over all of the sections. (See the table on page 25 for part 2 of this problem.)

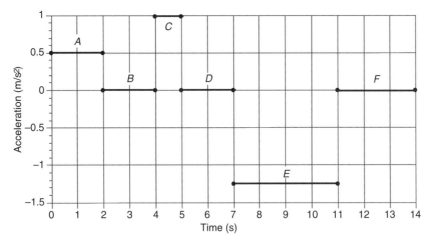

Note that the accelerations are drawn as straight line segments *within* each section, but they are not connected *between* sections. The reason is that the change in velocity between sections is so abrupt that the acceleration cannot be calculated accurately. (Contrast the velocity versus time graph shown in part 5 of this problem.)

QUESTIONS

1. A flashing light of constant 0.20-second period is situated on a lab cart. The diagram below represents a photograph of the light as the cart moves across a tabletop.

A   B

• •   •   •   •   •   •

How much time elapsed as the cart moved from position *A* to position *B*?
(1) 1.0 s   (2) 5.0 s   (3) 0.80 s   (4) 4.0 s

2. The average speed of a plane was 600 kilometers per hour. How long did it take the plane to travel 120 kilometers?
(1) 0.2 h   (2) 0.5 h   (3) 0.7 h   (4) 5 h

3. What is the total distance traveled by an object that moves with an average speed of 6.0 meters per second for 8.0 seconds?
(1) 0.75 m   (2) 1.3 m   (3) 14 m   (4) 48 m

4. An object travels for 8.00 seconds with an average speed of 160. meters per second. The distance traveled by the object is
(1) 20.0 m   (2) 200. m   (3) 1280 m   (4) 2,560 m

5. A blinking light of constant period is situated on a lab cart. Which diagram best represents a photograph of the light as the cart moves with constant velocity?
(1) • •   •   •   •   •
(2) •   •   •   •   •   •
(3) •   •   •   •   •   •   •
(4) • •   •   •   •   • •

6. A car is accelerated at 4.0 meters per second² from rest. The car will reach a speed of 28 meters per second at the end of
(1) 3.5 s   (2) 7.0 s   (3) 14 s   (4) 24 s

Base your answers to questions 7 through 9 on the information and diagram below. The diagram represents a block sliding along a frictionless surface between points *A* and *G*.

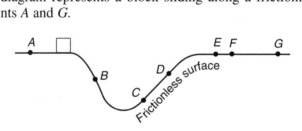

7. As the block moves from point *A* to point *B*, the speed of the block will be
(1) decreasing
(2) increasing
(3) constant, but not zero
(4) zero

8. Which expression represents the magnitude of the block's acceleration as it moves from point *C* to point *D?*
(1) $\dfrac{m}{F}$ (2) $\dfrac{\Delta v}{\Delta t}$ (3) $m\,\Delta v$ (4) $\dfrac{2\Delta s}{\Delta t}$

9. Which formula represents the velocity of the block as it moves along the horizontal surface from point *E* to point *F?*
(1) $\bar{v} = \dfrac{\Delta s}{\Delta t}$ (2) $\bar{v} = \dfrac{\Delta v}{2}$ (3) $v_f^2 = 2a\,\Delta s$ (4) $\Delta v = \dfrac{1}{2}a(\Delta t)^2$

10. An object that is originally moving at a speed of 20. meters per second accelerates uniformly for 5.0 seconds to a final speed of 50. meters per second. What is the acceleration of the object?
(1) 14 m/ s^2 (2) 10. m/ s^2 (3) 6.0 m/ s^2 (4) 4.0 m/s^2

11. If a car increases its speed from 15 meters/second to 30 meters/second in 15 seconds, the average acceleration during this time is
(1) 1.0 m/s^2 (2) 15 m/s^2 (3) 30. m/s^2 (4) 45 m/s^2

12. A block starting from rest slides down the length of an 18-meter plank with a uniform acceleration of 4.0 meters per second^2. How long does the block take to reach the bottom?
(1) 4.5 s   (2) 2.0 s   (3) 3.0 s   (4) 9.0 s

Base your answers to questions 13 through 17 on the information below.

A toy projectile is fired from the ground vertically upward with an initial velocity of +29 meters per second. The projectile arrives at its maximum altitude in 3.0 seconds. [Neglect air resistance.]

13. The greatest height the projectile reaches is approximately
(1) 23 m   (2) 44 m   (3) 87 m   (4) 260 m

14. What is the velocity of the projectile when it hits the ground?
(1) 0. m/s   (2) –9.8 m/s   (3) –29 m/s   (4) +29 m/s

15. What is the displacement of the projectile from the time it left the ground until it returned to the ground?
(1) 0. m   (2) 9.8 m   (3) 44 m   (4) 88 m

16. Which graph best represents the relationship between velocity (*v*) and time (*t*) for the projectile?

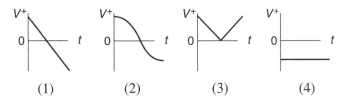

(1)   (2)   (3)   (4)

17. As the projectile rises and then falls back to the ground, its acceleration
(1) decreases, then increases   (3) increases, only
(2) increases, then decreases   (4) remains the same

18. A freely falling object near the Earth's surface has a constant
(1) velocity of –1.00 m/s   (3) acceleration of –1.00 m/s^2
(2) velocity of –9.81 m/s   (4) acceleration of –9.81 m/s^2

19. An experiment consists of throwing balls straight up with varying initial velocities. Which quantity will have the same value in all trials?
(1) initial momentum
(2) maximum height
(3) time of travel
(4) acceleration

20. A 2.0-kilogram stone that is dropped from the roof of a building takes 4.0 seconds to reach the ground. Neglecting air resistance, the maximum speed of the stone will be approximately
(1) 8.0 m/s (2) 9.8 m/s (3) 29 m/s (4) 39 m/s

Base your answers to questions 21 and 22 on the diagram below, which shows a 1-kilogram aluminum sphere and a 3-kilogram brass sphere, both having the same diameter and both at the same height above the ground. Both spheres are allowed to fall freely. [Neglect air resistance.]

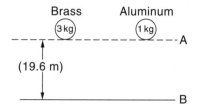

21. Both spheres are released at the same instant. They will reach the ground at
(1) the same time but with different speeds
(2) the same time with the same speeds
(3) different times but with the same speeds
(4) different times and with different speeds

22. If the spheres are 19.6 meters above the ground, the time required for the aluminum sphere to reach the ground is
(1) 1 s (2) 2 s (3) 8 s (4) 4 s

23. A softball is thrown straight up, reaching a maximum height of 20 meters. Neglecting air resistance, what is the ball's approximate vertical velocity when it hits the ground?
(1) –10 m/s (2) –20 m/s (3) –15 m/s (4) –40 m/s

24. An object is allowed to fall freely near the surface of a planet. The object has an acceleration due to gravity of 24 meters per second^2. How far will the object fall during the first second?
(1) 24 m (2) 12 m (3) 9.8 m (4) 4.9 m

25. Object *A* with a mass of 2 kilograms and object *B* with a mass of 4 kilograms are dropped simultaneously from rest near the surface of the Earth. At the end of 3 seconds, what is the ratio of the speed of object *A* to the speed of object *B?* [Neglect air resistance.]
(1) 1 : 1   (2) 2 : 1   (3) 1 : 2   (4) 1 : 4

26. Starting from rest, an object rolls freely down an incline that is 10 meters long in 2 seconds. The acceleration of the object is approximately
(1) 5 m/s   (2) 5 m/s^2 (3) 10 m/s   (4) 10 m/s^2

Base your answers to questions 27 and 28 on the information below.

An object starting from rest moves down an incline with an acceleration of 2 meters per second^2 for 2 seconds.

27. How far does the object move during the 2 seconds?
(1) 1 m   (2) 2 m   (3) 8 m   (4) 4 m

28. The final speed of the object after 2 seconds is
(1) 1 m/s   (2) 2 m/s   (3) 8 m/s   (4) 4 m/s

29. An object initially traveling in a straight line with a speed of 5.0 meters per second is accelerated at 2.0 meters per second^2 for 4.0 seconds. The total distance traveled by the object in the 4.0 seconds is
(1) 36 m   (2) 24 m   (3) 16 m   (4) 4.0 m

30. Starting from rest, object *A* falls freely for 2.0 seconds, and object *B* falls freely for 4.0 seconds. Compared with object *A,* object *B* falls
(1) one-half as far   (3) 3 times as far
(2) twice as far   (4) 4 times as far

31. An object starts from rest and falls freely. What is the speed of the object at the end of 3.00 seconds?
(1) 9.81 m/s   (2) 19.6 m/s   (3) 29.4 m/s   (4) 88.2 m/s

32. An object is allowed to fall freely near the surface of a planet. The object falls 54 meters in the first 3.0 seconds after it is released. The acceleration due to gravity on that planet is
(1) –6.0 m/s^2 (2) –12 m/s^2 (3) –27 m/s^2 (4) –108 m/s^2

33. An object, initially at rest, falls freely near the Earth's surface. How long does it take the object to attain a speed of 98 meters per second?
(1) 0.1 s   (2) 10 s   (3) 98 s   (4) 960 s

34. A car accelerates uniformly from rest at 3.2 meters per second^2. When the car has traveled a distance of 40. meters, its speed will be
    (1) 8.0 m/s   (2) 12.5 m/s   (3) 16 m/sec   (4) 128 m/sec

35. An object initially at rest accelerates at 5 meters per second^2 until it attains a speed of 30 meters per second. What distance does the object move while accelerating?
    (1) 30 m   (2) 90 m   (3) 3 m   (4) 600 m

Base your answers to questions 36 through 39 on the graph below, which is a velocity versus time graph for an object moving in a straight line.

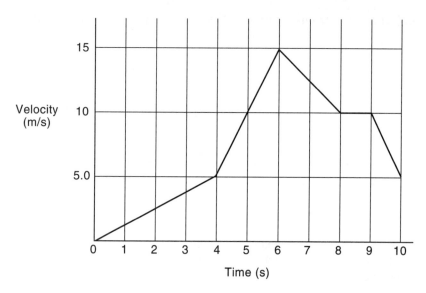

36. During which time interval did the object travel the greatest distance?
    (1) 0 s to 4 s   (2) 4 s to 6 s   (3) 6 s to 8 s   (4) 9 s to 10 s

37. During which time interval did the object have the greatest positive acceleration?
    (1) 0 s to 4 s   (2) 4 s to 6 s   (3) 6 s to 8 s   (4) 9 s to 10 s

38. What was the acceleration of the object during the time interval between 4 seconds and 6 seconds?
    (1) 20. m/s^2 (2) 10. m/s^2 (3) 5.0 m/s^2 (4) 0.20 m/s^2

39. What distance did the object travel during the time interval between 9 seconds and 10 seconds?
    (1) 5.0 m   (2) 7.5 m   (3) 10. m   (4) 20. m

Base your answers to questions 40 through 45 on the graph below, which represents the relationship between speed and time for an object in motion along a straight line.

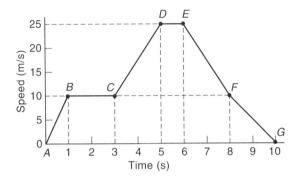

40. What is the acceleration of the object during the time interval $t = 3$ seconds to $t = 5$ seconds?
    (1) 5.0 m/s²   (2) 7.5 m/s²   (3) 12.5 m/s²   (4) 1 7.5 m/s²

41. What is the average speed of the object during the time interval $t = 6$ seconds to $t = 8$ seconds?
    (1) 7.5 m/s   (2) 10 m/s   (3) 15 m/s   (4) 17.5 m/s

42. What is the total distance traveled by the object during the first 3 seconds?
    (1) 15 m   (2) 20 m   (3) 25 m   (4) 30 m

43. During which interval is the object's acceleration the greatest?
    (1) *AB*   (2) *CD*   (3) *DE*   (4) *EF*

44. During the interval $t = 8$ seconds to $t = 10$ seconds, the speed of the object is
    (1) zero   (2) increasing   (3) decreasing   (4) constant, but not zero

45. What is the maximum speed reached by the object during the 10 seconds of travel?
    (1) 10 m/ s   (2) 25 m/s   (3) 150 m/s   (4) 250 m/s

46. An object is thrown vertically upward from the surface of the Earth. Which graph best represents the relationship between velocity and time for the object as it rises and then returns to the Earth?

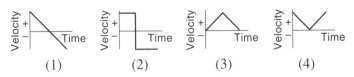

Base your answers to questions 47 through 49 on the graph and information below.

Cars A and B both start from rest at the same location at the same time.

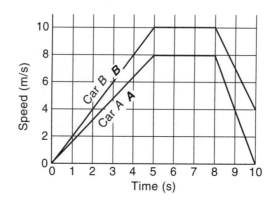

47. What is the magnitude of the acceleration of car A during the period between $t = 8$ seconds and $t = 10$ seconds?
(1) 20 m/s^2 (2) 16 m/s^2 (3) 18 m/s^2 (4) 4 m/s^2

48. Compared to the speed of car B at 6 seconds, the speed of car A at 6 seconds is
(1) less   (2) greater   (3) the same

49. Compared to the total distance traveled by car B during the 10 seconds, the total distance traveled by car A is
(1) less   (2) greater   (3) the same

50. The graph represents the acceleration acting on an object as a function of time. During which time interval is the velocity of the object constant?

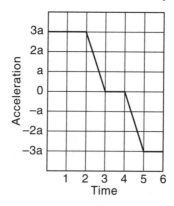

(1) 0 to 2   (2) 2 to 3   (3) 3 to 4   (4) 4 to 5

Base your answers to questions 51 through 55 on the accompanying graph, which represents the motions of four cars on a straight road.

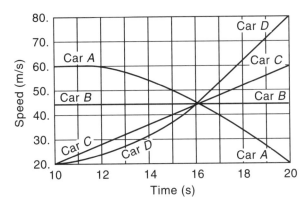

51. The speed of car C at time $t = 20$ seconds is closest to
   (1) 60 m/s   (2) 45 m/s   (3) 3.0 m/s   (4) 4.0 m/s

52. Which car has zero acceleration?
   (1) A (2) B (3) C (4) D

53. Which car has a negative acceleration?
   (1) A (2) B (3) C (4) D

54. Which car moves the greatest distance in the time interval $t = 10$ seconds to $t = 16$ seconds?
   (1) A (2) B (3) C (4) D

55. Which graph best represents the relationship between distance and time for car C?

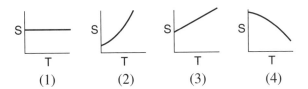

56. The motions of cars *A, B,* and *C* in a straight path are represented by the graph below.

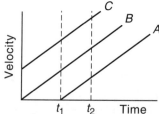

During the time interval from t_1 to t_2, the three cars travel
(1) the same distance
(2) with the same velocity, only
(3) with the same acceleration, only
(4) with both the same velocity and acceleration

Base your answers to questions 57 through 61 on the graph below, which represents velocity versus time for an object in linear motion. The object has a velocity of 20 meters per second at $t = 0$.

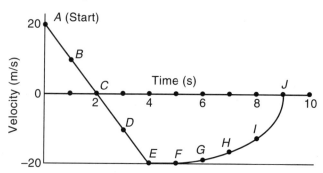

57. During which interval is the magnitude of the acceleration greatest?
(1) *EF*   (2) *FG*   (3) *GH*   (4) *IJ*

58. The acceleration of the object at point *D* on the curve is
(1) 0 m/s^2 (2) 5 m/s^2 (3) –10 m/s^2 (4) –20 m/s^2

59. During what interval does the object have zero acceleration?
(1) *BC*   (2) *EF*   (3) *GH*   (4) *HI*

60. At what point is the distance from start zero?
(1) *C*   (2) *E*   (3) *F*   (4) *J*

61. At what point is the distance from start a maximum?
(1) *C*   (2) *E*   (3) *G*   (4) *J*

62. Which combination of graphs best describes free-fall motion? [Neglect air resistance.]

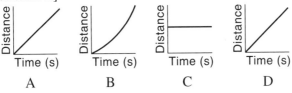

A   B   C   D

(1) *A* and *C*   (2) *B* and *D*   (3) *A* and *D*   (4) *B* and *C*

63. The graph below represents the relationship between velocity and time for an object moving in a straight line. What is the acceleration of the object?

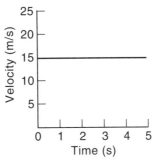

(1) 0 m/s^2 (2) 5 m/s^2 (3) 3 m/s^2 (4) 15 m/s^2

64. The graph below represents the velocity versus time relationship for a ball thrown vertically upward. Time zero represents the time of release.

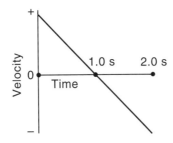

During the time interval between 1.0 second and 2.0 seconds, the displacement of the ball from the point where it was released
(1) decreases   (2) increases   (3) remains the same

65. The graph below represents the relationship between distance and time for an object moving in a straight line. According to the graph, the object is

(1) motionless   (3) decelerating
(2) moving at a constant speed   (4) accelerating

66. The area under a speed versus time curve is a measure of
     (1) acceleration   (2) distance   (3) momentum   (4) velocity

Chapter Three

FORCES AND NEWTON'S LAWS

KEY IDEAS

A *force* is a push or a pull in a given direction. If the force is unbalanced, an acceleration will result. The relationship between force and motion is given by Newton's first two laws of motion.

Special categories of forces include weight and friction. *Weight* is the force on an object as a result of gravitational attraction by a massive body such as a planet. Its direction is always toward the center of the body. *Friction* is a force resulting from the contact between two surfaces. Frictional forces are always directed so that they oppose any relative motion.

Forces always occur in pairs; if an object exerts a force on another object, the second object exerts an equal and opposite force on the first object. This relationship is known as Newton's third law of motion.

KEY OBJECTIVES
At the conclusion of this chapter you will be able to:
- Define the term *force* and state its SI unit.
- State Hooke's law, and use relevant data to measure a force.
- State Newton's first law of motion.
- State Newton's second law of motion, and use it to solve problems.
- Define the term *weight*, and relate it to Newton's second law of motion.
- Define the term *normal force*.
- Define the term *fricitional force*, and solve simple problems involving kinetic friction.
- Define the term *coefficient of kinetic friction*, and use it in the solution of problems.
- State Newton's third law of motion, and apply it to common situations.

3.1 INTRODUCTION

In Chapter 2 we examined the one-dimensional motion of an object in great detail, a study known as *kinematics*. Now we focus our attention on the relationship between forces and motion. This is the study of *dynamics*.

3.2 WHAT IS A FORCE?

A **force** is a push or a pull in a given direction. Since a force has both magnitude (the "strength" of the force) and direction, it is a vector quantity. We will use the letter **F** to represent force.

3.3 MEASURING FORCES USING HOOKE'S LAW

We can measure the magnitude of a force very simply by recognizing that an applied force will stretch or compress a spring. The English scientist Robert Hooke was able to show that the magnitude of a force (**F**) is directly proportional to the elongation (stretch) or compression of a spring (x) within certain limits. The relationship, known as *Hooke's law,* is given by the equation

---
PHYSICS CONCEPTS
$$F = kx$$

In the SI system of measurement, if x is measured in meters, **F** is measured in units called *newtons* (N). One newton is equivalent to approximately ¼ pound of force. The constant of proportionality, k, is known as the *spring constant*. Its unit is the newton per meter (N/m), and it is related to the stiffness of the spring; the greater the constant, the stiffer the spring. The following problem illustrates how Hooke's law can be used.

PROBLEM
The following set of data was obtained by a student while investigating Hooke's law in the laboratory:

Force (N)	Elongation of Spring (m)
0.0	0.0
3.0	0.9
6.0	2.2
9.0	2.8
12	4.1

(a) Graph this set of data.
(b) Using the graph, determine the elongation of the spring if a force of 7.0 newtons is applied to it.
(c) Calculate the spring constant *from the graph.*

SOLUTION

(a)

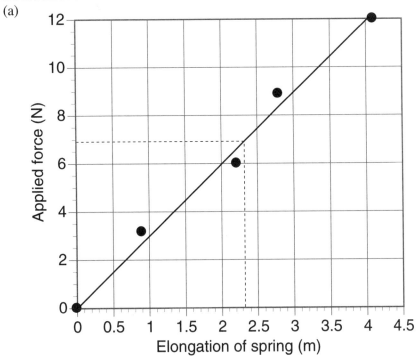

Note that the graph does not need to pass through the data points; it is a "best-fit" straight line.
(b) Reading directly from the graph, we see that an applied force of 7.0 N corresponds to an elongation of 2.3 m.
(c) The spring constant k can be found by calculating the *slope* of the best-fit straight line $\left(\dfrac{\Delta F}{\Delta x}\right)$. A careful calculation of the slope yields a value of 3.0 N/m for the spring constant.

3.4 NEWTON'S FIRST LAW OF MOTION

Originally, it was believed that forces were necessary to keep all objects in motion. The Italian astronomer Galileo Galilei, however, reasoned that a force *changed* the motion of an object as long as the force was not "balanced" by other forces. For example, if you push (lightly!) on a wall, the wall will not move appreciably because your force is balanced by a number of other forces present. However, if you apply an unbalanced force to a chair on a smooth floor, you may see a number of possible effects:

- If the chair is at rest, it will begin to move.
- If the chair is in motion, it may speed up, slow down, come to rest, or change its *direction* of motion.

If an object is at rest and no unbalanced force acts on it, its motion will not change; it will simply remain at rest. Also, if an object is traveling at constant velocity (constant speed in a straight line) and no unbalanced force acts on it, its motion will not change; it will continue to move with constant velocity. The English physicist Isaac Newton summarized these findings in a statement we now call *Newton's first law of motion*:

PHYSICS CONCEPTS

Every object persists in a state of uniform motion unless acted upon by an unbalanced force.

Newton's first law means that all material objects *resist* changes in motion. The quality of matter that is responsible for this property is known as *inertia*. The more inertia an object has, the more it resists such changes. The (*inertial*) mass of an object is a measure of the quantity of inertia it contains.

PROBLEM

In which of the following situations do unbalanced forces act?
(a) A car is traveling at 20 meters per second, and the driver steps on the brake.
(b) A chair is dragged across a floor at constant velocity.
(c) A plane makes a turn at constant speed.
(d) An object is dropped vertically toward the ground.

SOLUTION

(a) An unbalanced force acts since the speed of the car is lowered.
(b) No unbalanced force acts on the chair since its velocity is constant.
(c) An unbalanced force acts since the plane's direction is changing.
(d) An unbalanced force acts since the object's speed is increasing.

In every case, an unbalanced force gives rise to an *acceleration* (i.e., a change in velocity over time).

3.5 NEWTON'S SECOND LAW OF MOTION

Although an unbalanced force will give an object an acceleration, the magnitude of the acceleration is determined by two quantities: the magnitude of the force and the *mass* of the object. It has been verified countless times that the acceleration is *directly proportional* to the magnitude of the force and *inversely proportional* to the mass of the object. If we choose our units carefully, we can write this relationship, which we call *Newton's second law of motion,* as follows:

PHYSICS CONCEPTS

$$\mathbf{a} = \frac{\mathbf{F}_{net}}{m} \quad \text{or} \quad \mathbf{F}_{net} = m \cdot \mathbf{a}$$

We measure the mass (m) in kilograms, and the acceleration ($\mathbf{a}$) in meters per second^2. The unbalanced force ($\mathbf{F}_{net}$) is measured in newtons.

PROBLEM
What are the equivalent basic SI units for the newton?

SOLUTION
We can answer this question by using the second equation for Newton's second law with units instead of numbers:

$$\mathbf{F}_{net} = m \cdot \mathbf{a}$$

$$N = kg \cdot \frac{m}{s^2} = \frac{kg \cdot m}{s^2}$$

PROBLEM
Complete the following table (fill in the blank spaces) by applying Newton's second law of motion:

F_{net} (N)	m (kg)	a (m/s^2)
3.0	6.0	?
?	4.5	2.0
12	?	1.2

SOLUTION
We substitute the values for each row into the equation $F_{net} = m \cdot a$.

Row 1: $3.0 \text{ N} = 6.0 \text{ kg} \cdot \mathbf{a}$; $a = 0.50 \text{ m/s}^2$

Row 2: $\mathbf{F}_{net} = 4.5 \text{ kg} \cdot 2.0 \text{ m/s}^2$; $\mathbf{F}_{net} = 9.0 \text{ N}$

Row 3: $12 \text{ N} = m \cdot 1.2 \text{ m/s}^2$; $m = 10. \text{ kg}$

3.6   WEIGHT

When we hold an object, we feel its heaviness as it pushes into our hand. As we learned in Chapter 2, when the object drops to the ground, it falls with an acceleration of 9.8 meters per second^2 (if we ignore air resistance). These two observations lead us to conclude that a force is present on the object. We call this force **weight**, and its origin is the gravitational attraction between the object and the Earth.

How can we calculate the weight of an object? When an object falls freely, the force on it (its weight!) is unbalanced. We can apply Newton's second law to calculate this force:

PHYSICS CONCEPTS

$$\mathbf{F}_W = m \cdot \mathbf{g}$$

We use $\mathbf{F}_W$ to represent the weight of the object and $\mathbf{g}$ to represent the gravitational acceleration the object experiences in free fall.

PROBLEM

What is the weight of an object with a mass of 15 kilograms?

SOLUTION

$$\mathbf{F}_W = m \cdot \mathbf{g}$$

$$= (15 \text{ kg})(9.8 \text{ m/s}^2) = 149 \text{ N}$$

3.7   NORMAL FORCES

When an object is at rest on a horizontal surface, such as a table, it has weight but is not accelerating. In this situation, the weight of the object is not an unbalanced force. In fact, the net force on the object is zero! Therefore, another force, which serves to balance the effects of gravity, must be present on the object. The origin of this supporting force is the surface itself, and this force is responsible for the contact between the object and the surface.

Since this second force is perpendicular to both the object and the surface, it is called a **normal force**. (In mathematics, the word *normal* is used to describe lines that are perpendicular.) We represent the normal force by the symbol $\mathbf{F}_N$. The diagram below represents the relationship between the weight of an object and the normal force on it.

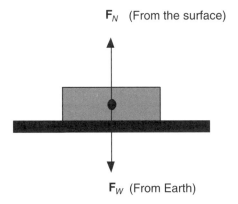

$\mathbf{F}_N$ (From the surface)

$\mathbf{F}_W$ (From Earth)

PROBLEM
How does the magnitude of the normal force compare with the weight of the object in the diagram shown above?

SOLUTION
Since the object is at rest, the net force on it is zero. Therefore, the magnitude of the normal force must be *equal* to the weight of the object.
(*Note*: This is not always the case, as we will see in Chapter 4.)

3.8 FRICTIONAL FORCES

Consider the following situation: A 5.0-kilogram object is pulled across a floor with an applied force of 20. newtons. The acceleration of the object is measured to be 3.0 meters per second^2. This situation is illustrated in the diagram below:

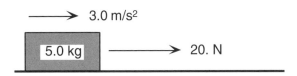

Something seems to be wrong here: $\mathbf{F} \neq ma$! But we know from Newton's second law that ma must equal the *unbalanced* force ($\mathbf{F}_{net}$). In this situation, $\mathbf{F}_{net}$ is equal to 15 newtons (5.0 kg · 3.0 m/s^2), *not* to the applied force of 20. newtons. Where has 5.0 newtons gone?

There is another force present that we have not considered: friction. **Frictional forces** are always present when two surfaces are in contact. The direction of a frictional force on an object is always *opposite* to the direction of the object's motion. We will represent a frictional force by the symbol $\mathbf{F}_f$. We now complete the diagram above by adding in the 5.0-newton frictional force:

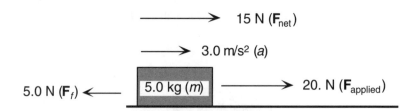

PROBLEM

A force of 50. newtons is used to drag a 10.-kilogram object across a horizontal table. If a frictional force of 15. newtons is present on the object, calculate (a) the unbalanced force on the object and (b) the acceleration of the object.

SOLUTION

We begin the solution by drawing a diagram of the situation presented in the problem:

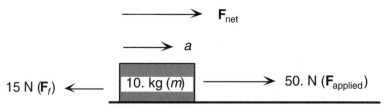

(a) We find the unbalanced force ($\mathbf{F}_{net}$) by *subtracting* the frictional force ($\mathbf{F}_f$) from the applied force ($\mathbf{F}_{applied}$). We subtract because the forces are in *opposite* directions:

$$\mathbf{F}_{net} = \mathbf{F}_{applied} - \mathbf{F}_f = 50. \text{ N} - 15 \text{ N} = 35 \text{ N}$$

(b) We calculate the acceleration of the object by applying Newton's second law:

$$a = \frac{\mathbf{F}_{net}}{m} = \frac{35 \text{ N}}{10. \text{ kg}} = 3.5 \text{ m/s}^2$$

PROBLEM

A student drags an object across a laboratory table at constant velocity using an applied force of 12 newtons. Calculate the frictional force present on the object.

SOLUTION

Since the object is moving with constant velocity, its acceleration is zero, as is the unbalanced force on it (why?). We can now apply this relationship:

$$0 \text{ N} = \mathbf{F}_{net} = \mathbf{F}_{applied} - \mathbf{F}_f = 12 \text{ N} - \mathbf{F}_f$$
$$\mathbf{F}_f = 12 \text{ N}$$

There are two ways in which we can view frictional forces. First, when an object is at rest on a surface, a certain minimum force is needed to get the object started. This frictional force is known as *static friction*. Second, there is the frictional force present on an object already in motion; this is known as *kinetic friction*. The force of static friction is usually larger than the force of kinetic friction because surfaces at rest tend to form stronger intermolecular attractions than surfaces in relative motion.

Friction is a contact force whose magnitude depends on two factors: the nature of the surfaces in contact and the magnitude of the contact force (i.e., the *normal force*) present on the object. To reduce friction, we generally try to reduce the amount of contact between the surfaces by using lubricants, such as oil or Teflon, or by employing rolling rather than sliding. Very smooth surfaces, such as glass on glass, may have extremely large frictional forces because of the higher degree of contact.

Friction is always with us. When we solve physics problems, however, we usually try to ignore frictional forces in order to make the problems simpler. For example, we assume that an object falling near the Earth's surface will have a constant acceleration of 9.8 meters per second per second, but this is not so. The friction between the air and the falling object will cause the object to reach a constant *terminal speed*.

3.9   THE COEFFICIENT OF KINETIC FRICTION

The **coefficient of kinetic friction**, represented by the symbol μ_k, is one way of predicting how much friction will be produced on an object because of its contact with another surface. Experimentally, it is found that the frictional force of an object in motion is *directly proportional* to the normal force present on the object. We can write this relationship as follows:

PHYSICS CONCEPTS

$$F_f = \mu_k \cdot F_N$$

We can calcuate μ_k by measuring the frictional force on an object (see the preceding problem) and then dividing this value by the normal force present on the object. The next problem demonstrates this procedure.

PROBLEM

As illustrated in the diagram below, a 5.0-kilogram object slides across a horizontal surface. If a 5.0-newton frictional force is present, calculate the coefficient of kinetic friction between the two surfaces.

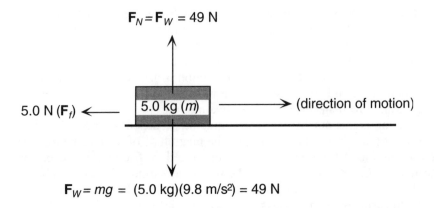

$$F_N = F_W = 49 \text{ N}$$

5.0 N (F_f) ⟵   5.0 kg (m) ⟶ (direction of motion)

$$F_W = mg = (5.0 \text{ kg})(9.8 \text{ m/s}^2) = 49 \text{ N}$$

SOLUTION

The first step is to calculate the weight of the object, as shown above, which is 49 N. Since this surface is horizontal and the object is not falling downward, the weight is balanced by the normal force, which is also 49 N.

We calculate μ_k from the relationship $\mu_k = \dfrac{F_f}{F_N} = \dfrac{5.0 \text{ N}}{49 \text{ N}} = 0.10$

Note that μ_k does not have units.

    The larger the value of μ_k, the greater the frictional force on the object. A table of possible values of μ_k is given below for reference:

Surface – Surface	μ_k
Steel – steel (nonlubricated)	0.6
Steel – steel (lubricated)	0.06
Rubber – concrete	0.8
Wood – wood	0.4
Waxed wood – dry snow	0.04

3.10 NEWTON'S THIRD LAW OF MOTION— ACTION AND REACTION

Consider the diagram below: An angry woman kicks a wall with her foot and is taken to the hospital to have her fractured toe set in a cast.

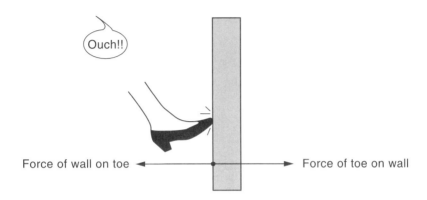

Force of wall on toe ← → Force of toe on wall

Why was the woman's toe broken? After all, it was she who exerted a force *on the wall*! As a consequence of her kick, however, the wall returned the favor and exerted a force back *on her toe*. Unfortunately, her toe was not as strong as the wall, and it broke under the stress.

This situation describes a very general principle in physics, one that has been verified countless times and is known as *Newton's third law of motion*:

PHYSICS CONCEPTS

If object A exerts a force on object B, then object B exerts an equal and opposite force back on object A.

The third law has also been stated equivalently in two other ways: (1) "Forces occur in pairs" and (2) "To every action, there is an equal and opposite reaction" (where the terms *action* and *reaction* refer to the two forces).

According to Newton's third law, *two objects* must always be present and only one force of the pair is present on each object. The two forces are always equal in magnitude and opposite in direction.

We can apply the third law to many situations:

- The recoil of all objects when they collide (e.g., a ball bouncing off a wall) is a consequence of this law.
- When we step on a scale, the scale exerts an *upward* force which supports us (the normal force). We, in turn, place a *downward* force on the scale. It is this force that registers our weight.

- We can walk because of Newton's third law! When we take a step, we push *backward* on the Earth. The Earth, in turn, pushes *forward* on us. (If you don't believe this, think of yourself stepping out of a boat onto the shore: Which way does the boat move as you move forward?)

QUESTIONS

1. The diagram below represents a block suspended from a spring The spring is stretched 0.200 meter. If the spring constant is 200 newtons per meter, what is the weight of the block?

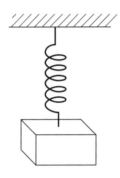

   (1) 40.0 N   (2) 20.0 N   (3) 8.00 N   (4) 4.00 N

2. Compared to the inertia of a 1-kilogram mass, the inertia of a 4-kilogram mass is
   (1) one-fourth as great   (3) 16 times as great
   (2) one-sixteenth as great   (4) 4 times as great

3. As the mass of an object decreases, its inertia will
   (1) decrease   (2) increase   (3) remain the same

4. If the mass of a moving object could be doubled, the inertia of the object would be
   (1) halved   (2) doubled   (3) unchanged   (4) quadrupled

5. The fundamental units for a force of 1 newton are
   (1) meters per second^2
   (2) kilograms
   (3) meters per second^2 per kilogram
   (4) kilogram · meters per second^2

6. An object with a mass of 2 kilograms is accelerated at 5 meters per second². The net force acting on the mass is
(1) 5 N   (2) 2 N   (3) 10 N   (4) 20 N

7. A cart is uniformly accelerating from rest. The net force acting on the cart is
(1) decreasing   (2) zero   (3) constant   (4) increasing

Base your answers to questions 8 and 9 on the information and diagram below.

A force is applied to a 5.0-kilogram object. The force is always applied in the same direction, but its magnitude varies with time according to the graph below. [Neglect friction.]

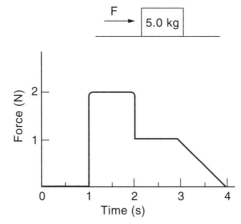

8. What is the acceleration of the object at time $t = 2.5$ seconds?
(1) 1.0 m/s²   (2) 0.2 m/s²   (3) 5.0 m/s²   (4) 9.8 m/s²

9. During which time interval did the object have a constant velocity?
(1) 0 s to 1 s   (2) 1 s to 2 s   (3) 2 s to 3 s   (4) 3 s to 4 s

10. The graph below shows the relationship between the acceleration of an object and the unbalanced force producing the acceleration. The ratio $(\Delta F/\Delta a)$ of the graph represents the object's

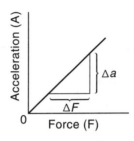

(1) mass   (3) kinetic energy
(2) momentum   (4) displacement

11. What force is needed to give an electron an acceleration of 1.00×10^{10} meters per second^2?
(1) 9.11×10^{-41} N   (3) 9.11×10^{-21} N
(2) 9.11×10^{-31} N   (4) 1.10×10^{43} N

12. When an unbalanced force of 10. newtons is applied to an object whose mass is 4.0 kilograms, the acceleration of the object will be
(1) 40. m/s^2 (2) 2.5 m/s^2 (3) 9.8 m/s^2 (4) 0.40 m/s^2

13. In the graph below the acceleration of an object is plotted against the unbalanced force on the object. The mass of the body is

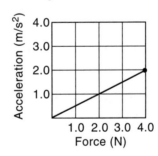

(1) 1.0 kg   (2) 2.0 kg   (3) 0.5 kg   (4) 8.0 kg

14. Assume that an object has no unbalanced force acting on it. Which statement about the object is true?
(1) The object may be in motion.
(2) The object must be slowing down.
(3) The object must be at rest.
(4) The object may be speeding up.

15. A cart is uniformly accelerating from rest. The net force acting on the cart is
(1) decreasing (2) zero (3) constant (4) increasing

16. If the mass of an object were doubled, its acceleration due to gravity would be
(1) halved (2) doubled (3) unchanged (4) quadrupled

17. What is the gravitational acceleration on a planet where a 2-kilogram mass has a weight of 16 newtons on the planet's surface?
(1) ⅛ m/s^2 (2) 8 m/s^2 (3) 10 m/s^2 (4) 32 m/s^2

18. On planet Gamma, a 4.0-kilogram mass experiences a gravitational force of 24 newtons. What is the acceleration due to gravity on planet Gamma?
(1) 0.17 m/s^2 (2) 6.0 m/s^2 (3) 9.8 m/s^2 (4) 96 m/s^2

19. What is the weight of a 5.0-kilogram object at the surface of the Earth?
(1) 0.5 N (2) 5.0 N (3) 49 N (4) 490 N

20. For a freely falling object, the ratio of the force of gravity to its acceleration is
(1) weight (2) momentum (3) kinetic energy (4) mass

21. The ratio of an object's weight to its mass is equal to the object's
(1) momentum (3) gravitational force
(2) inertia (4) gravitational acceleration

22. A force of 10. newtons applied to mass M accelerates the mass at 2.0 meters per second^2. The same force applied to a mass of $2M$ would produce an acceleration of
(1) 1.0 m/s^2 (2) 2.0 m/s^2 (3) 0.5 m/s^2 (4) 4.0 m/s^2

23. An object with a mass of 0.5 kilogram starts from rest and achieves a maximum speed of 20 meters per second in 0.01 second. What average unbalanced force accelerates this object?
(1) 1000 N (2) 10 N (3) 0.1 N (4) 0.001 N

24. An object is acted upon by a constant unbalanced force. Which graph best represents the motion of this object?

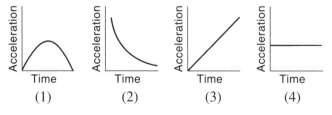

Time (1)   Time (2)   Time (3)   Time (4)

25. At a given location on the Earth's surface, which graph best represents the relationship between an object's mass (*M*) and weight (*W*)?

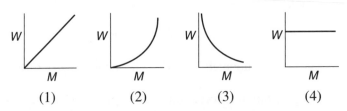

(1)   (2)   (3)   (4)

26. Two forces of 5 newtons and 10 newtons act on a 10-kilogram body as shown. The magnitude of the body's acceleration will be

(1) 50 m/s^2 (2) 2.0 m/s^2 (3) 1.5 m/s^2 (4) 0.5 m/s^2

27. If an object is moving north, the direction of the frictional force is
(1) north   (2) south   (3) east   (4) west

28. In order to keep an object weighing 20 newtons moving at constant speed along a horizontal surface, a force of 10 newtons is required. The force of friction between the surface and the object is
(1) 0 N   (2) 10 N   (3) 20 N   (4) 30 N

29. A box is sliding down an inclined plane as shown. The force of friction is directed toward point

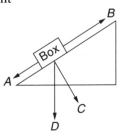

(1) *A*   (2) *B*   (3) *C*   (4) *D*

30. A horizontal force of 15 newtons pulls a 5-kilogram block along a horizontal surface. If the force produces an acceleration of 2 meters per second^2, the frictional force acting on the block is
(1) 1 N   (2) 2 N   (3) 5 N   (4) 15 N

31. A 2-kilogram mass moving along a horizontal surface at 10 meters per second is acted upon by a 5-newton force of friction. The time required to bring the mass to rest is
(1) 10 s   (2) 2 s   (3) 5 s   (4) 4 s

32. A force of 40 newtons applied horizontally is required to push a 20-kilogram box at constant velocity across the floor. The coefficient of friction between the box and the floor is
(1) 0.2   (2) 2   (3) 0.5   (4) 5

33. A 100-newton box is moving on a horizontal surface. A force of 10 newtons applied parallel to the surface is required to keep the box moving at constant velocity. What is the maximum coefficient of kinetic friction between the box and the surface?
(1) 0.1   (2) 10   (3) 0.5   (4) 1000

Base your answers to questions 34 and 35 on the diagram and information below.

The diagram shows a 2.0 kilogram mass suspended from a string that is attached to a 3.0 kilogram lab cart. (The system is kept at rest by means of hook *H*.) Assume that the system is frictionless, that the pulley has no mass, and that the table is level.

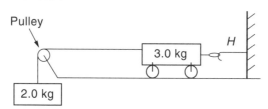

34. The force exerted on the cart by the hook is closest to
(1) 6.0 N   (2) 2.0 N   (3) 20. N   (4) 10. N

35. After the cart is released, the total mass being accelerated is
(1) 6.0 kg   (2) 2.0 kg   (3) 5.0 kg   (4) 4.0 kg

36. What is the net force on the block shown?

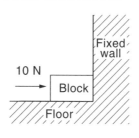

(1) 0 N   (2) 9.8 N   (3) 10 N   (4) 20 N

37. In the diagram below, surface *A* of the wooden block has twice the area of surface *B*. If a force of *F* newtons is required to keep the block moving at constant speed across the table when it slides on surface *A*, what force is needed to keep the block moving at constant speed when it slides on surface *B*?

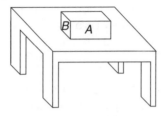

(1) *F*   (2) 2*F*   (3) $\frac{1}{2}F$ (4) 4*F*

38. A 1-kilogram bird stands on a limb. The force that the limb exerts on the bird is
(1) 1 N   (2) 2 N   (3) 0 N   (4) 9.8 N

39. A test booklet is sitting at rest on a desk. Compared to the magnitude of the force of the booklet on the desk, the magnitude of the force of the desk on the booklet is
(1) less   (2) greater   (3) the same

40. A 2.00-kilogram mass is at rest on a horizontal surface. The force exerted by the surface on the mass is approximately
(1) 0 N   (2) 2.00 N   (3) 9.80 N   (4) 19.6 N

41. As masses are added to a box resting on a table, the force exerted by the table on the box
(1) decreases   (2) increases   (3) remains the same

42. A baseball bat moving at high velocity strikes a feather. If air resistance is neglected, compared to the force exerted by the bat on the feather, the force exerted by the feather on the bat will be
(1) smaller   (2) larger   (3) the same

43. Which graph best represents the motion of an object on which the net force is zero?

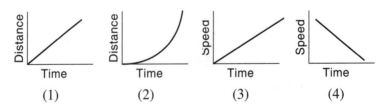

(1)   (2)   (3)   (4)

56

VECTOR QUANTITIES AND THEIR APPLICATIONS

Chapter Four

KEY IDEAS

Vectors are quantities that have both magnitude and direction. Forces and displacements are examples of vector quantities. A vector is best represented by an arrow: it points in the vector's direction, and its length is proportional to the vector's magnitude.

Vector quantities can be combined with one another. The process is known as vector addition, and the sum is equal to the combined result of the interaction of the individual vectors. The vector sum is also known as the resultant vector. Vectors can be added geometrically by using a measured scale or mathematically by using the laws of algebra and trigonometry.

The inverse process of vector addition is known as resolution. In resolution, a single vector is separated into a number of components. Resolution usually simplifies the solution of vector-related problems such as two-dimensional motion, static equilibrium, and motion on an inclined plane.

KEY OBJECTIVES

At the conclusion of this chapter you will be able to:
- Define the terms *scalar* and *vector* and list scalar and vector quantities.
- Represent a vector quantity by an arrow drawn to scale.
- Relate the direction of a vector to compass directions.
- Define the term *resultant vector*.
- Add vector quantities (1) graphically and (2) algebraically.
- Relate vector subtraction to vector addition.
- Define the term *vector resolution*, and resolve a vector into its *x*- and *y*-components.
- Add vector quantities by adding their *x*- and *y*-components.
- Define the term *static equilibrium.*
- Solve static equilibrium problems.
- Identify and calculate the parallel and perpendicular components of an object's weight when the object is on an inclined plane.
- Solve problems involving motion on an inclined plane.
- Solve problems involving the motion of an object in two dimensions.

4.1   INTRODUCTION

If we want to measure the mass of an object, it makes no difference whether we face in any particular direction while taking the measurement. Quantities such as mass, time, and temperature are called **scalar** quantities because they can be described solely in terms of their magnitudes (sizes).

If we are traveling in a car at 60 miles per hour, however, it makes a great deal of difference which way the car is facing: A car traveling east from Los Angeles might end up in New York, but if the car faced west it could end up in the Pacific! Quantities such as velocity, acceleration, and force are called vector quantities because they must be described in terms of their magnitudes *and* directions.

4.2   DISPLACEMENT, AND REPRESENTATION OF VECTOR QUANTITIES

In our study of motion in Chapter 2, we defined the simplest vector, called *displacement*, as a directed change in the position of an object. In other words, displacement is a distance (magnitude) in a given direction. For example, the quantity 30 meters [east] represents a displacement. (In this book, we will indicate the direction in brackets following the magnitude of the vector.)

We can easily represent a vector quantity by using an arrow. The length of the arrow represents the magnitude of the vector, and the direction of the vector points from the tail of the arrow toward its tip. The diagrams illustrate two displacement vectors.

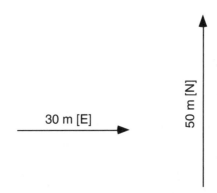

It should be obvious that the arrows shown above are not really 30 meters and 50 meters long. They are drawn *to scale*, as a map maker does when creating a map. In this instance, 1 inch of arrow length represents 20 meters of distance, and the arrows are 1.5 and 2.5 inches long, respectively.

PROBLEM

If 0.20 meter represents a displacement of 30. kilometers, what displacement is represented by a vector 0.80 meter long?

SOLUTION

$$0.80 \text{ m} \left(\frac{30. \text{ km}}{0.20 \text{ m}} \right) = 120 \text{ km}$$

There are many ways to represent the direction of a vector. We will use the method that we believe to be simplest in the long run. The direction is indicated by the angle that the vector makes with the x-axis (horizontal). We use the compass points to describe how many degrees the vector is north (N) or south (S) of the east-west (E-W) baseline. The diagram illustrates how we do this.

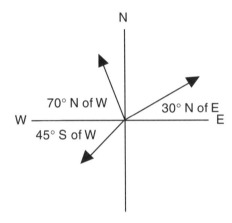

On occasion, it may be necessary to measure the angle that the vector makes with the y-axis (vertical). The conversion is an easy matter since the angles are complementary to one another. (In the diagram above, the angles would be 60° east of north, 20° west of north, and 45° west of south.)

4.3 VECTOR ADDITION

An airplane traveling at 300 meters per second [east] enters the jet stream, whose velocity is 100 meters per second [north]. What is the velocity of the airplane when measured by a person on the ground?

This problem, as well as a host of related problems, can be solved by a process called *vector addition*. When applied to vector quantities, the term *addition* means calculating the net effect of two or more vectors acting on the same object. It does not matter whether the vector quantities are velocities, displacements, forces, or others. The techniques of vector addition are the same in each case.

Suppose a person walks 3.0 meters [east] and then walks 4.0 meters [east]. What is the net displacement of the person? This is a simple problem in vector addition.

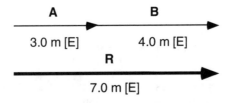

The addition is performed by placing the two vectors in a line, head to tail, as shown in the diagram. The sum of the vectors (indicated by the bold arrow) is called the **resultant vector (R)** and is determined by drawing an arrow beginning at the tail of the first vector and extending to the head of the second vector. The vector equation for this sum is:

$$\begin{array}{ccccc} \textbf{A} & + & \textbf{B} & = & \textbf{R} \\ \text{3.0 m [E]} & & \text{4.0 m [E]} & & \text{7.0 m [E]} \end{array}$$

Note that symbols representing vector quantities are set in boldface type (**A**). Another way of representing a vector quantity is to place an arrow above the symbol ($\vec{A}$). In this book we will use boldface type for vectors.

When vectors are oriented in opposite directions, their magnitudes are *subtracted*, as shown in the following problem.

PROBLEM
A bird flies north 3.0 kilometers and then south 4.0 kilometers. What is the resultant displacement of the bird?

SOLUTION

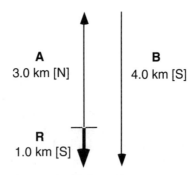

In this case, as the diagram shows, the addition yields a resultant of 1.0 kilometer south:

$$\begin{array}{ccccc} \textbf{A} & + & \textbf{B} & = & \textbf{R} \\ \text{3.0 m [N]} & & \text{4.0 km [S]} & & \text{1.0 km [S]} \end{array}$$

Now we will consider how vectors at right angles are added.

PROBLEM

A plane flies with a velocity of 300 meters per second [east] and enters the jet stream, whose velocity is 100 meters per second [north]. What is the resultant velocity of the plane?

SOLUTION

The solution to this problem also involves a diagram drawn to scale:

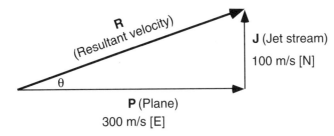

If we used a ruler to measure the magnitude of the resultant velocity **R**, we would find that the length of the resultant arrow translates to 320 m/s. Using a protractor, we see that the angle made by the resultant with the *x*-axis is approximately 18° N of E.

We could have solved this problem algebraically by recognizing that the resultant is the hypotenuse of a right triangle. Using the Pythagorean theorem, we have

$$R^2 = (300 \text{ m/s})^2 + (100 \text{ m/s})^2 = 100,000 \text{ m/s}^2$$
$$R = 316.2 \text{ m/s} = 320 \text{ m/s}$$

To calculate angle θ, we note that

$$\tan \theta = \frac{\text{opposite side}}{\text{adjacent side}} = \frac{100 \text{ m/s [N]}}{300 \text{ m/s [E]}} = 0.333$$

$$\theta = \tan^{-1} (0.33) = 18° \text{ [N of E]}$$

Finally, let's consider the effect of two forces that act *concurrently* (i.e., at the same point) at an angle other than 90°.

PROBLEM

As shown in the diagram, an object is subjected to two concurrent forces: **A** = 50 newtons [east] and **B** = 30 newtons [30° north of east]. What is the resultant force on the object?

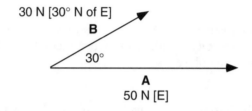

SOLUTION

We place the vectors head to tail by displacing vector **B** to the right, as shown in the diagram:

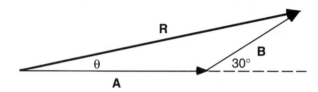

Then we draw **R** from the tail of **A** to the head of **B**. Using a ruler, a protractor, and a suitable scale factor, we find that the magnitude of **R** is 77 N and the angle θ is 11°.

This problem may also be solved algebraically by using the relationships known as the law of cosines and the law of sines. First we relabel the triangle as shown:

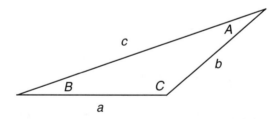

Then we have:

Law of cosines: $c^2 = a^2 + b^2 - 2ab \cos C$

Law of sines: $\dfrac{a}{\sin A} = \dfrac{b}{\sin B} = \dfrac{c}{\sin C}$

Using both diagrams and applying these two laws, we get:

$$R^2 = (50\ N)^2 + (30\ N)^2 - 2(50\ N)(30\ N)\ (\cos 150°)$$

$$R = 77\ N$$

$$\frac{77\ N}{\sin 150°} = \frac{30\ N}{\sin \theta} \Rightarrow \theta = 11°$$

This is a complicated procedure indeed! We will soon see that there is an easier—and more powerful—way to add vectors algebraically.

4.4   VECTOR SUBTRACTION

Subtraction is really a special case of addition. In mathematics, the expression $a - b$ is equivalent to the expression $a + (-b)$, where $-b$ is called the negative of b. Similarly, we subtract two vectors by adding one of them to the negative of the other:

$$\mathbf{A} - \mathbf{B} = \mathbf{A} + (-\mathbf{B})$$

Since vectors have both magnitude and direction, the negative of a vector $\mathbf{V}$ is equal to $\mathbf{V}$ in magnitude but opposite in direction, as illustrated below.

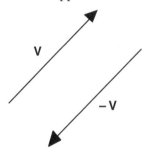

The following diagram compares vector addition and subtraction:

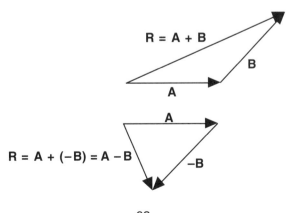

4.5 RESOLUTION OF VECTORS

If an automobile is traveling at 20 meters per second at 45° north of east, we know that the vehicle is traveling north and east at the same time. But *how fast* is it going in each direction? To solve this problem we use a technique known as **vector resolution**.

Resolving a vector means breaking it down into a number of components that will add to produce the original vector, as shown in the diagram.

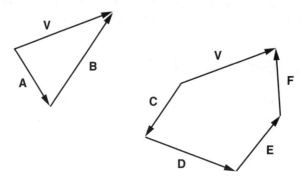

Here **A** and **B** are components of **V**, as are **C**, **D**, **E**, and **F**. As the diagram implies, a vector can be resolved into any number of components in virtually any direction. However, it is usually most helpful to resolve a vector into just *two* components that lie along the *x*- and the *y*-axis (i.e., the components are *perpendicular* to each other).

This resolution is illustrated in the following diagram:

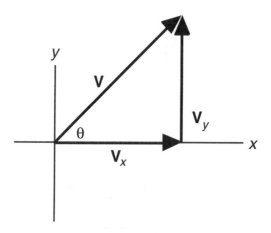

V_x and V_y are called the *x*- and *y*-components of **V**, respectively, and θ is the angle that **V** makes with the *x*-axis. Since the vector and its components form a right triangle, it follows that:

$$\frac{V_x}{V} = \cos \theta \implies V_x = V \cos \theta$$

$$\frac{V_y}{V} = \sin \theta \implies V_y = V \sin \theta$$

The x- and y-components of a vector have algebraic signs that depend on the quadrant in which the original vector lies, as shown in the diagram below:

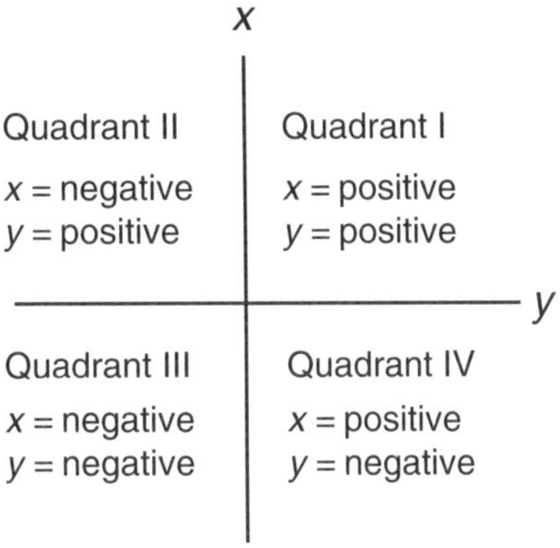

PROBLEM

Find the x- and y-components of the force $\mathbf{F} = 30$ newtons [30° north of east].

SOLUTION

Since the force lies in Quadrant I, its x- and y-components are positive:

$$\mathbf{F}_x = \mathbf{F} \cos \theta = (30 \text{ N}) (\cos 30°) = +26\text{N}$$
$$\mathbf{F}_y = \mathbf{F} \sin \theta = (30 \text{ N}) (\sin 30°) = +15\text{N}$$

PROBLEM

A plane is flying at 500 meters per second at 37° south of east. How fast is it flying east, and how fast is it flying south?

SOLUTION

To answer the questions we simply need to resolve the velocity vector into its x- and y-components. Since the velocity lies in Quadrant IV, its x-component is positive while its y-component is negative:

$$\mathbf{v}_x = \mathbf{v} \cos \theta = +(500 \text{ m/s}) \cdot (\cos 37°) = +400 \text{ m/s (i.e., 400 m/s [E])}$$
$$\mathbf{v}_y = \mathbf{v} \sin \theta = -(500 \text{ m/s}) \cdot (\sin 37°) = -300 \text{ m/s (i.e., 300 m/s [S])}$$

4.6 USING RESOLUTION TO ADD VECTORS

To solve a vector problem graphically (i.e., using a ruler and protractor) is usually fairly easy. However, we cannot achieve the precision of a mathematical solution. Unfortunately, mathematical solutions using the laws of cosines and sines are tedious and difficult.

We now develop a technique for vector addition that is relatively simple and yet gives us precision. The rules for this method are as follows:

1. Resolve each of the vectors to be added into its x- and y-components. Remember to include the proper sign (positive or negative), depending on the quadrant of each vector.
2. Be aware that, if a vector lies on the x-axis, its y-component is zero; if a vector lies on the y-axis, its x-component is zero.
3. Add the x-components together to produce the x-component of the resultant vector ($\mathbf{R}_x$).
4. Add the y-components together to produce the y-component of the resultant vector ($\mathbf{R}_y$).
5. Calculate the magnitude of the resultant vector by means of the Pythagorean theorem: $R = \sqrt{\mathbf{R}_x^2 + \mathbf{R}_y^2}$
6. Find the angle that the resultant vector makes with the x-axis from this relationship: $\tan \theta = \dfrac{\mathbf{R}_y}{\mathbf{R}_x}$.

Let's add three vectors using this method.

PROBLEM
Find the resultant of the force vectors illustrated below:

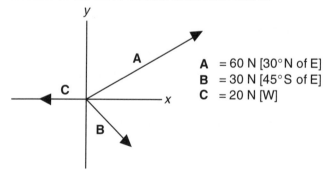

$$\mathbf{A} = 60 \text{ N } [30° \text{N of E}]$$
$$\mathbf{B} = 30 \text{ N } [45° \text{S of E}]$$
$$\mathbf{C} = 20 \text{ N } [W]$$

SOLUTION
We begin by calculating the x- and y-components of each force vector:
$$\mathbf{A}_x = A \cos \theta = + (60 \text{ N})(\cos 30°) = + 52 \text{ N}$$
$$\mathbf{A}_y = A \sin \theta = +(60 \text{ N})(\sin 30°) = + 30 \text{ N}$$
$$\mathbf{B}_x = B \cos \theta = + (30 \text{ N})(\cos 45°) = + 21 \text{ N}$$

$$B_y = B \sin \theta = -(30 \text{ N})(\sin 45°) = -21 \text{ N}$$
$$C_x = -20 \text{ N}$$
$$C_y = 0 \text{ N}$$

Next we add the x-components: $(+52 \text{ N}) + (+21 \text{ N}) + (-20 \text{ N}) = +53 \text{ N} = R_x$
and the y-components: $(+30 \text{ N}) + (-21 \text{ N}) + (0 \text{ N}) = +9 \text{ N} = R_y$

R_x and R_y are the respective x- and y-components of the resultant force. Since both values are positive, the resultant force lies in Quadrant I (N of E). The magnitude of **R** is calculated from the equation

$$R = \sqrt{R_x{}^2 + R_y{}^2} = \sqrt{(+53 \text{ N})^2 = (+9 \text{ N})^2} = \underline{54 \text{ N}}$$

and the angle θ is calculated from

$$\tan \theta = \frac{R_y}{R_x} = \frac{9 \, N \, [\text{N}]}{53 \, N \, [\text{E}]} = 0.17 \Rightarrow \theta = 9.6°[\text{N of E}]$$

4.7 STATIC EQUILIBRIUM

An object is at equilibrium if the vector sum of all the forces acting on it equals zero. If the object is at rest, we say that it is in **static equilibrium**.

The object in the diagram is in equilibrium because the vector sum of all forces acting on it (i.e., the net force) is zero.

PROBLEM
A 500-newton chandelier is hanging from a ceiling by a chain. What other force is acting on the chandelier?

SOLUTION

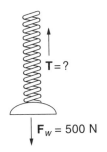

The term *tension* is used to describe the force that causes the chain to be taut. The direction of the tension (**T**) in this problem is upward because the tension supports the chandelier.

Since the chandelier is in equilibrium, the vector sum of the forces on it must equal zero. We can write the vector equation for the situation as follows:

$$\mathbf{T} + \mathbf{F}_W = 0$$
$$\mathbf{T} + (-500 \text{ N}) = 0$$
$$\mathbf{T} = +500 \text{ N}$$

(Note that the direction of $\mathbf{F}_W$, which is downward, is represented as a negative quantity. Since tension is a positive quantity, its direction is upward, as stated above.)

PROBLEM

Jesse, whose weight ($\mathbf{F}_W$) is 650 newtons, hangs by his arms from a horizontal bar in the gym. What is the tension in each of his arms?

SOLUTION

It is reasonable to assume that there is equal tension in each of Jesse's arms, and we use **T** to represent this tension. We can write the vector quantity as

$$\mathbf{T} + \mathbf{T} + \mathbf{F}_W = 0$$
$$\mathbf{T} + \mathbf{T} + (-650 \text{ N}) = 0$$
$$2\mathbf{T} = +650 \text{ N}$$
$$\mathbf{T} = +325 \text{ N}$$

The tension in each of Jesse's arms is equal to 325 N.

PROBLEM

Angela, whose weight ($\mathbf{F}_W$) is 400 newtons, sits on a tire swing in her backyard. Linda pulls Angela horizontally to the right so that the rope of the swing makes a 30° angle with the horizontal. If Angela is at rest in this position, what is the tension (**T**) in the rope and what force (**P**) does Linda exert on Angela?

SOLUTION

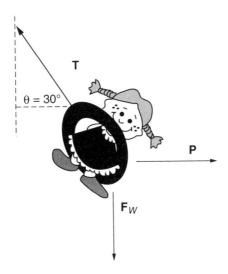

The vector equation for the forces on Angela is

$$\mathbf{T} + \mathbf{P} + \mathbf{F}_W = 0$$

The easiest way to solve this problem is to resolve the vectors into their x- and y-components. We will use the techniques developed earlier in this chapter.

Since the resultant is zero, both the x- and y-component of the resultant must also be zero. We can write these equations:

$$\mathbf{F}_{Wx} + \mathbf{P}_x + \mathbf{T}_x = \mathbf{R}_x = 0$$
$$\mathbf{F}_{Wy} + \mathbf{P}_y + \mathbf{T}_y = \mathbf{R}_y = 0$$

Force	x-component (N)	y-component (N)
$\mathbf{F}_W$	0	−400
$\mathbf{P}$	+$\mathbf{P}$	0
$\mathbf{T}$	−$\mathbf{T}$ cos 30°	+$\mathbf{T}$ sin 30°
$\mathbf{R}$	0	0

Substituting in the y-direction, we obtain

$$-400 \text{ N} + 0 + T \sin 30° = 0$$
$$T \sin 30° = 400 \text{ N}$$
$$T = \frac{400 \text{ N}}{\sin 30°} = 800 \text{ N}$$

Substituting in the *x*-direction, we obtain

$$0 + P + (-T \cos 30°) = 0$$
$$P = +T \cos 30° = (800)(\cos 30°)$$
$$= 693 \text{ N}$$

4.8   INCLINED PLANES

Let us now consider an object whose weight is $\mathbf{F}_W$, sliding down a friction-less inclined plane that makes an angle of θ with the ground. The diagram illustrates the situation.

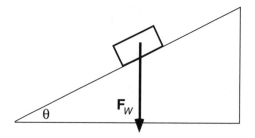

If we examined the object closely, we would find that it has an accelera-tion directly parallel to the surface of the plane. To analyze the forces acting on the object, we choose a set of axes that are parallel and perpendicular to the surface of the plane as shown in the diagram below.

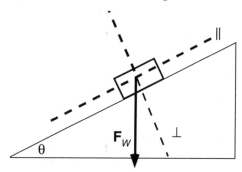

It is most convenient to resolve the weight of the object into components that are parallel and perpendicular to the inclined plane.

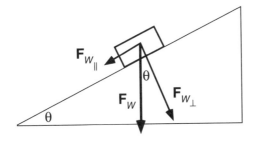

Simple trigonometry yields these relationships

$$\mathbf{F}_{W_\parallel} = \mathbf{F}_W \sin \theta \quad \text{and} \quad \mathbf{F}_{W_\perp} = \mathbf{F}_W \cos \theta$$

Since the object's acceleration is parallel to the plane, force $\mathbf{F}_{W_\parallel}$ is unbalanced. The object does not accelerate in the perpendicular direction, therefore, $\mathbf{F}_{W_\perp}$ is balanced by the normal force $\mathbf{F}_N$, which also acts perpendicularly to the surface of the plane and is equal to $\mathbf{F}_W \cos \theta$.

The diagram below illustrates all of the forces acting on the object as it travels down the plane.

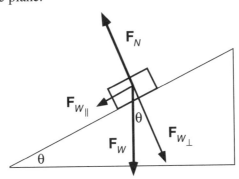

PROBLEM

A 5000-newton polar bear slides down an ice slide without friction. If the slide makes an angle of 37° with the ground, calculate the accelerating force and the normal force on the polar bear.

SOLUTION

$$F_{W_\parallel} = F_W \sin \theta$$
$$= (5000 \text{ N})(\sin 37°) = 3000 \text{ N}$$

$$F_N = F_{W_\perp} = F_W \cos \theta$$
$$= (5000 \text{ N})(\cos 37°) = 4000 \text{ N}$$

Suppose an object is sliding down an inclined plane at constant speed. In this situation, the forces in the parallel direction, as well as the forces in the perpendicular direction, are balanced. It is the force due to friction (F_f) that balances F_W, as shown in the diagram.

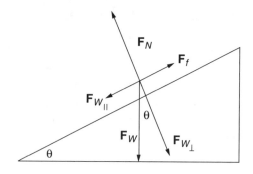

PROBLEM

A wooden crate whose weight (F_W) is 1200 newtons slides down a metal ramp at constant speed as it is unloaded from an airplane. The ramp makes an angle of 30° with the horizontal. Calculate the force of friction (F_f), the normal force (F_N), and the coefficient of kinetic friction (μ_k) between the crate and the ramp.

SOLUTION

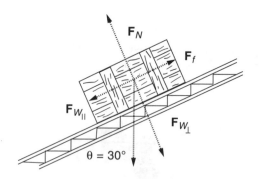

$$F_f = F_{W_\parallel} = F_W \sin \theta$$
$$= (1200 \text{ N})(\sin 30°) = 600 \text{ N}$$

$$\mathbf{F}_N = \mathbf{F}_{W\perp} = \mathbf{F}_W \cos \theta$$
$$= (1200 \text{ N})(\cos 30°) = 1040 \text{ N}$$

$$\mathbf{F}_f = \mu_k \mathbf{F}_N$$

$$\mu_k = \frac{\mathbf{F}_f}{\mathbf{F}_N}$$
$$= \frac{600 \text{ N}}{1040 \text{ N}} = 0.57$$

4.9   MOTION IN A PLANE (TWO-DIMENSIONAL MOTION)

Every student of physics is familiar with the following problem: An object is thrown horizontally away from a tall cliff at the same instant that an identical object is dropped vertically from the same height down the cliff. If air resistance is ignored, which object will reach the ground first? The diagram illustrates the problem.

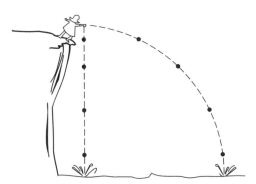

As unreasonable as it may seem, both objects hit the ground at exactly the same instant. Why does this happen? The explanation is based on the fact that the force due to gravity is the only factor that governs the downward motion of both objects. Since both objects fall at the same rate, they reach the ground at the same time. The difference is that the object thrown outward travels horizontally as well as vertically.

The following diagram represents a time lapse picture of this situation.

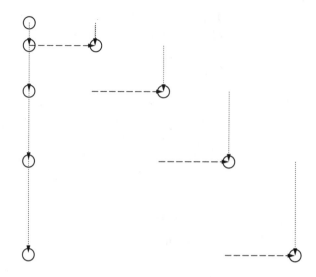

Notice that at any instant the vertical displacement of each object is the same. Also notice that in each time interval the horizontal displacement traveled by the object thrown outward is constant.

In the *y*-direction, both objects can be considered as falling freely from rest (with an acceleration of –9.8 meter per second per second) and are governed by the equations of motion developed in Chapter 2:

$$\Delta s_y = v_{iy}\, \Delta t + \frac{1}{2}\, a_y\, (\Delta t)^2$$

$$v_{fy}{}^2 = v_{iy}{}^2 + 2a_y\, \Delta s_y$$

$$a_y = \frac{\Delta v_y}{\Delta t} = \frac{v_{fy} - v_{iy}}{\Delta t}$$

In the *x*-direction, the thrown object travels at constant speed and its motion is described by

$$v_x = \frac{\Delta s_x}{\Delta t}$$

It is important to note that the *x*- and *y*-motions of the two objects are *completely independent* of one another.

PROBLEM

An object is thrown outward from a cliff with a horizontal velocity of 20. meters per second. With air resistance ignored, the object takes 15 seconds to reach the bottom of the cliff. Calculate (a) the height of the cliff and (b) the horizontal distance that was traveled by the object by the time it reaches the ground.

SOLUTION

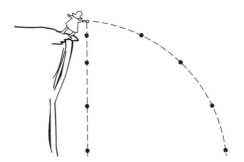

(a) We solve this part by considering the vertical motion of the object: its initial velocity is zero, and its acceleration is –9.8 m/s². Then:

$$\Delta s_y = v_{iy} \Delta t + \frac{1}{2} a_y (\Delta t)^2$$

$$= 0 + \frac{1}{2} a_y (\Delta t)^2$$

$$= \frac{1}{2} \left(-9.8 \ \frac{m}{s^2} \right) (15 \ s)^2$$

$$= -1102.5 \text{ m (i.e., } 1102.5 \text{ m } downward)$$

(b) Since the horizontal velocity is constant, we can apply this equation:

$$v_x = \frac{\Delta s_x}{\Delta t}$$

$$\Delta s_x = v_x \Delta t$$

$$= \left(20 \ \frac{m}{s} \right) (15 \ s) = 300 \text{ m}$$

Now let's consider an object such as a cannonball that is fired at an angle to the ground, as shown in the diagram.

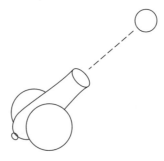

The shape of the path (known as the *trajectory*) is parabolic. If the initial velocity of the object is $\mathbf{v}_i$ and the angle is θ, we must resolve the velocity into its x- and y-components before we can solve any problem undergoing this type of motion:

$$\mathbf{v}_{ix} = \mathbf{v}_i \cos \theta$$
$$\mathbf{v}_{iy} = \mathbf{v}_i \sin \theta$$

In the x-direction, the object travels at constant speed. In the y-direction, we treat the object as if it were tossed directly upward and then returned to the ground: Its speed decreases as it rises and increases as it returns to the ground.

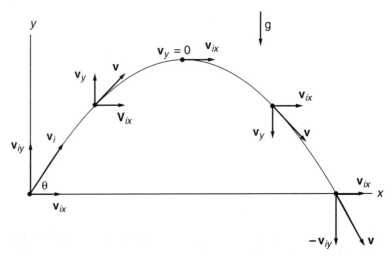

The motion equations developed in Chapter 2 can be used to solve situations involving this type of motion.

PROBLEM
An object is fired from the ground at 100 meters per second at an angle of 30° with the horizontal.

100 m/s

76

(a) Calculate the horizontal and vertical components of the initial velocity.
(b) After 2.0 seconds, how far has the object traveled in the horizontal direction?
(c) How high is the object at this point?

SOLUTION

(a) $\mathbf{v}_{ix} = \mathbf{v}_i \cos\theta = (100\ \text{m/s})(\cos 30°) = 87\ \text{m/s}$
$\mathbf{v}_{iy} = \mathbf{v}_i \sin\theta = (100\ \text{m/s})(\sin 30°) = 50\ \text{m/s}$

$(100\ m/s)$

(b) $\mathbf{v}_x = \dfrac{\Delta s_x}{\Delta t}$

$\Delta s_x = \mathbf{v}_x \cdot \Delta t$

$= \left(87\ \dfrac{\text{m}}{\text{s}}\right)(2.0\text{s}) = 174\text{m}$

✓ vertical acceleration
-9.8

(c) $\Delta s_y = \mathbf{v}_{iy}\,\Delta t + \dfrac{1}{2}\,\mathbf{a}_y\,(\Delta t)^2$

$\Delta s_y = \left(50\ \dfrac{\text{m}}{\text{s}}\right)(2.0\text{s}) + \dfrac{1}{2}\left(-9.8\ \dfrac{\text{m}}{\text{s}^2}\right)(2.0\text{s})^2$

$= 80\ \text{m}$

QUESTIONS

1. Which quantity has both magnitude and direction?
   (1) distance   (2) speed   (3) mass   (4) velocity

2. Which is a scalar quantity?
   (1) force   (2) energy   (3) displacement (4) velocity

3. Which is a vector quantity?
   (1) time   (2) work   (3) displacement (4) distance

4. Two concurrent forces of 6 newtons and 12 newtons could produce the same effect as a single force of
   (1) 5.0 N   (2) 15 N   (3) 20 N   (4) 72 N

5. A girl attempts to swim directly across a stream 15 meters wide. When she reaches the other side, she is 15 meters downstream. The magnitude of her displacement is closest to
   (1) 30 m   (2) 21 m   (3) 17 m   (4) 15 m

6. If a woman runs 100 meters north and then 70 meters south, her total displacement will be
(1) 30 m [N] (2) 30 m [S] (3) 170 m [N] (4) 170 m [S]

7. A 5-newton force directed north and a 5-newton force directed west both act on the same point. The resultant of these two forces is approximately
(1) 5 N [NW] (2) 7 N [NW] (3) 5 N [SW] (4) 7 N [SW]

8. Two forces of 5 newtons and 15 newtons acting concurrently could have a resultant with a magnitude of
(1) 5 N (2) 10 N (3) 25 N (4) 75 N

9. Two 10.0-newton forces act concurrently on a point at an angle of 180° to each other. The magnitude of the resultant of the two forces is
(1) 0 N (2) 10.0 N (3) 18.0 N (4) 20.0 N

10. The diagram below represents two forces acting concurrently on an object. The magnitude of the resultant force is closest to

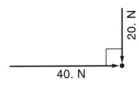

(1) 20 N (2) 40 N (3) 45 N (4) 60 N

11. Forces **A** and **B** have a resultant **R**. Force **A** and resultant **R** are shown in the diagram below.

Which vector below best represents force **B**?

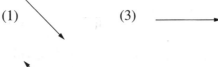

(1)   (3)

(2)   (4)

12. Which pair of concurrent forces could have a resultant of 5?
(1) 2 N and 2 N   (3) 2 N and 8 N
(2) 2 N and 4 N   (4) 2 N and 16 N

13. Two concurrent forces of 40 newtons and **X** newtons have a resultant of 100 newtons. Force **X** could be
(1) 20 N   (2) 40 N   (3) 80 N   (4) 150 N

14. Which vector best represents the resultant of the two vectors shown below?

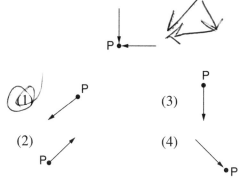

15. Two forces (**A** and **B**) act simultaneously at point P as shown in the diagram below.

The magnitude of the resultant force is closest to
(1) 8.0 N   (2) 11 N   (3) 15 N   (4) 16 N

16. Which vector best represents the resultant of forces F_1 and F_2 acting concurrently on point P as shown in the diagram below?

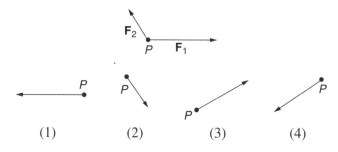

17. Four forces act on a point as shown.

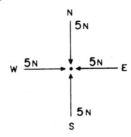

The resultant of the four forces is
(1) 0 N   (2) 5 N   (3) 14 N   (4) 20 N

18. The resultant of a 12-newton force and a 7-newton force is 5 newtons. The angle between the forces is
(1) 0°   (2) 45°   (3) 90°   (4) 180°

19. Three forces act concurrently on an object in equilibrium. These forces are 10 newtons, 8 newtons, and 6 newtons. The resultant of the 6-newton and 8-newton forces is
(1) 0   (3) 10 N
(2) between 0 and 10 N   (4) greater than 10 N

20. The resultant of two concurrent forces is minimum when the angle between them is
(1) 0°   (2) 45°   (3) 90°   (4) 180°

21. As the angle between two concurrent forces decreases from 180°, their resultant
(1) decreases   (2) increases   (3) remains the same

22. As the angle between two concurrent forces of 5.0 newtons and 7.0 newtons increases from 0° to 180°, the magnitude of their resultant changes from
(1) 0 N to 35 N   (3) 12 N to 2.0 N
(2) 2.0 N to 12 N   (4) 12 N to 0 N

23. The maximum number of components that a single force may be resolved into is
(1) one   (2) two   (3) three   (4) unlimited

24. A constant force is exerted on a box as shown in the diagram.

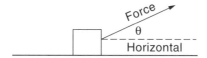

As angle θ decreases to 0°, the magnitude of the horizontal component of the force
(1) decreases (2) increases (3) remains the same

25. A force of 100. newtons is applied to an object at an angle of 30° from the horizontal as shown in the diagram below. What is the magnitude of the vertical component of this force?

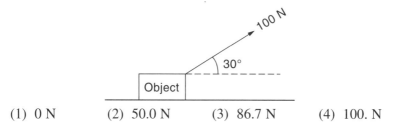

(1) 0 N (2) 50.0 N (3) 86.7 N (4) 100. N

26. Which diagram represents the vector with the largest downward component? [Assume each vector has the same magnitude.]

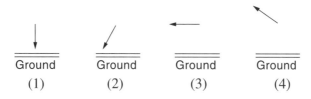

27. What is the magnitude of the vertical component of the velocity vector shown below?

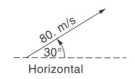

(1) 10. m/s (2) 69 m/s (3) 30. m/s (4) 40. m/s

28. In the diagram below, the numbers 1, 2, 3, and 4 represent possible directions in which a force could be applied to a cart. If the force applied in each direction has the same magnitude, in which direction will the vertical component of the force be the least?

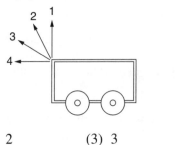

(1) 1   (2) 2   (3) 3   (4) 4

29. A resultant force of 10. newtons is made up of two component forces acting at right angles to each other. If the magnitude of one of the components is 6.0 newtons, the magnitude of the other component must be

(1) 16 N   (2) 8.0 N   (3) 6.0 N   (4) 4 N

30. A ball is fired vertically upward at 5.0 meters per second from a cart moving horizontally to the right at 2.0 meters per second. Which vector best represents the resultant velocity of the ball when fired?

(1)

(3)

(2)

(4)

31. An object weighing 24 newtons is placed on a 30° slope as shown. The component of the weight acting parallel to the slope is closest to

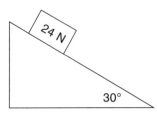

(1) 12 N   (2) 17 N   (3) 22 N   (4) 24 N

Base your answer to question 32 on the diagram below, which shows a cart held motionless by an external force **F** on a frictionless incline.

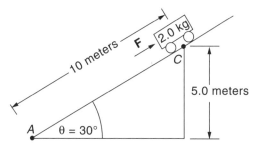

32. What is the magnitude of the external force **F** necessary to hold the cart motionless at point *C*?
(1) 4.9 N   (2) 2.0 N   (3) 9.8 N   (4) 19.6 N

33. Weight **W** is supported by a rod and a cable, as shown.

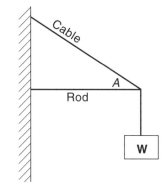

If angle *A* is made smaller, the tension in the cable will
(1) decrease   (2) increase   (3) remain the same

34. An object weighing 600 newtons is pulled up a frictionless incline at a constant speed. If the incline makes an angle of 30° with the horizontal, the force on the object parallel to the incline is
(1) 200 N   (2) 300 N   (3) 520 N   (4) 600 N

35. An object rests on an incline. As the angle between the incline and the horizontal increases, the force needed to prevent the object from sliding down the incline
(1) decreases   (2) increases   (3) remains the same

36. A projectile is fired with a velocity of 150. meters per second at an angle of 30° with the horizontal. What is the magnitude of the vertical component of the velocity at the time the projectile is fired?
(1) 75.0 m/s   (2) 130. m/s   (3) 150. m/s   (4) 225 m/s

37. A projectile is fired at an angle of 53° to the horizontal with a speed of 80. meters per second. What is the vertical component of the projectile's initial velocity?
(1) 130 m/s   (2) 100 m/s   (3) 64 m/s   (4) 48 m/s

38. A ball is fired with a velocity of 12 meters per second from a cannon pointing north, while the cannon is moving eastward at a velocity of 24 meters per second. Which vector best represents the resultant velocity of the ball as it leaves the cannon?

(1)   N   (3)   N

       E   E

(2)   N   (4)   N

       E   E

39. A batted softball leaves the bat with an initial velocity of 44 meters per second at an angle of 37° above the horizontal. What is the magnitude of the initial vertical component of the softball's velocity?
(1) 0 m/s   (2) 26 m/s   (3) 35 m/s   (4) 44 m/s

40. A ball rolls down a curved ramp as shown in the diagram below. Which dotted line best represents the path of the ball after leaving the ramp?

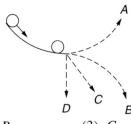

(1) *A*   (2) *B*   (3) *C*   (4) *D*

Base your answers to questions 41 and 42 on the diagram below, which shows a ball projected horizontally with an initial velocity of 20. meters per second east, off a cliff 100. meters high. [Neglect air resistance.]

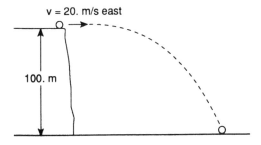

v = 20. m/s east

100. m

41. How many seconds does the ball take to reach the ground?
(1) 4.5   (2) 20.   (3) 9.8   (4) 2.0

42. During the flight of the ball, what is the direction of its acceleration?
(1) downward   (2) upward   (3) westward   (4) eastward

Base your answers to questions 43 and 44 on the diagram below, which shows a ball thrown toward the east and upward at an angle of 30° to the horizontal. Point X represents the ball's highest point.

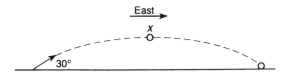

East

X

30°

43. What is the direction of the ball's velocity at point X? [Neglect friction.]
(1) down   (2) up   (3) west   (4) east

44. What is the direction of the ball's acceleration at point X? [Neglect friction.]
(1) down   (2) up   (3) west   (4) east

45. A bullet is fired horizontally from the roof of a building 100. meters tall with a speed of 850. meters per second. Neglecting air resistance, how far will the bullet drop in 3.00 seconds?
(1) 29.4 m   (2) 44.1 m   (3) 100. m   (4) 2,550 m

Base your answers to questions 46 and 47 on the diagram below, which represents a ball being kicked by a foot and rising at an angle of 30.° from the horizontal. The ball has an initial velocity of 5.0 meters per second. [Neglect friction.]

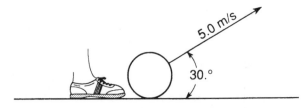

46. What is the magnitude of the horizontal component of the ball's initial velocity?
    (1) 2.5 m/s   (2) 4.3 m/s   (3) 5.0 m/s   (4) 8.7 m/s

47. As the ball rises, the vertical component of its velocity
    (1) decreases   (2) increases   (3) remains the same

Base your answers to questions 48 and 49 on the information below.

A rocket is launched at an angle of 60° to the horizontal. The initial velocity of the rocket is 500 meters per second. [Neglect friction.]

48. The vertical component of the initial velocity is
    (1) 250 m/s   (2) 433 m/s   (3) 500 m/s   (4) 1000 m/s

49. Compared to the horizontal component of the rocket's initial velocity, the horizontal component after 10 seconds would be
    (1) less   (2) greater   (3) the same

Base your answers to questions 50 through 52 on the information below.

An object is thrown horizontally off a cliff with an initial velocity of 5.0 meters per second. The object strikes the ground 3.0 seconds later.

50. What is the vertical speed of the object as it reaches the ground? [Neglect friction.]
    (1) 130 m/s   (2) 29 m/s   (3) 15 m/s   (4) 5.0 m/s

51. How far from the base of the cliff will the object strike the ground? [Neglect friction].
    (1) 2.9 m   (2) 9.8 m   (3) 15 m   (4) 44 m

52. What is the horizontal speed of the object 1.0 second after it is released? [Neglect friction.]
(1) 5.0 m/s   (2) 10. m/s   (3) 15 m/s   (4) 30. m/s

53. Which graph best represents the motion of an object sliding down a frictionless inclined plane?

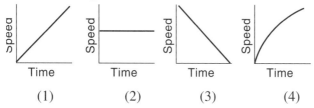

     (1)   (2)   (3)   (4)

54. Four forces are acting on an object as shown in the diagram below.

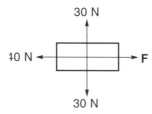

If the object is moving with a constant velocity, the magnitude of force **F** must be
(1) 0 N   (2) 20 N   (3) 100 N   (4) 40 N

55. An elevator containing a man weighing 800 newtons is rising at a constant speed. The force exerted by the man on the floor of the elevator is
(1) less than 80 N   (3) 800 N
(2) between 80 and 800 N   (4) more than 800 N

Base your answers to questions 56 through 58 on the information and diagram below.

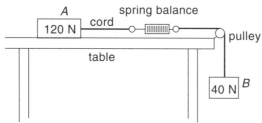

The reading of the spring balance is *t* newtons. [Neglect friction and the weight of the spring balance.]

56. What is the direction of the net force on object B?

    (1) ⟵   (3) ⟶

    (2) ↑   (4) ↓

57. The net force on object B is equal to
    (1) t N   (2) 40 N   (3) $(40 + t)$ N   (4) $(40 - t)$ N

58. The net force on object A is equal to
    (1) t N   (2) 120 N   (3) $(120 + t)$ N   (4) $(120 - t)$ N

CIRCULAR MOTION AND GRAVITATION

Chapter Five

KEY IDEAS

An object that travels in a circular path experiences an acceleration directed toward the center of the circle and known as a centripetal acceleration. The unbalanced force responsible for this acceleration is called a centripetal force and is also directed toward the center of the circle.

The study of planetary motion was greatly advanced by Kepler, who deduced three laws that provided a mathematical basis for this motion. Kepler's three laws were later proved by Newton when he developed his law of universal gravitation: two masses attract each other with a force proportional to their masses and inversely proportional to the square of the distance between them.

KEY OBJECTIVES

At the conclusion of this chapter you will be able to:

- Identify the direction of an object's velocity when it is undergoing uniform circular motion.
- Define the terms *centripetal acceleration* and *centripetal force*, and identify the directions of these quantities when an object undergoes uniform circular motion.
- State the equations for calculating centripetal force and centripetal acceleration.
- Solve problems involving uniform circular motion.
- Define the term *period of revolution* and relate it to the equations of uniform circular motion.
- State Kepler's three laws of planetary motion.
- State Newton's law of universal gravitation and solve problems related to it.
- Solve simple problems involving satellites in (a circular) orbit.
- Define the term *geosynchronous orbit*.
- Relate weight to gravitational force.
- Describe the field concept of gravitation.
- Relate the strength of a gravitational field with the acceleration due to gravity.

5.1 INTRODUCTION

In this chapter we study another aspect of motion in a plane, namely, motion in a circle. The subject of circular motion leads, in turn, to a study of gravitation because both natural and artificial satellites travel in nearly circular paths.

5.2 CIRCULAR MOTION

An object moves in a circular path at constant speed as indicated in the diagram below.

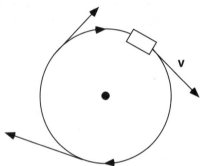

At various points in the path, the direction of the velocity, **v**, of the object is tangent to the circle, as shown in the diagram. Since the *direction* of the object's motion is changing, the object must be subjected to an unbalanced force and is, therefore, accelerating. Here is an example of an object that accelerates even though its speed does not change. This is due to the fact that acceleration is a change in the *velocity* of an object, and this change can be in the magnitude and/or direction of the velocity.

As the object moves in its circular path, its acceleration always points toward the center of the circle. For this reason we call the acceleration a **centripetal** ("center-seeking") acceleration (**a**$_c$).

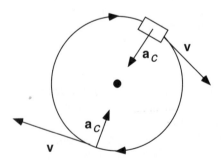

The centripetal acceleration of an object is calculated by means of this equation:

$$a_c = \frac{v^2}{r}$$

where **v** is the linear speed of the object and r is the radius of the circular path.

The unbalanced force associated with the centripetal acceleration is called the **centripetal force** (**F$_c$**). It also points toward the center of the circular path and is given by this relationship:

$$F_c = ma_c = \frac{mv^2}{r}$$

We can think of the centripetal force as the force needed to keep the object in its circular path. If the mass of the object were increased, or if the speed of the object were increased, more force would be needed to keep the object in its circular path. If, however, the radius of the path were increased, less force would be required for this purpose.

PROBLEM

A 5.0-kilogram object travels clockwise in a horizontal circle with a speed of 20. meters per second, as shown in the diagram. The radius of the circular path is 25 meters.

$M = 5\ Kg$
$V = 20\ m/s$
$r = 25 m$

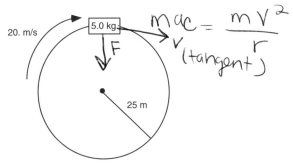

20. m/s   5.0 kg

$mac = \frac{mv^2}{r}$

V (tangent)

F

25 m

(a) Calculate the centripetal acceleration and the centripetal force on the object. $ac = 16$ $Fc = 80$
(b) In the position shown, indicate the *direction* of the velocity and of the centripetal force of the object.

SOLUTION

(a) $\quad a_c = \dfrac{v^2}{r}$

$$= \frac{\left(20 \frac{m}{s}\right)^2}{25 \text{ m}} = 16 \text{ m/s}^2$$

$$F_c = ma_c = (5.0 \text{ kg})\left(16 \frac{m}{s^2}\right) = 80. \text{ N}$$

(b)

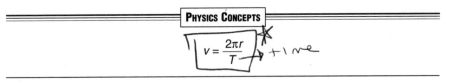

Measuring the speed of an object in circular motion is not always easy, but we can measure the speed of the object indirectly by making use of a quantity known as the period (*T*). The period of revolution is the time an object takes to complete one revolution in a circular path. In one revolution, the distance the object travels equals the circumference of the circle (2π*r*). Using the equation $v = \dfrac{\Delta s}{\Delta t}$, we substitute and obtain

PHYSICS CONCEPTS

$$v = \frac{2\pi r}{T}\ time$$

PROBLEM
An object traveling in a circular path makes 1200 revolutions in 1.0 hour. If the radius of the path is 10. meters, calculate the speed of the object.

SOLUTION
If the object makes 1200 revolutions in 1.0 h (3600 s), the time (*T*) for one revolution is

$$T = \frac{3600 \text{ s}}{1200 \text{ rev}} = 3\text{s} \qquad T = \frac{seconds}{Revolutions}$$

1200

The speed of the object is

$$v = \frac{2\pi r}{\Delta t} = \frac{(2)\,(3.1)\,(10.\ \text{m})}{3.0\ \text{s}} = 21\ \text{m/s}$$

The equation for the centripetal acceleration can be rewritten in terms of the period of revolution rather than the speed of the object:

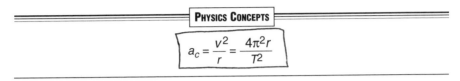

PHYSICS CONCEPTS

$$a_c = \frac{v^2}{r} = \frac{4\pi^2 r}{T^2}$$

5.3 KEPLER'S LAWS

The motions of the planets in the heavens were of great interest in the sixteenth century. The German astronomer Johannes Kepler summarized the laws of planetary motion; this work was based on data collected by the Danish astronomer Tycho Brahe. Kepler proposed three laws:

PHYSICS CONCEPTS

FIRST LAW

1. The path of each planet about the Sun is an ellipse with the Sun at one focus.

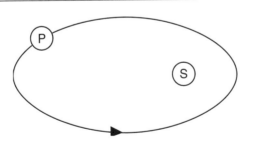

PHYSICS CONCEPTS

2. SECOND LAW

Each planet moves so that an imaginary line drawn from the Sun to the planet sweeps out equal areas of space in equal periods of time.

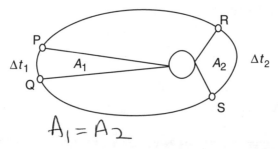

$$A_1 = A_2$$

In the diagram above, the time it takes for the planet to move between points P and Q (Δt_1) is equal to the time it takes the planet to move between points R and S (Δt_2). As a result of Kepler's second law, we can conclude that area A_1 is equal to Area A_2.

PHYSICS CONCEPTS

3. THIRD LAW

The cube of a planet's average distance from the Sun (r^3) divided by the square of its period (T^2) is a constant for all planets.

The table below illustrates the validity of Kepler's third law for our solar system.

Planet	Average Distance from Sun (r) ($/10^6$ km)	Period (T) (/earth-years)	r^3/T^2
Mercury	57.9	0.241	3.34
Venus	108.2	0.615	3.35
Earth	149.6	1.00	3.35
Mars	227.9	1.88	3.35
Jupiter	778.3	11.86	3.35
Saturn	1427	29.5	3.34
Uranus	2870	84.0	3.35
Neptune	4497	165	3.34
Pluto	5900	248	3.33

In fact, Kepler's three laws hold for *any* satellite orbiting a heavenly body.

5.4 NEWTON'S LAW OF UNIVERSAL GRAVITATION

More than half a century after Kepler proposed his laws, Isaac Newton was able to prove them mathematically by means of his law of universal gravitation.

Newton recognized that the force responsible for pulling an apple toward the Earth has the same origin as the force that keeps the Moon in its orbit about the Earth. This force, which we call *gravitation*, is present between all bodies of mass in the universe, from the largest galaxies to the smallest atoms.

Newton's law of universal gravitation can be explained as follows. Consider two point masses, m_1 and m_2, separated by a distance d as shown in the diagram.

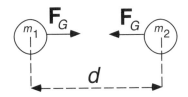

The force of gravitation is an *attractive* force. The force on either mass ($\mathbf{F}_G$) is given by this relationship:

PHYSICS CONCEPTS

$$\mathbf{F}_G = \frac{Gm_1m_2}{d^2}$$

According to *Newton's universal law*:

The gravitational force between two masses is directly proportional to each mass and inversely proportional to the square of the distance between the masses.

This is known as an *inverse square* law. The constant G is called the *universal gravitational constant*. Its measured value is 6.67×10^{-11} newton·meter^2 per kilogram^2. The table below illustrates how the relative gravitational force between two objects depends on their relative masses and the relative distance between them.

Relative Mass$_1$	Relative Mass$_2$	Relative Distance	Relative Force
m_1	m_2	d	F
$2m_1$	m_2	d	2F
m_1	$3m_2$	d	3F
$2m_1$	$3m_2$	d	6F
m_1	m_2	$2d$	F/4
m_1	m_2	$3d$	F/9
m_1	m_2	$d/4$	16F
m_1	m_2	$d/5$	25F

PROBLEM

Calculate the gravitational force of attraction between the Earth and the Moon, given that the mass of the Earth is 6.0×10^{24} kilograms, the mass of the Moon is 7.4×10^{22} kilograms, and the average Earth–Moon distance is 3.8×10^8 meters.

SOLUTION

$$F_G = \frac{Gm_{Earth}m_{Moon}}{(d_{Earth-Moon})}$$

$$= \frac{\left(6.67 \times 10^{-11} \dfrac{N \cdot m^2}{kg^2}\right)(6.0 \times 10^{24}\ kg)(7.4 \times 10^{22}\ kg)}{(3.8 \times 10^8\ m)^2}$$

$$= 2.1 \times 10^{20}\ N$$

This force is present on both the Earth and the Moon (in keeping with Newton's third law). It may seem like an outrageously large force, but consider the masses of the two objects involved! For "normal, Earth-sized" objects, on the other hand, the force of gravitational attraction is quite small.

5.5 SATELLITE MOTION

Suppose a satellite is traveling around the Earth in a circular orbit at a distance r from the center of the planet. The satellite is kept in orbit by a centripetal force whose magnitude is given by the relationship:

$$F_c = \frac{m_s v^2}{r}$$

where m_s is the mass of the satellite, v is its speed in orbit, and r is the distance from the center of the Earth. We know that the origin of this centripetal force is gravitational force; therefore, we can equate the centripetal force relationship with Newton's law of universal gravitation:

$$F_c = \frac{m_s v^2}{r} = \frac{Gm_{Earth}m_s}{r^2} = F_G$$

$$\frac{v^2}{r} = \frac{Gm_{Earth}}{r^2}$$

$$v = \sqrt{\frac{Gm_{Earth}}{r}}$$

We can use the relationship to find the speed necessary for the satellite to be in a given orbit. If the satellite were to move faster, its orbit would be closer to the Earth. If the satellite were to move slower, its orbit would be further from the Earth.

There is a specific distance at which the period of a satellite matches the period of the Earth's rotation (1 day). This type of orbit is known as a **geosynchronous orbit**. Satellites in geosynchronous orbits are especially useful for communication purposes because their positions with respect to the ground do not vary.

5.6 WEIGHT AND GRAVITATIONAL FORCE

We have measured an object's weight by the relationship $\mathbf{F}_w = m\mathbf{g}$, and we can measure an object's gravitational attraction to the Earth by the equation

$$\mathbf{F}_G = \frac{Gm_1 m_2}{d^2}$$

Weight and gravitational attraction, however, are one and the same force since an object's weight is gravitational in origin. Accordingly, we can set the two terms equal to each other:

$$\mathbf{F}_W = m\mathbf{g} = \frac{Gmm_{Earth}}{d^2} = \mathbf{F}_G$$

$$\mathbf{g} = \frac{Gm_{Earth}}{d^2}$$

We can now appreciate the fact that the gravitational acceleration of an object is independent of its mass; it depends only on the mass of the Earth and the distance of the object from the Earth's center.

PROBLEM
Calculate the gravitational acceleration of a satellite that is in orbit at a distance of 1.0×10^8 meters from the center of the Earth.

SOLUTION

$$\mathbf{g} = \frac{Gm_{Earth}}{d^2}$$

$$= \frac{\left(6.67 \times 10^{-11} \dfrac{N \cdot m^2}{kg^2}\right)(6.0 \times 10^{24}\ kg)}{(1 \times 10^8\ m)^2}$$

$$= 0.040\ m/s^2 \text{ (toward the center of the Earth)}$$

5.7 THE GRAVITATIONAL FIELD

A gravitational field is a region of space that attracts masses with a gravitational force. The gravitational-field concept assumes that one mass somehow changes the space around it and a second mass then interacts with the field. The result is that an attractive gravitational force is exerted on the second mass. One type of interaction is represented in the diagram below.

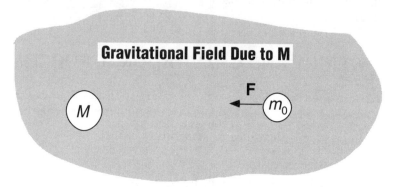

The gravitational field is a vector quantity: its direction is the direction of the force on the mass m_0, which we call the *test mass*. The test mass is assumed to be much smaller than the mass M, whose field is being determined. The strength of a gravitational field ($\mathbf{g}$) is defined to be the ratio of the gravitational force on the test mass ($\mathbf{F}$) to the test mass (m_0):

PHYSICS CONCEPTS

$$g = \frac{F}{m_0}$$

The unit of gravitational field strength is the *newton per kilogram* (N/kg). A little "fiddling" with units will show that this unit is equivalent to the unit *meter per second^2*—the unit of (gravitational) acceleration. The gravitational field is simply another way of viewing the action of gravity. If we drop an object near the surface of the Earth, it will accelerate at 9.8 meters per second^2; since it is in the Earth's gravitational field, it experiences a gravitational force of 9.8 newtons per kilogram of mass. Numerically, the gravitational acceleration and the gravitational field strength are always equal.

$$a_c = \frac{80\,v^2}{r\,m}$$

QUESTIONS

1. An object travels in a circular path of radius 5.0 meters at a uniform speed of 10. meters per second. What is the magnitude of the object's centripetal acceleration?
   (1) 10.0 m/s²   (2) 2.0 m/s²   (3) 5.0 m/s²   (4) 20. m/s²

2. The diagram below shows an object traveling clockwise in a horizontal, circular path at constant speed. Which arrow best shows the direction of the centripetal acceleration of the object at the instant shown?

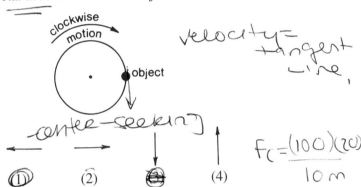

velocity = tangent line,

center-seeking

$F_c = (100)(20)$

10 m

    (1)   (2)   (3)   (4)

3. A motorcycle of mass 100 kilograms travels around a flat, circular track of radius 10 meters with a constant speed of 20 meters per second. What force is required to keep the motorcycle moving in a circular path at this speed?
   (1) 200 N   (2) 400 N   (3) 2000 N   (4) 4000 N

4. A mass of 10 kilograms is revolving at a linear speed of 5 meters per second in a circle with a radius of 10 meters. The centripetal force acting on the mass is
   (1) 5 N   (2) 10 N   (3) 20 N   (4) 25 N

5. Two masses, A and B, move in circular paths as shown in the diagram below.

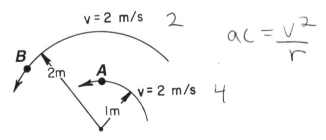

$a_c = \frac{v^2}{r}$

The centripetal acceleration of mass A, compared to that of mass B, is
   (1) the same   (3) one-half as great
   (2) twice as great   (4) 4 times as great

Base your answers to questions 6 through 10 on the diagram below, which represents a ball of mass M attached to a string. The ball moves at a constant speed around a flat horizontal circle of radius R.

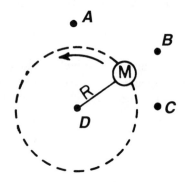

6. When the ball is in the position shown, the direction of the centripetal force is toward point
   (1) *A*   (2) *B*   (3) *C*   (4) *D*

7. The centripetal acceleration of the ball is
   (1) zero
   (2) constant in direction, but changing in magnitude
   (3) constant in magnitude, but changing in direction
   (4) changing in both magnitude and direction

8. If the velocity of the ball is doubled, the centripetal acceleration
   (1) is halved   (3) remains the same
   (2) is doubled   (4) is quadrupled

9. If the string is shortened while the speed of the ball remains the same, the centripetal acceleration will
   (1) decrease   (2) increase   (3) remain the same

10. If the mass of the ball is decreased, the centripetal force required to keep it moving in the same circle at the same speed will
    (1) decrease   (2) increase   (3) remain the same

Base your answers to questions 11 through 14 on the diagram below which represents a 2.0-kilogram mass moving in a circular path on the end of a string 0.50 meter long. The mass moves in a horizontal plane at a constant speed of 4.0 meters per second.

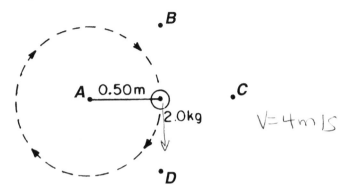

11. The force exerted on the mass by the string is
    (1) 8 N   (2) 16 N   (3) 32 N   (4) 64 N

12. In the position shown in the diagram, the velocity of the mass is directed toward point
    (1) A   (2) B   (3) C   (4) D

13. The centripetal force acting on the mass is directed toward point
    (1) A   (2) B   (3) C   (4) D

14. The speed of the mass is changed to 2.0 meters per second. Compared to the centripetal acceleration of the mass when moving at 4.0 meters per second, its centripetal acceleration when moving at 2.0 meters per second would be
    (1) half as great   (3) one-fourth as great $V = 2 m/s$
    (2) twice as great   (4) 4 times as great

    3 2 .

Base your answers to questions 15 and 16 on the information and diagram below.

A pendulum with a 10-kilogram bob is released at point *A* and allowed to swing without friction, as shown in the diagram.

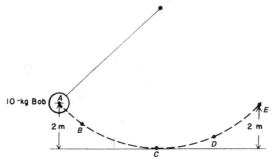

15. What is the weight of the bob?
  (1) 0.1 N   (2) 0.98 N   (3) 10 N   (4) 98 N

16. The centripetal acceleration of the bob is greatest at point
  (1) *E*   (2) *B*   (3) *C*   (4) *D*

Base your answers to questions 17 through 19 on the diagram below, which shows a section of a level road viewed from above. The road has one curve (*AB*) with a radius of 2*R* and a second curve (*CD*) with a radius of *R*. A car of mass *M* moves, with constant speed, *v*, along the road.

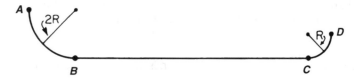

17. What is the magnitude of the net force acting on the car as it moves from *B* to *C*?
  (1) *Mv*

  (2) *Mv*^{2}

  (3) $\dfrac{Mv^2}{2}$

  (4) zero

18. What is the magnitude of the net force acting on the car when it is moving from *C* to *D*?
  (1) $\dfrac{Mv^2}{R}$

  (2) *Mv*^{2}*R*

  (3) *MvR*

  (4) zero

19. Compared to the centripetal acceleration of the car when it is moving from *C* to *D*, the centripetal acceleration when it is moving from *A* to *B* is

(1) less   (2) greater   (3) the same

Base your answers to questions 20 through 23 on the diagram and information below.

At an amusement park, a passenger whose mass is 50 kilograms rides in a cage. The cage has a constant speed of 10 meters per second in a vertical circular path of radius *R* equal to 10 meters.

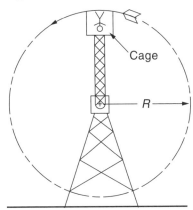

20. What is the magnitude of the centripetal acceleration of the passenger?

(1) 1.0 m/s^2 (3) $5.0 \times 10^2 \text{ m/s}^2$
(2) $2.0 \times 10^3 \text{ m/s}^2$ (4) $10. \text{ m/s}^2$

21. What is the direction of the centripetal acceleration of the passenger at the instant the cage reaches the highest point in the circle?

(1) to the left   (2) to the right   (3) up   (4) down

22. What does the 50.-kilogram passenger weigh at rest?

(1) 1600 N   (2) 490 N   (3) 50 N   (4) 0 N

23. What is the magnitude of the centripetal force acting on the passenger?

(1) 0 N   (2) 50 N   (3) $4.9 \times 10^2 \text{ N}$ (4) $5.0 \times 10^2 \text{ N}$

Base your answers to questions 24 through 27 on the diagram below, which shows a 10-newton centripetal force causing a 1-kilogram mass to move in a circular path with a speed of 10 meters per second.

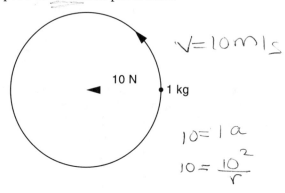

24. What is the radius of the circular path?
    (1) 1 m   (2) $\sqrt{10}$ m   (3) 10 m   (4) 100 m

25. The centripetal acceleration of the object is equal to
    (1) 100 m/s²   (2) 1000 m/s²   (3) $\dfrac{100}{r}$ m/s²   (4) $\dfrac{1000}{r}$ m/s²

26. If the speed of the mass increases and the radius of its path remains constant, the centripetal force will
    (1) decrease   (2) increase   (3) remain the same

27. If the radius of the path decreases and the speed of the mass remains constant, the centripetal acceleration of the mass will
    (1) decrease   (2) increase   (3) remain the same

Base your answers to questions 28 through 32 on the diagram below which shows the path of an object moving counterclockwise in a circle of radius 2.0 meters. The speed of the object is 6.0 meters per second, and the mass of the object is 0.2 kilogram.

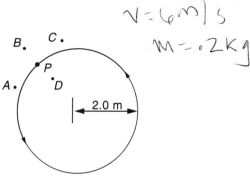

28. When the object is at point P, the direction of the velocity of the object is toward point
    (1) A　　　　(2) B　　　　(3) C　　　　(4) D

29. When the object is at point P, the direction of the acceleration of the object is toward point
    (1) A　　　　(2) B　　　　(3) C　　　　(4) D

30. What is the magnitude of the acceleration of the object?
    (1) 6.0 m/s²　　(2) 12 m/s²　　(3) 3.0 m/s²　　(4) 18 m/s²

31. What is the magnitude of the centripetal force on the object?
    (1) 2.4 N　　　(2) 3.6 N　　　(3) 6.0 N　　　(4) 12 N

32. If the speed of the object increases, the magnitude of the centripetal force needed to keep the object moving in the same circle will
    (1) decrease　　(2) increase　　(3) remain the same

Base your answers to questions 33 through 36 on the diagram below, which represents a car of mass 1000 kilograms traveling around a horizontal circular track of radius 200 meters at a constant speed of 20 meters per second.

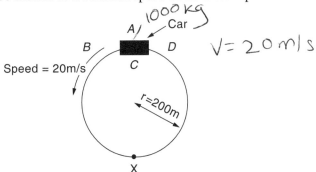

33. When the car is in the position shown, the direction of its centripetal acceleration is toward point
    (1) A　　　　(2) B　　　　(3) C　　　　(4) D

34. If the speed of the car were doubled, the centripetal acceleration of the car would be
    (1) the same　　　　　　(3) one-half as great
    (2) doubled　　　　　　 (4) 4 times as great

35. The magnitude of the centripetal force acting on the car is closest to
    (1) 100 N　　(2) 1000 N　　(3) 2000 N　　(4) 4000 N

36. If additional passengers were riding in the car, at the original speed, the car's centripetal acceleration would be
(1) less   (2) greater   (3) the same

37. If the mass of one of two particles is doubled and the distance between them is doubled, the force of attraction between the two particles will
(1) decrease   (2) increase   (3) remain the same

38. If the mass of an object near the surface of the Earth is increased from M to $3M$, the acceleration of the object due to gravity will be
(1) one-third as great   (3) 3 times as great
(2) 9 times as great   (4) unchanged

39. Gravitational force of attraction F exists between two point masses, A and B, when they are separated by a fixed distance. After mass A is tripled and mass B is halved, the gravitational attraction between the two masses is
(1) 1/6 F   (2) 2/3 F   (3) 3/2 F   (4) 6F

40. A rocket weighs 10,000 newtons at the Earth's surface. If the rocket rises to a height equal to the Earth's radius, its weight will be
(1) 2500 N   (2) 5000 N   (3) 10000 N   (4) 40000 N

41. Compared to the mass of a 10-newton object on the Earth, the mass of the same object on the Moon is
(1) less   (2) greater   (3) the same

42. As a space ship from Earth goes toward the Moon, the force it exerts on the Earth
(1) decreases   (2) increases   (3) remains the same

43. Two masses of 10.0 kilograms and 1.0 kilogram, respectively, are located 1.0 meter apart. The gravitational force that each mass exerts on the other is
(1) 6.7×10^{-9} N   (3) 6.7×10^{-11} N
(2) 6.7×10^{-10} N   (4) 6.7×10^{-12} N

44. The diagram below shows spheres A and B with masses of M and $3M$, respectively.

Mass M Mass $3M$

If the gravitational force of attraction of sphere A on sphere B is 2 newtons, then the gravitational force of attraction of sphere B on sphere A is
(1) 9 N   (2) 2 N   (3) 3 N   (4) 4 N

45. Which graph best represents the gravitational force between two point masses as a function of the distance between the masses?

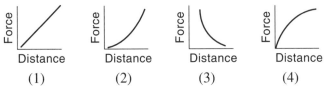

(1)   (2)   (3)   (4)

46. Which graph best represents the relationship between the mass of an object and its distance from the center of the Earth?

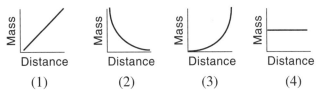

(1)   (2)   (3)   (4)

47. Two objects of fixed mass are moved apart so that they are separated by 3 times their original distance. Compared to the original gravitational force between them, the new gravitational force is
(1) one-third as great   (3) 3 times greater
(2) one-ninth as great   (4) 9 times greater

48. Three equal masses, A, B, and C, are arranged as shown in the diagram.

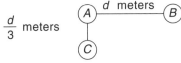

If the gravitational force between A and B is 3 newtons, then the gravitational force between A and C is
(1) 1 N   (2) 9 N   (3) 3 N   (4) 27 N

49. An object has a weight W at the surface of the Earth. At a distance of 3 Earth radii from the center of the Earth, the weight of the object will be
(1) $W/9$ (2) $W/3$ (3) $3W$ (4) $9W$

50. The acceleration due to gravity at a point near the surface of the Moon is 1/6 that near the surface of the Earth. The weight of a 2.0-kilogram mass at that same point near the surface of the Moon is approximately
(1) 1.6 N   (2) 2.0 N   (3) 3.3 N   (4) 0.33 N

51. The gravitational field at a given location is 9.73 newtons per kilogram. The gravitational acceleration at this location is
(1) 4.90 m/s^2 (2) 9.73 m/s^2 (3) 9.81 m/s^2 (4) 19.6 m/s^2

52. A satellite is moving at constant speed in a circular orbit about the Earth, as shown in the diagram below.

The net force acting on the satellite is directed toward point
(1) *A*　　　　　(2) *B*　　　　　(3) *C*　　　　　(4) *D*

53. As the distance between the Moon and Earth increases, the Moon's orbital speed
(1) decreases　　(2) increases　　(3) remains the same

54. The diagram below shows the movement of a planet around the Sun. Area 1 equals area 2.

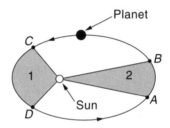

Compared to the time the planet takes to move from *C* to *D*, the time it takes to move from *A* to *B* is
(1) less　　　　(2) greater　　　　(3) the same

55. The path of a satellite orbiting the Earth is best described as
(1) linear　　(2) hyperbolic　　(3) parabolic　　(4) elliptical

56. As the distance of a satellite from the Earth's surface increases, the time the satellite takes to make one revolution around the Earth
(1) decreases　　(2) increases　　(3) remains the same

Base your answers to questions 57 through 60 on the diagram below, which represents a satellite orbiting the Earth. The satellite's distance from the center of the Earth equals 4 Earth radii.

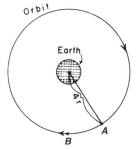

57. Which vector best represents the velocity of the satellite at point *B*?

(1) ↑

(3) ←

(2) ↓

(4) →

58. The original satellite is replaced by one with twice the mass, and the orbit speed and radius are unchanged. Compared to the magnitude of the acceleration of the original satellite, the magnitude of the acceleration of the new satellite is

(1) one-half as great

(3) twice as great

(2) the same

(4) 4 times as great

59. Which vector best represents the acceleration of the satellite at point *A* in its orbit?

(1) ↗

(3) ↘

(2) ↙

(4) ↖

60. If the satellite's distance from the center of the Earth were increased to 5 Earth radii, the centripetal force on the satellite would

(1) decrease (2) increase (3) remain the same

61. The orbital period for a satellite in geosynchronous orbit around the Earth is

(1) 1 hour (2) 1 day (3) 1 month (4) 1 year

62. The shapes of the paths of the planets about the Sun are all

(1) circles with the Sun at the center

(2) circles with the Sun off center

(3) ellipses with the Sun at the center

(4) ellipses with the Sun at one focus

109

63. The diagram below represents the motion of a planet around the Sun, S. The time that the planet takes to go from point 1 to point 2 is identical to the time the planet talem to go from point 3 to point 4. Which statement must be true?

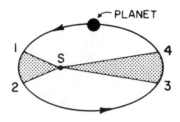

(1) The two shaded regions of the diagram have equal areas.
(2) The centripetal acceleration of the planet is constant.
(3) The planet moves at a constant speed.
(4) The planet moves faster when it is farthest from the Sun.

64. Which diagram best represents the orbit of the planet Pluto around the Sun?

Key:
C = Center of orbit
F = Focus   (not drawn to scale)

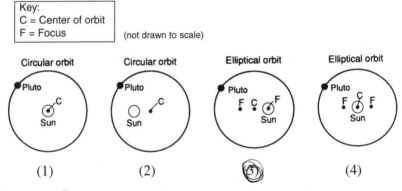

65. Two identical planets are orbiting the Sun. The mean radius of the orbit of the second planet is 4 times the mean radius of the orbit of the first planet. Compared to the orbital period of the first planet, the period of the second planet will be
(1) one-fourth as great
(2) one-half as great
(3) 8 times greater
(4) 4 times greater

66. A satellite is in geosynchronous orbit around the Earth. The period of the satellite's orbit is closest to
(1) 6 h   (2) 12 h   (3) 24 h   (4) 48 h

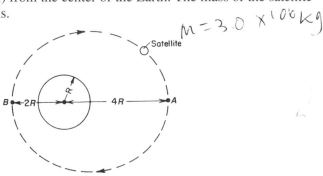

67. For planets orbiting the Sun, the ratio of the mean radius of the orbit cubed to the orbital period of motion squared is
(1) greatest for the most massive planet
(2) greatest for the least massive planet
(3) constantly changing as a planet rotates
(4) the same for all planets

68. A satellite orbits the Earth in a circular orbit. Which statement best explains why the satellite does not move closer to the center of the Earth?
(1) The gravitational field of the Earth does not reach the satellite's orbit.
(2) The Earth's gravity keeps the satellite moving with constant velocity.
(3) The satellite is always moving perpendicularly to the force due to gravity.
(4) The satellite does not have any weight.

69. Which condition is required for a satellite to be in a geosynchronous orbit about the Earth?
(1) The period of revolution of the satellite must be the same as the rotational period of the Earth.
(2) The altitude of the satellite must be equal to the radius of the Earth.
(3) The orbital speed of the satellite around the Earth must be the same as the orbital speed of the Earth around the Sun.
(4) The daily distance traveled by the satellite must be equal to the circumference of the Earth.

Base your answers to questions 70 through 73 on the diagram below which represents a satellite in an elliptical orbit about the Earth. The highest point, A, is four Earth radii ($4R$) from the center of the Earth. The lowest point, B, is two Earth radii ($2R$) from the center of the Earth. The mass of the satellite is 3.0×10^6 kilograms.

70. Which vector represents the direction of the satellite's velocity at point *A*?

71. Which vector represents the direction of the centripetal force on the satellite at point *B*?

72. As the satellite moves from point *A* toward point *B*, the velocity of the satellite
(1) decreases   (2) increases   (3) remains the same

73. Compared to the magnitude of the force of the satellite on the Earth, the magnitude of the force of the Earth on the satellite is
(1) less   (2) greater   (3) the same

Base your answers to questions 74 through 77 on the diagram and information below.

Two satellites having masses m_1 and m_2 are moving in circular orbits around the Earth. The Earth has a mass of m_E.

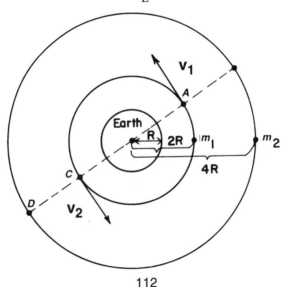

[Assume g to be the acceleration due to gravity at the surface of the Earth.]

74. The gravitational force acting on mass m_1 is best represented by

(1) m_1G 　　 (2) $\dfrac{Gm_Em_1}{R^2}$ 　　 (3) $\dfrac{Gm_E}{4\,R^2}$ 　　 (4) $\dfrac{Gm_Em_1}{4\,R^2}$

75. The acceleration due to gravity of mass m_2 is best represented by

(1) $\dfrac{Gm_2}{16\,R^2}$ 　　 (2) $\dfrac{Gm_E}{R^2}$ 　　 (3) $\dfrac{Gm_E}{16\,R^2}$ 　　 (4) $\dfrac{Gm_2}{4\,R^2}$

76. What is the total change in the velocity of mass m_1 as it moves from point A to point C?

(1) 0 　　 (2) $\mathbf{v}_1$ 　　 (3) $2\mathbf{v}_1$ 　　 (4) $2\mathbf{v}_2$

77. The speed of mass m_1 is best expressed by

(1) Gt 　　 (2) $\sqrt{\dfrac{Gm_E}{2\,R}}$ 　　 (3) $\dfrac{Gm_E}{4\,R^2}$ 　　 (4) $\dfrac{G}{4}t$

Chapter Six

MOMENTUM AND ITS CONSERVATION

KEY IDEAS

Momentum is the product of mass and velocity. It is a vector quantity that measures the tendency of an object to remain in motion. An *impulse* is the product of force and time and is responsible for changing the momentum of an object. It is also a vector quantity.

If two objects interact (and are not subject to external forces) the momentum of the system (i.e., the two objects) is conserved. This basic law of physics can be used to describe collision and explosion phenomena.

KEY OBJECTIVES

At the conclusion of this chapter you will be able to:
- Define the term *momentum*, and state its SI unit.
- Solve problems involving mass, velocity, and momentum.
- Define the term *impulse*, and state its SI unit.
- Relate impulse to change in momentum.
- Solve impulse-momentum problems.
- State the law of conservation of momentum, and solve problems based on this law.
- Relate the law of conservation of momentum to Newton's third law of motion.

6.1 MOMENTUM

Suppose we wanted to measure the "tendency" of an object to remain in motion. We would find that two factors are necessary to describe this tendency: the mass and the velocity of the object. The more mass an object has, the more force is required to bring it to rest. Similarly, the greater the velocity of the object, the more force is necessary to bring it to rest. These two factors, mass and velocity, are combined into a single quantity that we call **momentum**. We define momentum, symbolized as **p**, as the product of mass and velocity:

$$\mathbf{p} = m\mathbf{v}$$

Momentum is a vector quantity and its direction is the direction of the velocity of the object. Its unit is the *kilogram · meter per second* (kg · m/s).

PROBLEM

An object whose mass is 3.5 kilograms is traveling at 20 meters per second [east]. Calculate the momentum of the object.

SOLUTION

$$\mathbf{p} = m\mathbf{v}$$
$$= (3.5 \text{ kg})(20\text{m/s [E]})$$
$$= 70. \text{ kg} \cdot \text{m/s [E]}$$

6.2 NEWTON'S SECOND LAW AND MOMENTUM

When Newton developed his second law of motion, he recognized that the unbalanced force on an object caused a change in the object's momentum:

$$\mathbf{F}_{net} = \frac{\Delta \mathbf{p}}{\Delta t} = \frac{\Delta(m\mathbf{v})}{\Delta t}$$

Since mass is usually constant:

$$\mathbf{F}_{net} = \frac{m\Delta \mathbf{v}}{\Delta t} = m\mathbf{a}$$

We can rewrite Newton's second law in the following form:

$$\mathbf{F}_{net} \Delta t = \Delta \mathbf{p}$$

We call the quantity F Δt, the **impulse** delivered to the object. Impulse is symbolized by the letter **J**, and its unit is the *newton · second* (N · s). Impulse is a vector quantity, and its direction is the direction of the net force.

From Newton's second law it follows that the impulse delivered to the object changes its momentum, and the unit newton · second is equivalent to the unit kilogram · meter per second.

PROBLEM

A 5-kilogram object traveling at 3 meters per second [east] is subjected to a force that increases its velocity to 7 meters per second [east]. Calculate: (a) the initial momentum of the object, (b) the final momentum of the object, (c) the change in momentum of the object, and (d) the impulse delivered to the object. (e) If the force acts for 0.2 second, what are its magnitude and its direction?

SOLUTION

(a) $\mathbf{p}_i = m\mathbf{v}_i$
$= (5 \text{ kg})(3 \text{ m/s [E]})$
$= 15 \text{ kg} \cdot \text{m/s [E]}$

(b) $\mathbf{p}_f = m\mathbf{v}_f$
$= (5 \text{ kg}) (7 \text{ m/s [E]})$
$= 35 \text{ kg} \cdot \text{m/s [E]}$

(c) $\Delta\mathbf{p} = \mathbf{p}_f - \mathbf{p}_i$
$= 35 \text{ kg} \cdot \text{m/s [E]} - 15 \text{ kg} \cdot \text{m/s [E]}$
$= 20 \text{ kg} \cdot \text{m/s [E]}$

(d) $\mathbf{J} = \mathbf{F} \Delta t = \Delta\mathbf{p}$
$= 20 \text{ N} \cdot \text{s [E]}$

(e) $\mathbf{J} = \mathbf{F} \Delta t$

$\mathbf{F} = \dfrac{\mathbf{J}}{\Delta t}$

$= \dfrac{20 \text{ N} \cdot \text{s [E]}}{0.2 \text{ s}}$

$= 100 \text{ N [E]}$

6.3   CONSERVATION OF MOMENTUM

If two objects that are not subjected to any external forces interact (e.g., they collide), the total momentum of the objects before the interaction is equal to their total momentum after the interaction:

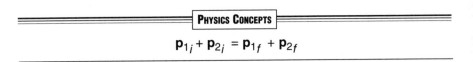

PHYSICS CONCEPTS

$$\mathbf{p}_{1i} + \mathbf{p}_{2i} = \mathbf{p}_{1f} + \mathbf{p}_{2f}$$

This relationship, known as *conservation of momentum*, is a fundamental law of physics. The diagram below illustrates the conservation of momentum.

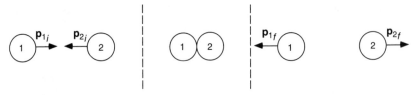

PROBLEM

A 0.5-kilogram object traveling at 2 meters per second [east] collides with a 0.3-kilogram object traveling at 4 meters per second [west]. After the collision, the 0.3-kilogram object is traveling at 2 meters per second [east]. What are the magnitude and the direction of the velocity of the first object?

SOLUTION

$$\mathbf{p}_{1i} + \mathbf{p}_{2i} = \mathbf{p}_{1f} + \mathbf{p}_{2f}$$

$$m_1\mathbf{v}_{1i} + m_2\mathbf{v}_{2i} = m_1\mathbf{v}_{1f} + m_2\mathbf{v}_{2f}$$

$$\mathbf{v}_{1f} = \frac{m_1\mathbf{v}_{1i} + m_2\mathbf{v}_{2i} - m_2\mathbf{v}_{2f}}{m_1}$$

$$= \frac{(0.5 \text{ kg})(2 \text{ m/s}) + (0.3 \text{ kg})(-4 \text{ m/s}) - (0.3 \text{ kg})(2 \text{ m/s})}{0.5 \text{ kg}}$$

$$= -1.6 \text{ m/s} = 1.6 \text{ m/s [W]}$$

Consider a gun about to fire a bullet. The initial momentum of the gun-bullet system is zero. Since the bullet moves in one direction, the gun recoils in the opposite direction, so that the momentum after the "explosion" is also zero.

PROBLEM

A 5.0-kilogram gun fires a 0.002-kilogram bullet. If the bullet exits the gun at 800 meters per second [east], calculate the recoil velocity of the gun.

SOLUTION

$$\mathbf{p}_{1i} + \mathbf{p}_{2i} = \mathbf{p}_{1f} + \mathbf{p}_{2f}$$

$$0 = \mathbf{p}_{1f} + \mathbf{p}_{2f}$$

$$0 = m_1\mathbf{v}_{1f} + m_2\mathbf{v}_{2f}$$

$$\mathbf{v}_{1f} = \frac{0 - m_2\mathbf{v}_{2f}}{m_1}$$

$$= \frac{0 - (0.0020 \text{ kg})(+800. \text{ m/s})}{5.0 \text{ kg}}$$

$$= -0.32 \text{ m/s} = 0.32 \text{ m/s [W]}$$

6.4 CONSERVATION OF MOMENTUM AND NEWTON'S THIRD LAW

We can rewrite the law of conservation of momentum as follows:

$$\mathbf{p}_{1i} + \mathbf{p}_{2i} = \mathbf{p}_{1f} + \mathbf{p}_{2f}$$

$$-(\mathbf{p}_{2f} - \mathbf{p}_{2i}) = (\mathbf{p}_{1f} - \mathbf{p}_{1i})$$

$$-\Delta\mathbf{p}_2 = \Delta\mathbf{p}_1$$

The new equation means that the objects undergo equal but opposite changes in momentum. Because momentum change is equal to impulse, the two objects deliver equal but opposite impulses to one another in the same time interval. *Therefore, they must exert equal and opposite forces on one another.*

Does this seem familiar? We have just shown that Newton's third law is a direct consequence of the law of conservation of momentum!

QUESTIONS

1. If the direction of the momentum of an object is west, the direction of the velocity of the object is
   (1) north   (2) south   (3) east   (4) west

2. A 2-newton force acts on a mass. If the momentum of the mass changes by 120 kilogram · meters per second, the force acts for a time of
   (1) 8 s   (2) 30 s   (3) 60 s   (4) 120 s

3. As an object falls freely toward the Earth, its momentum
   (1) decreases   (2) increases   (3) remains the same

4. A 50-kilogram mass has a momentum of 100 kilogram · meters per second. What is the velocity of the mass?
   (1) 0.5 m/s   (2) 2 m/s   (3) 2500 m/s   (4) 5000 m/s

5. An impulse $\mathbf{J}$ is applied to an object. The change in the momentum of the object is
   (1) $\mathbf{J}$ (2) $2\mathbf{J}$ (3) $\dfrac{\mathbf{J}}{2}$ (4) $4\mathbf{J}$

6. The momentum of an object is the product of its
   (1) mass and acceleration
   (2) mass and velocity
   (3) force and displacement
   (4) force and distance

7. Momentum may be expressed in
   (1) joules
   (2) watts
   (3) kilogram · meters per second^2
   (4) Newton · seconds

8. If a 3.0-kilogram object moves 10. meters in 2.0 seconds, its average momentum is
   (1) 60. kg · m/s
   (2) 30. kg · m/s
   (3) 15 kg · m/s
   (4) 10. kg · m/s

9. The direction of an object's momentum is always the same as the direction of the object's
   (1) inertia
   (2) potential energy
   (3) velocity
   (4) weight

10. A mass is moving with a uniform velocity. If the time required to stop the mass increases, the force required to stop it must
    (1) decrease   (2) increase   (3) remain the same

11. A force of 3.0 newtons applied to an object produces a change in velocity of 12 meters per second in 0.40 second. The mass of the object is
    (1) 1.0 kg   (2) 0.10 kg   (3) 10. kg   (4) 0.31 kg

12. An object traveling at 4.0 meters per second has a momentum of 16 kilogram · meters per second. What is the mass of the object?
    (1) 64 kg   (2) 20 kg   (3) 12 kg   (4) 4.0 kg

13. What is the magnitude of the velocity of a 25-kilogram mass that is moving with a momentum of 100. kilogram · meters per second?
    (1) 0.25 m/s   (2) 2500 m/s   (3) 40. m/s   (4) 4.0 m/s

14. A 15-newton force acts on an object in a direction due east for 3.0 seconds. What will be the change in momentum of the object?
    (1) 45 kg · m/s [E]
    (2) 45 kg · m/s [W]
    (3) 5.0 kg · m/s [E]
    (4) 0.20 kg · m/s [W]

15. An unbalanced 6.0-newton force acts eastward on an object for 3.0 seconds. The impulse produced by the force is
    (1) 18 N · s [E]
    (2) 2.0 N · s [E]
    (3) 18 N · s [W]
    (4) 2.0 N · s [W]

16. As the unbalanced force applied to an object increases, the time rate of change of the object's momentum
    (1) decreases   (2) increases   (3) remains the same

17. A rocket with a mass of 1000 kilograms is moving at a speed of 20 meters per second. The magnitude of the momentum is
    (1) 50 kg · m/s   (3) 20,000 kg · m/s
    (2) 200 kg · m/s   (4) 400,000 kg · m/s

18. An impulse of 30.0 newton · seconds is applied to a 5.00-kilogram mass. If the mass had a speed of 100. meters per second before the impulse, its speed after the impulse could be
    (1) 250. m/s   (2) 106 m/s   (3) 6.00 m/s   (4) 0 m/s

19. What is the magnitude of the change in momentum produced when a force of 5.0 newtons acts on a 10-kilogram object for 3.0 seconds?
    (1) 1.5 kg · m/s   (3) 10 kg · m/s
    (2) 5.0 kg · m/s   (4) 15 kg · m/s

20. An object is brought to rest by a constant force. Which factor other than the mass and velocity of the object must be known in order to determine the magnitude of the force required to stop the object?
    (1) the time that the force acts on the object
    (2) the gravitational potential energy of the object
    (3) the density of the object
    (4) the weight of the object

21. A constant unbalanced force acts on an object for 3.0 seconds, producing an impulse of 6.0 newton · seconds. What is the magnitude of the force?
    (1) 6.0 N   (2) 2.0 N   (3) 3.0 N   (4) 18 N

22. A car with a mass of 1.0×10^3 kilograms is moving with a speed of 1.4×10^2 meters per second. The impulse required to bring the car to rest is
    (1) 1.4×10^2 N · s   (3) 7.0×10^4 N · s
    (2) 1.4×10^4 N · s.   (4) 1.4×10^5 N · s.

Base your answers to questions 23 through 26 on the information and graph below.

A force acts on a 3.0-kg. cart that is on a frictionless horizontal surface. The resulting motion is shown by the graph.

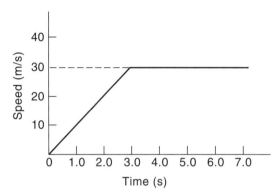

23. How long does the force act on the cart?
 (1) 1.0 s (2) 2.0 s (3) 3.0 s (4) 7.0 s

24. What is the acceleration of the cart during the first 2 seconds?
 (1) 1.0 m/s^2 (2) 0 m/s^2 (3) 4.5 m/s^2 (4) 10. m/s^2

25. What is the momentum of the cart at the end of 6 seconds?
 (1) 10. N · s (2) 90. N · s (3) 210 N · s (4) 630 N · s

26. What is the total distance traveled by the cart during 7 seconds?
 (1) 45 m (2) 105 m (3) 165 m (4) 210 m

27. When two objects collide, there will be no net change in the
 (1) velocity of each object
 (2) displacement of each object
 (3) kinetic energy of each object
 (4) total momentum of the objects

28. An 80.-kilogram skater and a 60.-kilogram skater stand at rest in the center of a skating rink. The two skaters push each other apart. The 60.-kilogram skater moves with a velocity of 10. meters per second east. What is the velocity of the 80.-kilogram skater? [Neglect any frictional effects.]
 (1) 0.13 m/s [W] (3) 10. m/s [E]
 (2) 7.5 m/s [W] (4) 13. m/s [E]

29. A 1.0-kilogram object moving east with a velocity of 10 meters per second collides with a 0.50-kilogram object that is at rest. Neglecting friction, what is the momentum of the system after the collision?
(1) 15 kg · m/s [E]   (3) 10 kg · m/s [E]
(2) 15 kg · m/s [W]   (4) 10 kg · m/s [W]

30. Two carts having masses of 5.0 kilograms and 1.0 kilogram, respectively, are pushed apart by a compressed spring. If the 5.0-kilogram cart moves westward at 2.0 meters per second, the magnitude of the velocity of the 1.0-kilogram cart will be
(1) 2.0 kg · m/s   (3) 10. kg · m/s
(2) 2.0 m/s   (4) 10. m/s

31. A 20.-kilogram cart traveling east with a speed of 6.0 meters per second collides with a 30.kilogram cart traveling west. If both carts come to rest after the collision, what was the speed of the westbound cart before the collision?
(1) 0 m/s   (2) 9.0 m/s   (3) 3.0 m/s   (4) 4.0 m/s

32. As a ball falls freely toward the Earth, the momentum of the Earth-ball system
(1) decreases   (2) increases   (3) remains the same

33. A cart has a momentum of 10. kilogram · meters per second east before a collision and a momentum of 5.0 kilogram · meters per second west after the collision. The net change in momentum of this cart is
(1) 5.0 kg · m/s [E]   (3) 15. kg · m/s [E]
(2) 5.0 kg · m/s [W]   (4) 15. kg · m/s [W]

34. A projectile with a mass of 0.01 kilogram has a muzzle velocity of 1000 meters per second when fired from a rifle weighing 5 kilograms. The recoil velocity of the rifle is
(1) 0.1 m/s   (2) 2 m/s   (3) 5 m/s   (4) 50 m/s

Base your answers to questions 35 through 39 on the diagram below which represents carts *A* and *B* being pushed apart by a spring that exerts an average force of 50. newtons for a period of 0.20 second. [Assume frictionless conditions.]

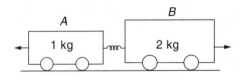

35. What is the magnitude of the impulse applied by the spring on cart *A*?
(1) 5.0 N · s   (2) 10. N · s   (3) 50. N · s   (4) 100 N · s

122

36. Compared to the magnitude of the impulse acting on cart A, the magnitude of the impulse acting on cart B is
(1) one-half as great   (3) the same
(2) twice as great   (4) 4 times as great

37. Compared to the velocity of cart B at the end of the 0.20-second interaction, the velocity of cart A is
(1) one-half as great   (3) the same
(2) twice as great   (4) 4 times as great

38. What is the average acceleration of cart B during the 0.20-second interaction?
(1) 0 m/s^2 (2) 10. m/s^2 (3) 25 m/s^2 (4) 50. m/s^2

39. Compared to the total momentum of the carts before the spring is released, the total momentum of the carts after the spring is released is
(1) one-half as great   (3) the same
(2) twice as great   (4) 4 times as great

Base your answers to questions 40 and 41 on the information and diagram below.

A 0.4-kilogram toy cannon is at rest on a horizontal, frictionless surface. When a 0.1-kilogram projectile is fired horizontally from the barrel of the cannon, the cannon recoils with a speed of 2.5 meters per second.

40. The speed of the projectile as it leaves the barrel is
(1) 0.125 m/s   (2) 2.5 m/s   (3) 10.0 m/s   (4) 4.0 m/s

41. The total change in the momentum of the cannon as it is fired is
(1) 1.0 N · s   (2) 0.25 N · s   (3) 0.6 N · s   (4) 2.9 N · s

42. A 2-kilogram car and a 3-kilogram car are originally at rest on a horizontal, frictionless surface as shown in the diagram below. A compressed spring is released, causing the cars to separate. The 3-kilogram car reaches a maximum speed of 2 meters per second. What is the maximum speed of the 2-kilogram car?

   (1) 1 m/s   (2) 2 m/s   (3) 3 m/s   (4) 6 m/s

43. A 2.0-kilogram cart traveling east with a speed of 6 meters per second collides with a 3.0-kilogram cart traveling west. If both carts come to rest immediately after the collision, what was the speed of the westbound cart before the collision?
   (1) 6 m/s   (2) 2 m/s   (3) 3 m/s   (4) 4 m/s

44. Two carts resting on a frictionless surface are forced apart by a spring. One cart has a mass of 6 kilograms and moves to the left at a speed of 3 meters per second. If the second cart has a mass of 9 kilograms, it will move to the right at a speed of
   (1) 1 m/s   (2) 2 m/s   (3) 3 m/s   (4) 6 m/s

45. A 2.0-kilogram rifle initially at rest fires a 0.002-kilogram bullet. As the bullet leaves the rifle with a velocity of 500 meters per second, what is the momentum of the rifle-bullet system?
   (1) 2.5 kg · m/s   (3) 0.5 kg · m/s
   (2) 2.0 kg · m/s   (4) 0 kg · m/s

46. A 4.0-kilogram mass is moving at 3.0 meters per second toward the right, and a 6.0-kilogram mass is moving at 2.0 meters per second toward the left, on a horizontal, frictionless table. If the two masses collide and remain together after the collision, their final momentum is
   (1) 1.0 kg · m/s   (3) 12 kg · m/s
   (2) 24 kg · m/s   (4) 0 kg · m/s

WORK AND ENERGY

Chapter
Seven

KEY IDEAS

Work is the product of force and the component of displacement in the direction of the force; work is a scalar quantity. Without motion there can be no work.

Power is the rate at which work is done, and it is also a scalar quantity. If work is done on an object, the work may be used to change the object's kinetic energy (the energy associated with its motion), its potential energy (the energy associated with its position), or its internal energy (the energy associated with its atoms and molecules).

An elastic collision is one in which momentum and kinetic energy are conserved. When gas molecules collide with the walls of a container, these collisions are very nearly elastic.

Simple machines provide examples of how work can be used to a person's advantage in performing such chores as lifting heavy objects and exerting large forces that can accomplish a task such as cutting through steel.

KEY OBJECTIVES

At the conclusion of this chapter you will be able to:

- Define the following terms: *kinetic energy; gravitational potential energy; elastic potential energy; partially inelastic collision; totally inelastic collision; elastic collision; simple machine; ideal mechanical advantage, actual mechanical advantage;* and *efficiency.*
- Define the term *work*, and state its SI unit.
- Solve problems involving force, displacement, and work.
- Define the term *power*, and state its SI unit.
- Solve problems involving power and work.
- State the equation for calculating kinetic energy, and solve problems using this equation.
- State the equation for calculating gravitational potential energy, and solve problems using this equation.
- Solve problems that relate changes in kinetic energy to changes in gravitational potential energy.

- State the equation for calculating the elastic potential energy of a spring, and solve problems using this equation.
- Solve problems involving the inclined plane as a simple machine.
- List other types of simple machines.

7.1 WORK

Would you pay a person $5 an hour to do this? The typical answer is "No!" because the person isn't doing any work. What exactly do we mean by the term *work*? In physics, **work** is defined as the product of force and displacement, assuming that both are in the *same direction*.

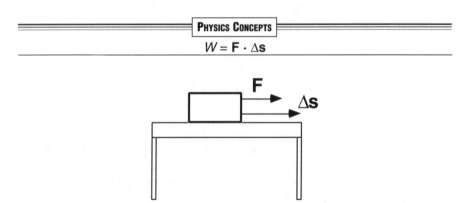

PHYSICS CONCEPTS
$W = \mathbf{F} \cdot \Delta \mathbf{s}$

Even though force and displacement are vector quantities, work is a scalar quantity. The unit of work is the *newton · meter*, which is called a joule (**J**) in honor of English scientist James Prescott Joule.

PROBLEM
How much work is done on an object if a force of 30 newtons [south] displaces the object 200 meters [south]?

SOLUTION
$$W = \mathbf{F} \cdot \Delta\mathbf{s}$$
$$= (30 \text{ N[S]})(200 \text{ m [S]})$$
$$= 6000 \text{ J}$$

Now we will suppose that the force and the displacement are not in the same direction. We define work to be the product of the component of the force in the direction of the displacement and the displacement.
    The diagram illustrates this relationship.

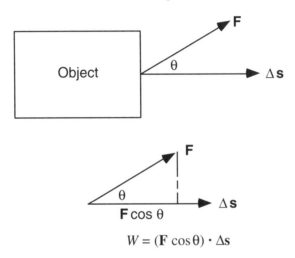

$$W = (\mathbf{F} \cos\theta) \cdot \Delta\mathbf{s}$$

PROBLEM
As Alex pulls his red wagon down the sidewalk, the handle of the wagon makes an angle of 60° with the pavement. If Alex exerts a force of 100 newtons along the direction of the handle, how much work is done when the displacement of the wagon is 20 meters along the ground?

SOLUTION

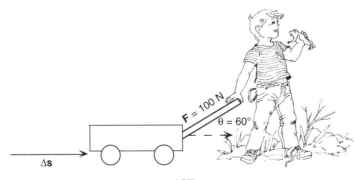

127

$$W = (\mathbf{F} \cos\theta) \cdot \Delta\mathbf{s}$$
$$= (100 \text{ N})(\cos 60°)(20 \text{ m})$$
$$= (50 \text{ N})(20 \text{ m})$$
$$= 1000 \text{ J}$$

PROBLEM

A constant force of 50 newtons is applied over a distance of 10 meters.
(a) Prepare a graph of force versus displacement.
(b) Calculate the area underneath the curve.
(c) Calculate the work done.

SOLUTION

(a)

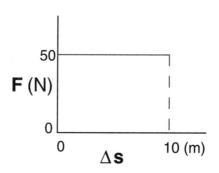

(b) *Area = base · height*
    $= (10 \text{ m})(50 \text{ N})$
    $= 500 \text{ J}$

(c) $W = \mathbf{F} \cdot \Delta\mathbf{s}$
    $= (50 \text{ N})(10 \text{ m})$
    $= 500 \text{ J}$

It is evident that the area under the force-versus-displacement curve, shown below, is equal to the work done on the object.

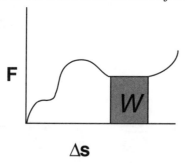

In general, the area under *any* force-versus-displacement curve is equal to the work done.

7.2   POWER

Power is a term that is frequently misused. We define **power** as the *rate* at which work is done:

PHYSICS CONCEPTS

$$P = \frac{W}{\Delta t}$$

Power is also a scalar quantity, and its unit is joules per second (J/s), also known as the watt (W).

PROBLEM
If 3000 joules of work is performed on an object in 1.0 minute, what is the power expended on the object?

SOLUTION

$$P = \frac{W}{\Delta t}$$

$$= \frac{3000 \text{ J}}{60 \text{ s}}$$

$$= 50 \text{ W}$$

PROBLEM
A 200-newton force is applied to an object that moves in the direction of the force. If the object travels with a constant velocity of 10 meters per second, calculate the power expended on the object.

SOLUTION

$$P = \frac{W}{\Delta t} = \frac{\mathbf{F} \cdot \Delta \mathbf{s}}{\Delta t} = \mathbf{F} \cdot \bar{\mathbf{v}}$$

$$= (200 \text{ N})(10 \text{ m/s})$$

$$= 2000 \text{ W}$$

7.3   ENERGY

When work is done on an object, the "energy" of the object is changed. *Energy* is a very broad term related to work, and it has a variety of forms. In this chapter we will consider two of these forms, kinetic and potential energy. Together the kinetic energy and the potential (gravitational and elastic) energies are called the *mechanical energy* of the object.

Kinetic Energy

An object is traveling along a frictionless horizontal surface. A constant force is applied to the object in the direction of its displacement. What is the result of the work done on the object?

We know from Newton's second law that $F = ma$, therefore $W = ma \cdot \Delta s$. If the force on the object is also constant, its acceleration is constant and we can write

$$v_f^2 = v_i^2 + 2a \cdot \Delta s$$

If we solve this relationship for $a \cdot \Delta s$, we find that

$$a \cdot \Delta s = \frac{v_f^2 - v_i^2}{2}$$

We can substitute this solution for $\mathbf{a} \cdot \Delta \mathbf{s}$ into our work relationship:

$$W = ma \cdot \Delta s$$

$$= m \left(\frac{v_f^2 - v_i^2}{2} \right)$$

$$= \frac{1}{2} m v_f^2 - \frac{1}{2} m v_i^2$$

The work done on the object changes a quantity called kinetic energy that is related to the mass and the square of the speed of the object. We define kinetic energy (KE) to be:

PHYSICS CONCEPTS

$$KE = \frac{1}{2} mv^2$$

Therefore, $W = KE_f - KE_i = \Delta KE$.

PROBLEM

A 10-kilogram object subjected to a 20-newton force moves across a horizontal, frictionless surface in the direction of the force. Before the force was applied, the speed of the object was 2 meters per second. When the force is removed, the object is traveling at 6 meters per second. Calculate the following quantities: (a) KE_i, (b) KE_f, (c) ΔKE, (d) W, and (e) Δs.

SOLUTION

(a) $KE_i = \frac{1}{2} m v_i^2$

$= \frac{1}{2}(10 \text{ kg})(2 \text{ m/s})^2$

$= 20 \text{ J}$

(b) $KE_f = \frac{1}{2} m v_f^2$

$= \frac{1}{2}(10 \text{ kg})(6 \text{ m/s})^2$

$= 180 \text{ J}$

(c) $\Delta KE = KE_f - KE_i$

$\qquad = 180\ \text{J} - 20\ \text{J}$

$\qquad = 160\ \text{J}$

(d) $W = \Delta KE$

$\qquad = 160\ \text{J}$

(e) $\qquad W = F \cdot \Delta s$

$$\Delta s = \frac{W}{F}$$

$$= \frac{160\ \text{J}}{20\ \text{N}}$$

$$= 8\ \text{m}$$

Gravitational Potential Energy

Darya lifts a textbook vertically off a desk and holds it above her head. It is clear that Darya has done work on the book because she has applied a force through a distance. However, the change in the kinetic energy of the book is zero. What did Darya's work accomplish?

In this case, the work overcame the attraction of the gravitational field, and as a result the position of the book with respect to the Earth changed. We relate this change to a quantity we call **gravitational potential energy**.

To calculate the change in the gravitational potential energy (PE) of the object, we measure the work done on the object. The force needed to overcome gravity is simply F_W, which is equal to mg. Therefore, since $W = F_W \cdot \Delta s$, we define the change in the gravitational potential energy (ΔPE) as:

PHYSICS CONCEPTS

$$\Delta PE = mg\Delta h$$

Here we use Δh, rather than Δs, to represent the change in vertical displacement above the Earth. Since we are dealing with a scalar quantity, we will not consider the algebraic signs of the quantities involved. We will simply agree that an object decreases its gravitational potential energy as it moves closer to the Earth (and vice versa).

PROBLEM

A 2.0-kilogram mass is lifted to a height of 10. meters above the surface of the Earth. Calculate the change in the gravitational potential energy of the object.

131

SOLUTION

$$\Delta PE = mg\,\Delta h$$
$$= (2.0\ kg)\,(9.8\ \text{m/s}^2)(10\ m)$$
$$= 196\ J$$

Since the object has moved *away* from the Earth's surface, its gravitational potential energy has *increased* by 196 joules.

For a change in gravitational energy to occur, there must be a change in the *vertical* displacement of an object; if it is moved only *horizontally*, its gravitational potential energy change is zero. If an object is moved up an inclined plane, its potential energy change is measured by calculating only its *vertical* displacement; the horizontal part of its displacement does not change its potential energy.

Interaction of Gravitational Potential and Kinetic Energy

Ketan tosses an object upward, and it returns to the Earth. Let's analyze the motion of this object using an "energy" point of view. For simplicity we will ignore air resistance.

As the object rises, we observe that its speed decreases to zero. As a result, the kinetic energy of the object is decreasing while its potential energy is increasing. This represents a tranformation of energy from kinetic to potential.

On the downward trip, the speed of the object increases. As the potential energy of the object decreases, its kinetic energy increases. This represents a transformation from potential to kinetic energy.

In the system, the sum of potential energy and kinetic energy has been conserved; a change in one is accompanied by an opposite change in the other.

$$\Delta PE = -\Delta KE$$
$$PE_i + KE_i = PE_f + KE_f$$

PROBLEM
A 0.5-kilogram ball is projected vertically and rises to a height of 2.0 meters above the ground. Calculate: (a) the increase in the ball's potential energy, (b) the decrease in the ball's kinetic energy, (c) the initial kinetic energy, and (d) the intitial speed of the ball.

SOLUTION

(a) $\Delta PE = mg\Delta h$
$\quad = (0.5\ \text{kg})(9.8\ \text{m/s}^2)(2.0\ \text{m})$
$\quad = 9.8\ \text{J}$

(b) $\Delta KE = -\Delta PE$
$\quad = -9.8\ \text{J}$

(c) At the highest point, the speed of the ball is zero; therefore its kinetic energy is zero. As a result, the initial kinetic energy represents the change in the kinetic energy of the object.

$$\Delta KE = KE_f - KE_i$$
$$-9.8\ \text{J} = 0 - KE_i$$
$$KE_i = 9.8\ \text{J}$$

(d) $KE_i = \dfrac{1}{2}mv_i{}^2$

Solving for v_i gives

$$v_i = \sqrt{\frac{2KE_i}{m}}$$

$$\quad = \sqrt{\frac{2(9.8\ \text{J})}{0.5\ \text{kg}}}$$

$$\quad = 6.3\ \text{m/s}$$

An amusement-park roller coaster is an example of the interchange of the kinetic and potential energies of the coaster car. As the car falls, its kinetic energy increases; as it rises, its kinetic energy decreases, as shown in the diagram.

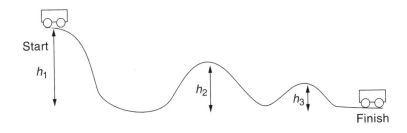

Subsequent hills are made shorter and shorter so that the car will continue to have kinetic energy as it moves along the track.

A simple pendulum, shown in the following diagram, is another device that illustrates the transformation between kinetic and potential energies.

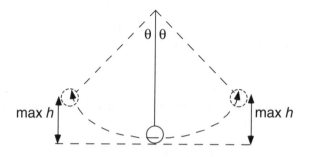

In the absence of friction, the swing of a pendulum back and forth will go on continuously. This type of motion, known as *simple harmonic motion* (SHM), occurs often in nature. For example, the oscillation of a spring and the vibration of a tuning fork are examples of SHM.

The time needed to complete one full swing of the pendulum is known as its period (*T*). The period of a pendulum (for small angles) is related to the length of the pendulum (ℓ) and to gravitational acceleration (*g*) according to the relationship

$$T = 2\pi \sqrt{\frac{\ell}{g}}$$

→ length not string
→ gravity

Note that the period of a simple pendulum is independent of the mass of the bob.

PROBLEM

A pendulum whose bob weighs 12 newtons is lifted a vertical height of 0.4 meter from its equilibrium position. Calculate: (a) the change in potential energy between maximum height and equilibrium height, (b) the gain in kinetic energy, and (c) the velocity at the equilibrium point.

SOLUTION

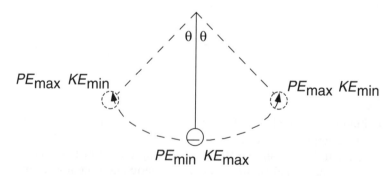

(a) We take the potential energy at the lowest point to be zero. Then:

$$\Delta PE = mg\Delta h = F_w\Delta h$$
$$= (12\ N)(-0.4\ m)$$
$$= -4.8\ J$$

(b) $\Delta KE = -\Delta PE$
$$= -(-4.8\ J)$$
$$= +4.8\ J$$

(c) First we calculate the mass of the bob:
$$F_w = mg$$

$$m = \frac{F_w}{g}$$

$$= \frac{12\ N}{9.8\ m/s^2}$$

$$= 1.2\ kg$$

Then we calculate the velocity:

$$KE = \frac{1}{2}mv^2$$

$$v = \sqrt{\frac{2KE}{m}}$$

$$= \sqrt{\frac{2(4.8\ J)}{1.2\ kg}}$$

$$= 2.8\ m/s^2$$

Elastic Potential Energy and Springs

Consider the arrangement shown in the diagram:

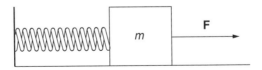

Here a spring is attached to a wall and to a mass resting on a horizontal, frictionless table. If we apply a force on the mass and displace it to the right, we have done work. This work has been converted into the spring's potential energy, a quantity we call **elastic potential energy**.

We can calculate the work done in stretching the spring as follows. We know that springs obey Hooke's law, a graph of which is shown below.

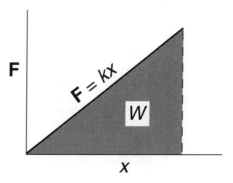

Since the area under this graph equals the work done, and we know that the area of a triangle is equal to 1/2 (base · height), we use the equations

$$W = \frac{1}{2} x \cdot F \text{ and } F = kx$$

Then we have

$$W = \frac{1}{2} x \cdot kx = \frac{1}{2} kx^2$$

and the potential energy of the spring (PE_s) is given by:

PHYSICS CONCEPTS

$$PE_s = \frac{1}{2} kx^2$$

PROBLEM

A spring whose constant is 2.0 newtons per meter is stretched 0.40 meter from its equilibrium position. What is the increase in the elastic potential energy of the spring?

SOLUTION

$$\Delta PE_s = \frac{1}{2} kx^2$$

$$= \frac{1}{2} \left(2.0 \frac{N}{m} \right) (0.40 \text{ m})^2$$

$$= 0.16 \text{ J}$$

What would happen if we released the spring after stretching it? The force exerted on the mass by the spring (the restoring force) would displace the mass toward the wall (to the left). The mass would then overshoot its equilibrium position and compress the spring. In turn, the spring would exert a force on the mass away from the wall (to the right). Since friction is absent, this back-and-forth motion, namely, SHM would continue indefinitely. As the spring moved back and forth, there would be a continual exchange between kinetic and potential energies, as shown in the diagram.

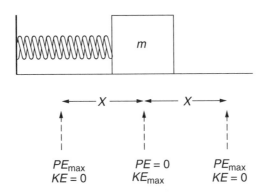

Elastic and Inelastic Collisions

Imagine a ball bouncing repeatedly on a sidewalk, as illustrated below.

After each successive bounce, the ball's height above the ground diminishes; eventually the ball comes to rest on the ground.

When the ball is on the ground, part of its kinetic energy is lost and there is an incomplete conversion to potential energy, a phenomenon known as a **partially inelastic collision**. If the ball stuck to the ground after its first bounce, the collision would be termed **totally inelastic**.

If, however, the ball rose repeatedly to the same height, the collisions would be termed **elastic**. In an elastic collision both kinetic energy and momentum are conserved, as outlined below:

$$\mathbf{p}_{1_i} + \mathbf{p}_{2_i} = \mathbf{p}_{1_f} + \mathbf{p}_{2_f}$$
$$KE_{1_i} + KE_{2_i} = KE_{1_f} + KE_{2_f}$$

137

7.4 SIMPLE MACHINES AND WORK

A **simple machine** is a device that allows work to be done and offers an advantage to the user. We need to use machines because the force available to us is not always adequate. For example, we cannot lift a 1000-newton box using our muscles alone. However, a simple machine such as a pulley or an inclined plane may allow us to accomplish this task.

Suppose we used a system of pulleys and exerted a downward force of 250 newtons to raise the 1000-newton box. We say that the pulley offers us a mechanical advantage of 4. We calculate the mechanical advantage from this relationship: (F_{out}/F_{in}). In nature, however, one never gets something for nothing and we pay for this mechanical advantage by pulling the rope of the pulley *four* times as far as the box is lifted. Since energy must be conserved, we could have deduced this fact from the ideal relationship that work input must equal work output:

$$W_{in} = W_{out}$$

$$F_{in} \cdot \Delta s_{in} = F_{out} \cdot \Delta s_{out}$$

$$\frac{F_{out}}{F_{in}} = \frac{\Delta s_{in}}{\Delta s_{out}}$$

The relationship given above assumes that no friction is present in the machine, and the mechanical advantage is called the **ideal mechanical advantage** (IMA). In reality, all machines have friction, so the **actual mechanical advantage** (AMA) is always less than the IMA. The IMA is based on the relative distances involved in the operation of the machine $(\Delta s_{in}/\Delta s_{out})$, while the AMA is based on the ratio F_{out}/F_{in}. The **efficiency** of the machine is the AMA/IMA ratio; this value is always less than 100%.

PROBLEM

A 100-newton object is moved 2 meters up an inclined plane whose end is lifted 0.5 meter from the floor. If a force of 50 newtons is needed to accomplish this task, calculate the (a) IMA, (b) AMA, and (c) efficiency of the inclined plane.

SOLUTION

A diagram will help us to identify the relevant quantities:

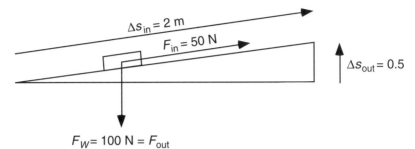

$\Delta s_{in} = 2\ m$

$F_{in} = 50\ N$

$\Delta s_{out} = 0.5$

$F_W = 100\ N = F_{out}$

The input force (F_{in}) is the force needed to move the object along the plane (50 N). The output force (F_{out}) is the force that has been lifted (100 N). The input distance (Δs_{in}) is the distance the object is moved along the plane (2 m). The output distance (Δs_{out}) is the distance to which the object has been raised (0.5 m). Then we have

(a)
$$IMA = \frac{\Delta s_{in}}{\Delta s_{out}} = \frac{2\,m}{0.5\,m} = 4$$

(b)
$$AMA = \frac{F_{out}}{F_{in}} = \frac{100\ N}{50\ N} = 2$$

(c)
$$Efficiency = \frac{AMA}{IMA} = \frac{2}{4} = 0.5\ (50\%)$$

Examples of simple machines include inclined planes, pulleys, levers, wheels and axles, and screwdrivers.

QUESTIONS

1. Which represents a scalar quantity?
   (1) acceleration
   (2) momentum
   (3) energy
   (4) displacement

2. Work is measured in the same units as
   (1) force   (2) momentum   (3) mass   (4) energy

3. Which terms represent scalar quantities?
   (1) power and force
   (2) work and displacement
   (3) time and energy
   (4) distance and velocity

4. Which term is a unit of power?
   (1) joule   (2) newton   (3) watt   (4) hertz

5. Which symbolic expression shows how the energy unit (joule) is related to the fundamental units of kilogram, meter, and second?
   (1) $kg \cdot m^2/s^2$ (2) $N \cdot m$ (3) $kg \cdot m/s$ (4) $kg \cdot m^2 \cdot s^2$

6. A 2.2-kilogram mass is pulled by a 30.-newton force through a distance of 5.0 meters, as shown in the diagram below.

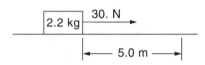

What amount of work is done?
   (1) 11 J   (2) 66 J   (3) 150 J   (4) 330 J

7. Work is being done when a force
   (1) acts vertically on a cart that can move only horizontally
   (2) is exerted by one team in a tug of war when there is no movement
   (3) is exerted on a wagon while pulling it up a hill
   (4) of gravitational attraction acts on a person standing on the surface of the Earth

8. In the diagram below, 55 joules of work is needed to raise a 10.-newton weight 5.0 meters at a constant speed.

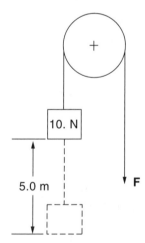

How much work is done to overcome friction as the weight is raised?
   (1) 5 J   (2) 5.5 J   (3) 11 J   (4) 50. J

9. A force of 80. newtons pushes a 50.-kilogram object across a level floor for 8.0 meters. The work done is

(1) 10 J   (2) 400 J   (3) 640 J   (4) 3920 J

10. How much work is done by a force of 8 newtons acting through a distance of 6 meters?

(1) 0 J   (2) 12 J   (3) 48 J   (4) 192 J

Base your answers to questions 11 and 12 on the graph below, which represents the relationship between the force applied to an object and the distance the object moves along a frictionless, horizontal surface.

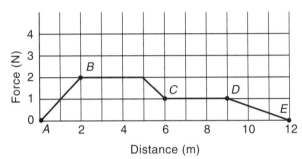

11. The work done in moving the object from *A* to *B* is

(1) 12 J   (2) 2 J   (3) 6 J   (4) 4 J

12. The largest amount of work is done when the object is moved from

(1) *A* to *B*   (2) *B* to *C*   (3) *C* to *D*   (4) *D* to *E*

Base your answers to questions 13 and 14 on the information below.

A 2.0-kilogram mass is pushed along a horizontal, frictionless surface by a 3.0-newton force that is parallel to the surface.

13. How much work is done in moving the mass 1.5 meters horizontally?

(1) 4.5 J   (2) 2.0 J   (3) 3.0 J   (4) 30 J

14. How much gravitational potential energy would be gained by the mass if it is moved 2 meters horizontally?

(1) 0 J   (2) 6 J   (3) 40 J   (4) 4 J

15. An object has a mass of 8.0 kilograms. A 2.0-newton force displaces the object a distance of 3.0 meters to the east, and then 4.0 meters to the north. What is the total work done on the object?

(1) 10. J   (2) 14 J   (3) 28 J   (4) 56 J

16. If a 2.0-kilogram mass is raised 0.05 meter vertically, the work done on the mass is approximately
(1) 0.10 J   (2) 0.98 J   (3) 40. J   (4) 100 J

17. A 20-newton block is at rest at the bottom of a frictionless incline as shown in the diagram below.

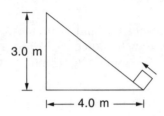

3.0 m

|◄——— 4.0 m ———►|

How much work must be done against gravity to move the block to the top of the incline?
(1) 10 J   (2) 60 J   (3) 80 J   (4) 100 J

18. The work done in raising an object must result in an increase in the object's
(1) gravitational potential energy   (3) internal energy
(2) kinetic energy   (4) heat energy

19. A 1.0-kilogram mass falls a distance of 0.50 meter, causing a 2.0-kilogram mass to slide the same distance along a table top, as represented in the diagram below.

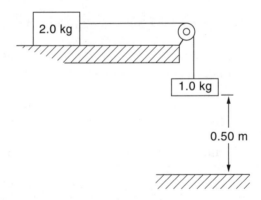

2.0 kg

1.0 kg

0.50 m

How much work is done by the falling mass?
(1) 1.5 J   (2) 4.9 J   (3) 9.8 J   (4) 14.7 J

20. Which is *not* a unit of power?
(1) joules/second   (3) watts
(2) newton-seconds   (4) calories/second

21. As the time required to lift a mass the same vertical distance is increased, the power developed
(1) decreases   (2) increases   (3) remains the same

22. As the power of a machine is increased, the time required to move an object a fixed distance
(1) decreases   (2) increases   (3) remains the same

23. The rate at which a force does work may be measured in
(1) watts   (2) newtons   (3) joules   (4) kilocalories

24. An electric motor lifts a 10-kilogram mass 100 meters in 10 seconds. The power developed by the motor is
(1) 9.8 W   (2) 98 W   (3) 980 W   (4) 9800 W

25. A horizontal force of 40 newtons pushes a block along a level table at a constant speed of 2 meters per second. How much work is done on the block in 6 seconds?
(1) 80 J   (2) 120 J   (3) 240 J   (4) 480 J

26. If an engine rated at 5.0×10^4 watts exerts a constant force of 2.5×10^3 newtons on a vehicle, the velocity of the vehicle is
(1) 0.050 m/s   (3) 20. m/s
(2) $2\sqrt{10}$ m/s   (4) 1.25×10^8 m/s

27. A motor rated at 100. watts accelerates an object along a horizontal, frictionless surface with an average speed of 4.0 meters per second. What force is supplied by the motor in the direction of motion? [Assume 100% efficiency.]
(1) 0.04 N   (2) 25 N   (3) 100 N   (4) 400 N

28. Car *A* and car *B* are of equal mass and travel up a hill. Car *A* moves up the hill at a constant speed that is twice the constant speed of car *B*. Compared to the power developed by car *B*, the power developed by car *A* is
(1) the same   (3) half as great
(2) twice as great   (4) 4 times as great

29. A weightlifter lifts a 2000-newton weight a vertical distance of 0.5 meter in 0.1 second. What is the power output?
(1) 1×10^{-4} W   (2) 4×10^{-4} W   (3) 1×10^4 W   (4) 4×10^4 W

30. One elevator lifts a mass a given height in 10 seconds, and a second elevator does the same work in 5 seconds. Compared to the power developed by the first elevator, the power developed by the second elevator is
    (1) one-half as great   (3) the same
    (2) twice as great   (4) 4 times as great

31. What is the maximum distance that a 60.-watt motor may vertically lift a 90.-newton weight in 7.5 seconds?
    (1) 2.3 m   (2) 5.0 m   (3) 140 m   (4) 1100 m

32. If the velocity of an automobile is doubled, its kinetic energy
    (1) decreases to one-half   (3) decreases to one-fourth
    (2) doubles   (4) quadruples

33. If the kinetic energy of a 10-kilogram object is 2000 joules, its velocity is
    (1) 10 m/s   (2) 20 m/s   (3) 100 m/s   (4) 400 m/s

34. Which graph best represents the relationship between the kinetic energy (*KE*) of a moving object as a function of its velocity (**v**)?

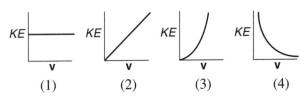

    (1)   (2)   (3)   (4)

35. Compared to the kinetic energy of an object that has fallen freely 1 meter, the kinetic energy of the object after falling 2 meters is
    (1) one-half as great   (3) the same
    (2) twice as great   (4) 4 times as great

36. The kinetic energy of a 10.0-kilogram mass moving at a speed of 5.00 meters per second is
    (1) 50.0 J   (2) 2.00 J   (3) 125 J   (4) 250. J

37. Which cart shown below has the greatest kinetic energy?

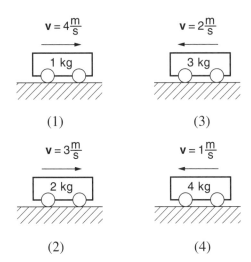

(1) (3)

(2) (4)

38. What is the kinetic energy of a 0.15-kg. baseball if its speed is 10 meters per second?
(1) 0.75 J (2) 1.5 J (3) 7.5 J (4) 15 J

39. The diagram below represents a cart traveling from left to right along a frictionless surface with an initial speed of **v**.

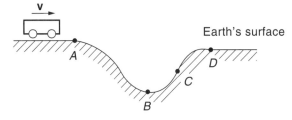

At which point is the gravitational potential energy of the cart least?
(1) *A* (2) *B* (3) *C* (4) *D*

40. A block raised through a given distance acquires a potential energy of 9.8 joules. A second block with twice the mass is raised through the same distance. The potential energy acquired by the second block is
(1) 4.9 J (2) 9.8 J (3) 19.6 J (4) 39.2 J

41. A mass resting on a shelf 10.0 meters above the floor has a gravitational potential energy of 980. joules with respect to the floor. The mass is moved to a shelf 8.00 meters above the floor. What is the new gravitational potential energy of the mass?
(1) 960. J (2) 784 J (3) 490. J (4) 196 J

42. A 2.0-kilogram rock is raised 5.0 meters above the ground. What is its change in potential energy with respect to the ground?
(1) 7.0 J   (2) 10 J   (3) 98 J   (4) 320 J

43. A 20.-newton block falls freely from rest from a point 3.0 meters above the surface of the Earth. With respect to the surface of the Earth, what is the gravitational potential energy of the block-Earth system after the block has fallen 1.5 meters?
(1) 20. J   (2) 30. J   (3) 60. J   (4) 120 J

44. An object gains 10. joules of potential energy as it is lifted vertically 2.0 meters. If a second object with one-half the mass is lifted vertically 2.0 meters, the potential energy gained by the second object will be
(1) 10. J   (2) 20. J   (3) 5.0 J   (4) 2.5 J

45. As a mass is displaced parallel to the surface of the Earth, its gravitational potential energy
(1) decreases   (2) increases   (3) remains the same

46. A ball is thrown upward from the Earth's surface. While the ball is rising, its gravitational potential energy will
(1) decrease   (2) increase   (3) remain the same

47. As the velocity of an object falling toward the Earth increases, the gravitational potential energy of the object with respect to the Earth
(1) decreases   (2) increases   (3) remains the same

48. A 5.0-newton object starts from rest and slides down a frictionless incline that is 5.0 meters long and 3.0 meters high.

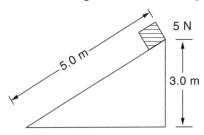

The kinetic energy of the object at the bottom of the incline is approximately
(1) 1.0 J   (2) 9.8 J   (3) 15 J   (4) 25 J

49. A 2.0-newton book falls from a table 1.0 meter high. After falling 0.5 meter, the book's kinetic energy is
(1) 1.0 J   (2) 2.0 J   (3) 10 J   (4) 20 J

50. As a satellite in orbit moves from a distance of 300 kilometers to a distance of 160 kilometers above the Earth, the kinetic energy of the satellite
(1) decreases   (2) increases   (3) remains the same

51. Ten joules of work are done in accelerating a 2.0-kilogram mass from rest across a horizontal, frictionless table. The total kinetic energy gained by the mass is
(1) 3.2 J   (2) 5.0 J   (3) 10. J   (4) 20. J

52. A 10.-kilogram mass falls freely a distance of 6.0 meters near the Earth's surface. The total kinetic energy gained by the mass as it falls is approximately
(1) 60. J   (2) 590 J   (3) 720 J   (4) 1200 J

53. As the pendulum swings from position *A* to position *C* as shown in the diagram below, what is the relationship of kinetic energy to potential energy? [Neglect friction.]

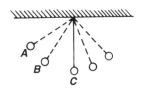

(1) The kinetic energy decreases more than the potential energy increases.
(2) The kinetic energy increases more than the potential energy decreases.
(3) The kinetic energy decrease is equal to the potential energy increase.
(4) The kinetic energy increase is equal to the potential energy decrease.

54. Mass *m* is raised to a vertical height *h* above the ground and then released. After the mass has fallen three-fourths of the way to the ground, its kinetic energy will be equal to
(1) $\dfrac{mgh}{4}$ (2) $\dfrac{3\,mgh}{4}$ (3) mgh (4) $\dfrac{4}{3\,mg}$

55. The diagram shows the path of a satellite in an elliptical orbit around the Earth.

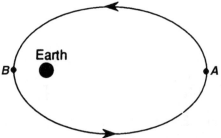

As the satellite moves from point A to point B, what changes occur in its potential and kinetic energies?
(1) Both potential energy and kinetic energy increase.
(2) Both potential energy and kinetic energy decrease.
(3) Potential energy increases and kinetic energy decreases.
(4) Potential energy decreases and kinetic energy increases.

56. A basketball player who weighs 600 newtons jumps 0.5 meter vertically off the floor. What is her kinetic energy just before hitting the floor?
(1) 30 J   (2) 60 J   (3) 300 J   (4) 600 J

57. In the diagram below, an object starting from rest at A slides down a frictionless hill and then up to B. The object's potential energy at A is 30. joules; its kinetic energy at B is

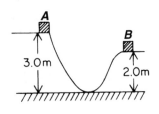

(1) 3.3 J   (2) 10. J   (3) 13 J   (4) 20. J

58. When a rising baseball encounters air resistance, its total mechanical energy
(1) decreases   (2) increases   (3) remains the same

59. An object is lifted at constant speed a distance h above the surface of the Earth in a time t. The total potential energy gained by the object is equal to the
(1) average force applied to the object
(2) total weight of the object
(3) total work done on the object
(4) total momentum gained by the object

60. As an object falls freely in a vacuum, its total energy
(1) decreases (2) increases (3) remains the same

61. As a pendulum swings from position A to position B as shown in the diagram, its total mechanical energy (neglecting friction)

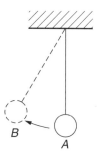

(1) decreases (2) increases (3) remains the same

62. A 2-kilogram mass is thrown vertically upward from the Earth's surface with an initial kinetic energy of 400 newton-meters. The mass will rise to a height of approximately
(1) 10 m (2) 20 m (3) 400 m (4) 800 m

Base your answers to questions 63 through 67 on the diagram below which represents a block with initial velocity v_1 sliding along a frictionless track from point A through point E.

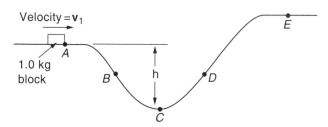

63. At which two points will the kinetic energy of the block be the same?
(1) A and B (2) A and C (3) B and D (4) B and E

64. At which two points will the gravitational potential energies of the block-Earth system be the same?
(1) A and B (2) A and C (3) B and D (4) B and E

65. The kinetic energy of the block will be greatest when it reaches point
(1) A (2) B (3) C (4) D

66. Which expression best represents the kinetic energy of the block at point *C*?

(1) mgh

(3) $\dfrac{mv_1^2}{2} - mgh$

(2) $mgh - \dfrac{mv_1^2}{2}$

(4) $\dfrac{mv_1^2}{2} + mgh$

67. The velocity of the block will be least at point
    (1) *A*   (2) *B*   (3) *C*   (4) *E*

Base your answers to questions 68 and 69 on the diagram below, which represents a 1.00-kilogram object being held at rest on a frictionless incline.

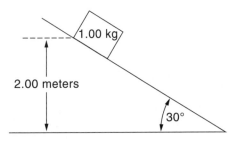

2.00 meters

1.00 kg

30°

68. The object is released and slides the length of the incline. When it reaches the bottom of the incline, the object's kinetic energy will be closest to
    (1) 19.6 J   (2) 2.00 J   (3) 9.81 J   (4) 4.00 J

69. As the object slides down the incline, the sum of the gravitational potential energy and kinetic energy of the object will
    (1) decrease   (2) increase   (3) remain the same

Base your answers to questions 70 through 72 on the diagram below, which represents a 3.0-kilogram mass being moved at a constant speed by a force of 6.0 newtons.

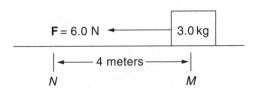

F = 6.0 N   3.0 kg

4 meters

N   *M*

70. What is the change in the kinetic energy of the mass as it moves from point *M* to point *N*?
    (1) 24 J   (2) 18 J   (3) 6 J   (4) 0 J

71. If energy is supplied at the rate of 10 watts, how much work is done during 2 seconds?
(1) 20 J　　　(2) 15 J　　　(3) 10 J　　　(4) 5 J

72. If the 3.0-kilogram mass were raised 4 meters from the surface, its gravitational potential energy would increase by approximately
(1) 120 J　　　(2) 40 J　　　(3) 30 J　　　(4) 12 J

Base your answers to questions 73 through 76 on the diagram below, which shows an object at A that moves over a frictionless surface from A to E. The object has a mass of m.

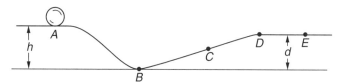

73. As the object moves from point A to point D, the sum of its gravitational potential and kinetic energies
(1) decreases, only　　　　　　(3) increases and then decreases
(2) decreases and then increases　(4) remains the same

74. The object will have a minimum gravitational potential energy at point
(1) A　　　(2) B　　　(3) C　　　(4) D

75. The object's kinetic energy at point C is less than its kinetic energy at point
(1) A　　　(2) B　　　(3) D　　　(4) E

76. The object's kinetic energy at point D is equal to
(1) mgd　　　(2) $mg(d + h)$　　　(3) mgh　　　(4) $mg(h - d)$

77. A spring of negligible mass with a spring constant of 200 newtons per meter is stretched 0.2 meter. How much potential energy is stored in the spring?
(1) 40 J　　　(2) 20 J　　　(3) 8 J　　　(4) 4 J

78. A 0.10-meter spring is stretched from equilibrium to position *A* and then to position *B* as shown in the diagram below.

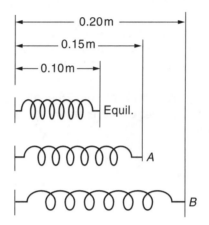

Compared to the spring's potential energy at *A*, what is its potential energy at *B*?
(1) the same   (3) one-half as great
(2) twice as great   (4) 4 times as great

79. What is the spring constant of a spring of negligible mass that gained 8 joules of potential energy as a result of being compressed 0.4 meter?
(1) 100 N/m   (2) 50 N/m   (3) 0.3 N/m   (4) 40 N/m

Base your answers to questions 80 and 81 on the figure below. A block of mass *M* sits on a frictionless surface and is pushed, compressing a spring 0.1 meter. The spring constant is 200 newtons per meter.

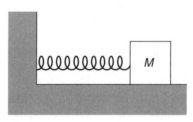

80. What force is required to compress the spring 10 meters?
(1) 0.5 N   (2) 20 N   (3) 200 N   (4) 2000 N

81. How much potential energy is stored in the spring?
(1) 1 J   (2) 10 J   (3) 1000 J   (4) 10,000 J

Chapter Eight

INTERNAL ENERGY AND THE PROPERTIES OF MATTER

KEY IDEAS

Temperature is the quantity used to measure the "hotness" of a body. A thermometer and an appropriate scale are needed to make this measurement.

Heat energy is the energy associated with changes in internal energy. If no phase changes occur, the transfer of heat energy is accompanied by a change in temperature. The specific heat of a substance indicates how much heat energy is needed to cause a given change in the substance's temperature. When two objects are brought into contact, heat energy will be exchanged until the temperatures of the objects are equal.

The gas phase is, in many respects, the simplest phase of matter. Scientists such as Boyle and Charles developed experimental laws that demonstrate gas behavior under different conditions. These laws relate pressure, volume, and temperature. The universal gas law combines Boyle's and Charles's laws and relates these quantities to the number of gas particles present.

Gas behavior can be explained by constructing a model of an ideal gas and applying the laws of physics to the model. This construction is known as the kinetic-molecular theory. With this theory, the pressure, temperature, and volume of a gas are explained in molecular terms and the universal gas law is derived from physical principles.

The liquid and solid phases of matter result from the presence of attractive forces between molecules. Changes of phase occur at specific temperatures and pressures, and heat energy is absorbed or released during these changes.

The laws of thermodynamics relate internal energy to heat energy and work. The first law is equivalent to the law of conservation of energy. The second law expresses the impossibility of converting heat energy completely into useful work. It also introduces the concept of entropy, the disorder of factor. The third law states the impossibility of attaining a temperature of absolute zero.

KEY OBJECTIVES

At the conclusion of this chapter you will be able to:

- Define the following terms: *internal energy* and *temperature, absolute zero, heat energy* and *specific heat,* and *pressure.*
- State the fixed points on the Celsius temperature scale and the fixed point on the Kelvin temperature scale.
- Relate Kelvin and Celsius temperatures, and solve problems involving this relationship.
- State the equation that relates heat energy, specific heat, and temperature changes, and solve problems using this equation.
- Define the term *thermal equilibrium,* and solve thermal equilibrium problems.
- State the equations for Boyle's and Charles's laws, and solve problems using them.
- Interpret graphs that illustrate Boyle's and Charles's laws.
- State the equation for the combined gas law, and solve problems using it.
- State the relationship for the universal gas law.
- Define the term *ideal gas,* and list the properties of an ideal gas according to the kinetic-molecular theory (KMT) of gas behavior.
- Define pressure, volume, and temperature as they are explained by the KMT.
- State the conditions under which real gases exhibit ideal behavior.
- Define the terms *melting (fusion), freezing, boiling (vaporization), condensation, sublimation,* and *deposition* as they relate to phase changes.
- Define the terms *heat of fusion* and *heat of vaporization,* and solve problems involving these quantities.
- State the factors that affect the boiling and freezing points of a liquid.
- Define the term *thermodynamics,* and state the three laws of thermodynamics.

8.1   INTERNAL ENERGY AND WORK

Suppose we use a force to move an object along a horizontal table at constant speed. We know from Chapter 7 that we have done work on the object, but what has this work accomplished? The kinetic energy has not changed because the speed has been kept constant. The gravitational potential energy has not changed because the table is horizontal. Our work has been used to overcome the friction between the object and the table, and, therefore we say that the *internal energy* of the object-table system has been increased by the work we have done.

Roughly speaking, the **internal energy** of a system is the total kinetic and potential energies of the atoms and molecules that make up the system. A change in the internal energy of an object is usually accompanied by a change in its *temperature*.

8.2 TEMPERATURE AND TEMPERATURE SCALES

Temperature is a measure of the "hotness" of an object with respect to some predefined standard. It is a scalar quantity. We will see in Section 8.4 that the temperature of an object is related to the average kinetic energy of its molecules.

To measure temperature, we need a property that changes regularly with changes in temperature. One such property is the volume of a liquid such as mercury. When mercury is placed in a thin tube (whose diameter is uniform), the length of the mercury column increases with rising temperature. Such a device is called a *thermometer*.

We also need to establish a scale of measurement. To do this, one or more fixed reference points must be set. On the *Celsius* scale, the freezing and boiling points of water (at 1 atmosphere of pressure) are assigned the respective temperatures of 0°C and 100°C. (The *Fahrenheit* scale sets these points at 32°F and 212°F.)

Another scale is the *Kelvin* or absolute temperature scale. Its single reference point is the temperature at which water exists simultaneously as a gas, liquid, and solid. The temperature assigned to this "triple point" is 273.16 K. On the Kelvin scale, zero kelvin (0 K) is the lowest temperature possible and is known as **absolute zero**. At this temperature, molecular motion is at a minimum.

The relationship between the Celsius (T_C) and Kelvin (T_K) temperature scales is as follows:

PHYSICS CONCEPTS

$$T_K = T_C + 273$$

PROBLEM
(a) Convert 37°C to the Kelvin scale.
(b) Convert 0 K to the Celsius scale.

SOLUTION
(c) $T_K = T_C + 273 = 37°C + 273 = 310$ K
(d) $T_K = T_C + 273$

Therefore,
$$T_C = T_K - 273 = 0 \text{ K} - 273 = -273°C$$

Frequently, scientists refer to *standard temperature*, which has a defined value of 273 K (the freezing point of water at 1 atmosphere of pressure).

8.3 HEAT AND INTERNAL ENERGY

An object can change its internal energy by absorbing or releasing **heat energy**. If no phase changes (e.g., freezing, boiling) are occurring, the absorption or release of heat energy is accompanied by a change in temperature.

The amount of heat energy absorbed or released is directly related to three factors: the substance itself, its mass, and the size of the temperature change. These three quantities can be combined into a single equation:

PHYSICS CONCEPTS

$$Q = cm \, \Delta T_C$$

---

where Q stands for the amount of heat energy in joules, m is the mass of the substance in kilograms, and ΔT_C is the temperature change of the substance in Celsius degrees (C°).

The symbol c stands for the **specific heat** of the substance. The value of the specific heat identifies the substance; its units are kilojoules per kilogram per Celsius degree (kJ/kg · C°). The lower the specific heat of a substance, the more readily it changes its temperature in response to heat loss or heat gain. For example, dry land has a much lower specific heat than water. For this reason, landlocked areas tend to have much larger changes in temperature (warmer summers and colder winters) than areas that are near large bodies of water.

A table of specific heats for various substances is given below:

Substance	Specific Heat (kJ/kg·C°)
Aluminum	0.90
Copper	0.39
Ice	2.05
Iron	0.45
Lead	0.13
Silver	0.24
Steam	2.01
Water	4.19

PROBLEM

How much heat energy is needed to raise the temperature of 20 kilograms of liquid water from 5°C to 20°C?

SOLUTION

$$Q = cm \cdot \Delta T_C$$
$$= (4.19 \text{ kJ/kg} \cdot \text{C}°)(20 \text{ kg})(20°\text{C} - 5°\text{C})$$
$$= +1257 \text{ kJ}$$

The positive sign associated with Q is a result of the sign of ΔT_C and indicates that heat has been *absorbed*. If the temperature had decreased, ΔT_C would have been negative, as would Q. This would mean that heat had been *released*.

Suppose two objects with different temperatures are brought into contact. Experience tells us that heat energy is always transferred from the hotter object to the colder one until *they both reach the same final temperature*. At this point, we say that the objects have reached **thermal equilibrium**; the final temperature is known as the *equilibrium temperature* This will occur, regardless of the nature of the objects or their masses. (The identity of the objects and their masses will determine the *value* of the final temperature, however.)

Problems involving heat-energy transfer utilize the relationship $Q = cm \Delta T_C$. However, this relationship must be used *twice*—once for the object that releases ("loses") heat energy, and once for the object that absorbs ("gains") it. We assume that the total internal energy is *conserved*; therefore, the heat energy "lost" by the hotter object is set equal to the heat energy "gained" by the colder object. The equations look like this:

$$|Q|_{\text{lost}} = |Q|_{\text{gained}}$$
$$(cm \, |\Delta T_C|)_{\text{lost}} = (cm \, |\Delta T_C|)_{\text{gained}}$$

The vertical bars (| |) indicate that we are interested only in the (absolute) value of the quantity, not whether it is positive or negative. The quantity $|\Delta T_C|$ is evaluated by *subtracting the smaller temperature from the larger temperature*.

PROBLEM

A 10-kilogram block of copper at 60°C is placed in contact with an identical 10-kilogram block of copper at 20°C. What is the equilibrium temperature of both blocks?

SOLUTION

We could guess at the answer (40°C, the midpoint temperature between 20°C and 60°C), and we would be correct! But let's see why this is so.

Both blocks have the same mass (10 kg), and both are composed of the same substance (copper). The change in heat energy should therefore affect the temperature of each block equally (but in opposite directions). If both blocks must reach the same final temperature, the midpoint temperature of 40°C is the only temperature that satisfies these requirements.

Now let's *prove* that this is the case by solving the pair of equations given above:

$$|Q|_{lost} = |Q|_{gained}$$

$$(cm \, |\Delta T_C|)_{lost} = (cm \, |\Delta T_C|)_{gained}$$

$$\left(0.39 \frac{kJ}{kg \cdot C°}\right)(10 \text{ kg}) (60°C - T_{final}) = \left(0.39 \frac{kJ}{kg \cdot C°}\right)(10 \text{ kg}) (T_{final} - 20°C)$$

$$60°C - T_{final} = T_{final} - 20°C$$

$$T_{final} = 40°C$$

PROBLEM

Calculate the equilibrium temperature when a 5.0-kilogram block of platinum (specific heat = 0.13 kJ/kg · C°) at 20°C is placed in contact with a 10.-kilogram block of silver (specific heat = 0.24 kJ/kg · C°) at 40°C.

SOLUTION

$$|Q|_{lost} = |Q|_{gained}$$

$$(cm \, |\Delta T_C|)_{silver} = (cm \, |\Delta T_C|)_{platinum}$$

$$\left(0.24 \frac{kJ}{kg \cdot C°}\right)(10 \text{ kg}) (40°C - T_{final}) = \left(0.13 \frac{kJ}{kg \cdot C°}\right)(5.0 \text{ kg}) (T_{final} - 20°C)$$

$$T_{final} = 35.7°C$$

We see that the equilibrium temperature is much closer to the initial temperature of the silver. This is a direct result of the fact that the mass and specific heat of the silver are larger than the mass and specific heat of the platinum.

8.4   THE GAS PHASE

Matter commonly exists in the solid, liquid, or gas phase. The phase of a substance is usually recognized by the characteristics of the substance's shape and volume. What phase a particular sample of matter is in depends on the nature of the sample, its temperature, and the pressure exerted on it.

Gases have neither definite shape nor definite volume. In our study of gases, we will assume that gases behave *ideally*. (On page 164 we will indicate exactly what we mean by the term *ideal behavior*.) Although no gas is truly ideal, many samples of gases exhibit ideal behavior under appropriate conditions. We can make this assumption because the "laws" that govern ideal gas behavior are very simple.

To describe the behavior of a sample of an ideal gas, we need to know four characteristics of the sample: the pressure it exerts, its volume, its temperature, and the number of particles (i.e., the mass) it contains.

We have already learned how volume and temperature are measured. The number of particles contained in a substances of any type (including a gas) is measured in terms of a unit known as the *mole* (mol). One mole stands for a specific number (Avogadro's number = 6.02×10^{23}), just as the term *dozen* stands for the specific number 12. For the present, however, we will use mass to describe gas behavior since it is directly related to the number of particles present.

Pressure is defined as the force per unit area of surface:

PHYSICS CONCEPTS

$$P = \frac{F}{A}$$

Pressure gives us a means of describing how a force is distributed over an entire surface. In the SI metric system, the unit of pressure is the *newton per square meter* (N/m^2), also known as the *pascal* (Pa). The pascal is a very small unit, and under normal atmospheric conditions air exerts a pressure of approximately 1.01×10^5 pascals. This value is known as *standard pressure.*

Boyle's Law

Boyle's law states the relationship between the volume of a gas and its pressure (at constant temperature and mass). The table below gives the measurements obtained in one volume-pressure experiment carried out on an ideal gas at constant temperature:

Pressure (Pa)	Volume (m³)	Pressure × Volume
1,000	0.08	80
2,000	0.04	80
4,000	0.02	80
8,000	0.01	80
20,000	(We will calculate this value on pages 160–161.)	80

According to the first four pairs of data given in the table, the volume varies *inversely* with the pressure at constant temperature, assuming that the mass is also constant. In other words, an *increase* in pressure is accompanied by a *decrease* in volume. Moreover, *the product of pressure and volume is constant under these conditions.* This relationship is known as *Boyle's law*:

Pressure · Volume = constant
(at contant temperature and mass)
In symbolic form: $PV = k$

If we graph the data given in the table, we produce a curve (shown below) known as a *rectangular hyperbola*. This curve is characteristic of inverse relationships. Each point on the graph obeys the Boyle's law equation: PV = constant.

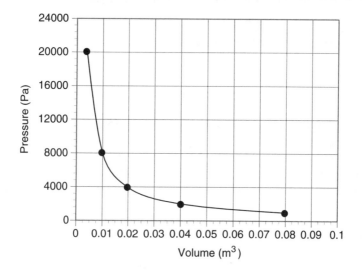

The alternative form for expressing Boyle's law:

$$P_1 V_1 = P_2 V_2$$

---

is very useful for solving mathematical problems involving pressure and volume at constant temperature and mass.

PROBLEM
Calculate the volume of the gas in the table on page 159 when the pressure is 20,000 pascals.

SOLUTION
Boyle's law indicates that we can use any other set of values in the table to solve this problem. We will use the third set. Then:

$$P_1V_1 = P_2V_2$$

$$(4000 \text{ Pa})(0.02 \text{ m}^3) = (20{,}000 \text{ Pa})V_2$$

$$V_2 = 0.004 \text{ m}^3$$

Charles's Law

To examine the relationship between the (Kelvin) temperature and the volume of an ideally behaving gas, we must hold the pressure and the mass constant. The table below provides data for such an investigation.

Temperature (K)	Volume (m³)
200	0.06
400	0.12
600	0.18
800	0.24

From data in the table, we can draw the following conclusion: The volume of the gas at constant pressure and mass is *directly proportional* to the Kelvin temperature. The graph illustrates this relationship.

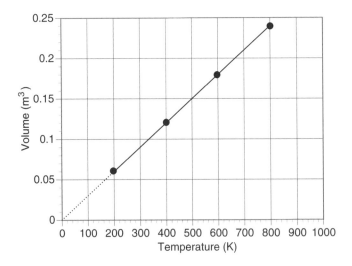

The mathematical form of this relationship is as follows:

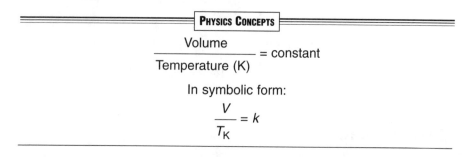

We can write this relationship, known as Charles's law, in its equivalent form:

PHYSICS CONCEPTS

$$\frac{V_1}{T_{K_1}} = \frac{V_2}{T_{K_2}}$$

PROBLEM

The volume of an ideally behaving gas is 0.03 meter^3 at 500 K. What volume will the gas occupy at 200 K, pressure and mass remaining constant?

SOLUTION

$$\frac{V_1}{T_{K_1}} = \frac{V_2}{T_{K_2}}$$

$$\frac{0.03 \text{ m}^3}{500 \text{ K}} = \frac{V_2}{200 \text{ K}}$$

$$V_2 = 0.012 \text{ m}^3$$

Combined Gas Law

It is possible to combine Boyle's and Charles's laws into a single relationship that holds as long as the mass is constant:

PHYSICS CONCEPTS

$$\frac{\text{Pressure} \cdot \text{Volume}}{\text{Temperature (K)}} = \text{constant}$$

In symbolic form:

$$\frac{PV}{T_K} = k$$

We can write this relationship, known as the *combined gas* law, in its equivalent form:

PHYSICS CONCEPTS

$$\frac{P_1 V_1}{T_{K_1}} = \frac{P_2 V_2}{T_{K_2}}$$

PROBLEM

A gas occupies 0.6 cubic meter at 5×10^4 pascals and 400 K. What volume does it occupy at 4×10^4 pascals and 200 K, mass remaining constant?

SOLUTION

$$\frac{P_1 V_1}{T_{K_1}} = \frac{P_2 V_2}{T_{K_2}}$$

$$\frac{(5 \times 10^4 \text{ Pa})(0.6 \text{ m}^3)}{400 \text{ K}} = \frac{(4 \times 10^4 \text{ Pa})(V_2)}{200 \text{ K}}$$

$$V_2 = 0.375 \text{ m}^3 \text{ (0.4 m}^3 \text{ to one significant digit)}$$

Universal Gas Law

We now turn our attention to how the combined gas law relates to the number of gas particles. Since the mass of a substance is directly related to the number of molecules present, we can reformulate the combined gas law to include the number of particles. We need only realize that adding more gas molecules at constant temperature and pressure will increase the volume proportionally, a fact known as *Avogadro's law*. The combined law now takes this form:

PHYSICS CONCEPTS

$$\frac{\text{Pressure} \cdot \text{Volume}}{\text{Number of particles} \cdot \text{Temperature (K)}} = \text{constant}$$

In symbolic form:

$$\frac{PV}{nT_K} = R$$

The number of particles (n) is measured in moles. The constant (R) is known as the *universal gas constant*. In the SI metric system, its value is 8.315 joules per mole per kelvin (J/mol · K). (You should convince yourself, as an exercise, that the units of pressure times volume are equivalent to a joule.)

The universal gas law is usually written in its equivalent form:

================= PHYSICS CONCEPTS =================

$$PV = nRT_K$$

Problems dealing with the universal gas law are usually covered in chemistry courses, so we will not encroach on their territory!

The Kinetic-Molecular Theory (KMT) of Gas Behavior

So far, we have described gas behavior in terms of *experimental laws*, that is, laws that have been investigated in the laboratory. We have assumed that all of the gases behaved ideally, but we never actually defined the term *ideal behavior*.

Now we take a different approach: We propose a model of an ideal gas, and we examine the model according to the basic laws of physics.

An **ideal gas** is a collection of particles that:

- have mass but negligible volume;
- move randomly in straight lines;
- are not subject to any attractive or repulsive forces (except during collisions with each other or with the walls of their container);
- collide in a *perfectly elastic* fashion, that is, both linear momentum and kinetic energy are conserved during collisions.

From the KMT model of gas behavior, the following facts are obtained:

- The volume occupied by an ideal gas is essentially the volume of its container.
- The pressure exerted by an ideal gas is related to the number of collisions that the particles make with the walls of the container in a given amount of time.
- The absolute (Kelvin) temperature of a gas is directly proportional to the *average* kinetic energy of the gas particles. (In other words, if the absolute temperature is doubled, the average kinetic energy of the particles will also be doubled.)

All of the experimental gas laws developed in this chapter can be *derived* mathematically as a direct result of the KMT model. Therefore, we can conclude that the KMT is a well-constructed model for explaining the behavior of ideal gases.

Real gases, such as oxygen, hydrogen, helium, and carbon dioxide, may or may not exhibit ideal behavior, depending on the conditions of the environment. For a real gas to behave ideally, its particles must be relatively far apart and must be moving at relatively high speeds. Under these conditions, the particles will exert only negligible forces on one another and the volume of the particles will be negligible in relation to the volume of the container. The behavior of real gases is most nearly ideal under conditions of high temperatures and low pressures.

8.5   LIQUIDS AND SOLIDS

Liquids and solids exist because particles of matter do *not* exhibit ideal behavior; they have measurable volumes, and they exert forces on one another. When the temperature of a gas is sufficiently low and the pressure is sufficiently high, the gas condenses to a liquid. The motion of the liquid particles becomes more restricted (the particles cannot move freely through space). As a result, the volume of the liquid becomes definite but it continues to take the shape of its container.

As the temperature of a liquid is lowered, the forces of attraction between the particles become stronger and the particles begin to arrange themselves in an orderly fashion. The motion of the particles becomes severely restricted (largely vibrations), and the substance is said to be in the solid phase. All true solids have crystalline structure.

8.6   CHANGE OF PHASE

We now take a closer look at how substances change their phases. Phase change is dependent on the nature of the substance and on the pressure and temperature of its environment. Various terms are associated with phase changes:

Phase Change	Term
Solid → liquid	**Melting (fusion)**
Liquid → solid	**Freezing**
Liquid → gas	**Boiling (vaporization)**
Gas → liquid	**Condensation**
Solid → gas	**Sublimation**
Gas → solid	**Deposition**

Reverse phase changes, melting and freezing for example, occur at the same temperature. Therefore, the melting point and the freezing point of a substance refer to the same temperature. If heat energy is being added to the system, melting occurs; if it is being removed, freezing results. Otherwise, the phases are said to coexist in *dynamic equilibrium*.

165

During a phase change, the temperature of the substance remains constant until all of the substance has been converted from one phase to another. This fact means that the average kinetic energy of the particles is not changing. The energy that is absorbed—or released—during a phase change is associated with changes in the *potential energy* of the molecules of the substance.

Heat Energy and Change of Phase

During melting, a solid absorbs a specific amount of heat energy in order to be converted to a liquid. This energy, known as the **heat of fusion** (H_f), is sometimes called a *latent* (hidden) heat because its addition is not accompanied by a temperature change. Heats of fusion are measured in kilojoules per kilogram (kJ/kg). During freezing, a liquid releases an amount of energy equal to its heat of fusion.

Similarly, during boiling, a liquid absorbs a specific amount of heat energy in order to be converted to a gas. This energy, known as the **heat of vaporization** (H_v), is also measured in kilojoules per kilogram. During condensation, a liquid releases an amount of energy equal to its heat of vaporization.

A table of heats of fusion and vaporization is given below.

Substance	Heat of Fusion (kJ/kg)	Heat of Vaporization (kJ/kg)
Aluminum	396	10,500
Copper	205	4,790
Ice	334	—
Iron	267	6,290
Tungsten	192	4,350
Water	—	2,260
Zinc	113	1,770

The table indicates that the heat of vaporization of a substance is significantly larger than its heat of fusion. For this reason it is more difficult to convert a liquid into a gas than to convert a solid into a liquid. The heat of vaporization for ice and the heat of fusion of water are not listed: we do not vaporize ice, nor do we melt water. (We vaporize water, and we melt ice!)

We can solve problems involving heats of fusion and vaporization by using the following relationships:

PHYSICS CONCEPTS

$$Q_f = mH_f$$
$$Q_v = mH_v$$

PROBLEM

How much heat energy is needed to melt 5.0 kilograms of iron at its melting point?

SOLUTION

$$Q_f = mH_f$$
$$= (5.0 \text{ kg})(267 \text{ kJ/kg})$$
$$= 1335 \text{ kJ}$$

PROBLEM

How many kilograms of steam will release 1130 kilojoules when the steam condenses at the boiling point of water?

SOLUTION

We need to use the heat of vaporization of water, since the steam will release this amount of energy when it condenses.

$$Q_v = mH_v$$
$$1130 \text{ kJ} = m(2260 \text{ kJ/kg})$$
$$m = 0.50 \text{ kg}$$

Factors Affecting Boiling and Freezing

The boiling and freezing temperatures of a substance are affected by pressure. An increase in pressure raises the boiling point of a substance. Generally, the freezing point is also raised by an increase in pressure. Water, however, is a substance that behaves irregularly because the density of liquid water is greater than the density of ice. Increasing the pressure *lowers* the freezing point of water.

When a substance such as salt is dissolved in water, the salt interferes with the water's ability to freeze and boil. As a consequence, the water-salt solution has a higher boiling point and a lower freezing point than pure water has. The degree to which the boiling point is elevated and the freezing point is depressed depends on the number of salt particles dissolved in the water.

8.7   THE LAWS OF THERMODYNAMICS

Thermodynamics is the study of the relationships among heat, work, and energy in the universe. The laws of thermodynamics are based on our experiences in observing nature. Thermodynamics has many applications in disciplines ranging from physics and engineering to biology and medicine.

The First Law

The *first law* of thermodynamics is a restatement of the law of conservation of energy. It states that the change in the internal energy of a system (ΔU) is equal to the heat (Q) that the system absorbs (or releases) minus the work (W) it does (or has done on it). In symbolic form the first law is written as follows:

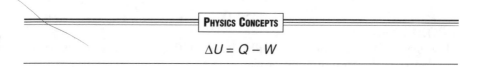

PHYSICS CONCEPTS

$$\Delta U = Q - W$$

The Second Law

The *second law* is a result of the work of the French physicist Nicolas Carnot with heat engines. This law states that heat cannot flow from a colder object to a warmer one without work being done on the system. For example, refrigerators must be run by motors in order to withdraw heat from objects placed inside them.

Another consequence of the second law is that heat can never be converted *completely* into work. In other words, no heat engine can be 100 percent efficient. Some of the heat absorbed by the engine must be lost in the random motions of its molecules. The quantity known as *entropy* is a measure of this disorder.

The Third Law

The *third law* is tied to the second law in the following way: The efficiency of a heat engine depends on its operating temperatures; the engine would reach 100 percent efficiency only if its "cold" temperature were absolute zero (0 K). Since such an engine *cannot* be completely efficient, it follows that a temperature of absolute zero cannot be reached. Although temperatures extremely close to absolute zero have been achieved, the remaining gap will never be bridged.

QUESTIONS

1. When an object moves at a constant speed against friction on a horizontal tabletop, there is an increase in the object's
   (1) temperature
   (2) momentum
   (3) potential energy
   (4) acceleration

2. Equal masses of aluminum and copper, both at 0°C, are placed in the same insulated can of hot water. Which statement describes this system at equilibrium (the net exchange of internal energy is zero)?
   (1) The water has a higher temperature than the aluminum and copper.
   (2) The aluminum has a higher temperature than the copper and water.
   (3) The copper has a higher temperature than the aluminum and water.
   (4) The aluminum, copper, and water have the same temperature.

3. Block A, at 100°C, and block B, at 50°C, are brought together in a well-insulated container. The internal energy of block A will
   (1) decrease, and the internal energy of block B will decrease
   (2) decrease, and the internal energy of block B will increase
   (3) increase, and the internal energy of block B will decrease
   (4) increase, and the internal energy of block B will increase

4. After a hot object is placed in an insulated container with a cold object, the hot object changes temperature and the cold object changes phase. The total amount of internal energy in the system will
   (1) decrease   (2) increase   (3) remain the same

5. If only the temperature of two objects is known, it is always possible to determine the
   (1) total internal kinetic energy
   (2) total internal potential energy
   (3) direction of heat flow between them
   (4) phase of the objects

6. As a rising baseball is acted on by air friction, its thermal energy
   (1) decreases   (2) increases   (3) remains the same

7. Heat will always flow from object A to object B if object B has a lower
   (1) mass   (2) total energy   (3) temperature   (4) specific heat

8. The direction of exchange of internal energy between objects is determined by their relative
   (1) inertias   (2) momentums   (3) temperatures   (4) masses

9. The internal energy of a solid is equal to the
   (1) absolute temperature of the solid
   (2) total kinetic energy of its molecules
   (3) total potential energy of its molecules
   (4) sum of the kinetic and potential energy of its molecules

10. At which temperature is the internal energy of a substance at a minimum?
    (1) 0 K   (2) 0° C   (3) 273 K   (4) 373 K

11. The ratio of a kelvin unit to a Celsius degree is
(1) 1 : 1   (2) 5 : 9   (3) 1 : 273   (4) 273 : 1

12. What is the boiling point of ammonia in kelvins?
(1) –306   (2) –33   (3) 240   (4) 316

13. A temperature change of 51 Celsius degrees is equivalent to a temperature change of
(1) 51 K   (2) 324 K   (3) 222 K   (4) –222 K

14. What temperature reading on the Kelvin scale is equivalent to a reading of zero 0°C?
(1) –273 K   (2) –100 K   (3) 100 K   (4) 273 K

15. A Celsius temperature reading may be converted to the corresponding Kelvin temperature reading by
(1) adding 273   (3) adding 180
(2) subtracting 273   (4) subtracting 180

16. Under standard conditions (a pressure of 1.0 atmosphere), water boils at a temperature of
(1) 0 K   (2) 100 K   (3) 273 K   (4) 373 K

17. What is the normal melting point of lead?
(1) 600°C   (2) 327°C   (3) 273°C   (4) 0°C

Base your answers to questions 18 through 22 on the information below:

Physical Constants of Substance X
Melting point: –20°C
Boiling point: 160°C
Heat of fusion: 60 kJ/kg
Heat of vaporization: 240 kJ/kg
Specific heat of solid: 0.2 kJ/kg · C°
Specific heat of liquid: 0.4 kJ/kg · C°

18. How many Celsius degrees are there between the melting point and the boiling point of substance X?
(1) 20   (2) 140   (3) 160   (4) 180

19. How many kilojoules of heat energy are needed to change the temperature of 1.0 kilogram of X from 120°C to 140°C?
(1) 0   (2) 8   (3) 20   (4) 4

20. How many kilojoules of heat energy are liberated when 1 kilogram of X solidifies at its freezing point?
(1) 60 (2) 240 (3) 600 (4) 2,400

21. If 1 kilogram of solid X at –20°C gains 40 kilojoules of energy, its temperature will be
(1) –120°C (2) –80°C (3) –20°C (4) 80°C

22. How many kilojoules of heat energy must be added to 1 kilogram of X to change its temperature from –40°C to –20°C?
(1) 0.8 (2) 0.4 (3) 8.0 (4) 4.0

Base your answers to questions 23 and 24 on the information below.

A 0.20-kilogram sample of lead at 120°C is put into a calorimeter containing 0.30 kilogram of water at 18°C. The equilibrium temperature of the mixture is 20°C. [Assume that the calorimeter does not absorb heat.]

23. Compared to the heat lost by the lead, the heat gained by the water is
(1) less (2) more (3) the same

24. The experiment above is repeated using 0.20 kilogram of zinc instead of lead. All other initial conditions remain the same. Compared to the water temperature change when lead was used, the water temperature change when zinc is used is
(1) smaller (2) larger (3) the same

Base your answers to questions 25 through 27 on the graph below, which represents the temperature of 2.0 kilograms of ammonia as heat is liberated.

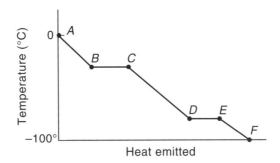

25. What is the phase of the ammonia at –15°C?
(1) solid (2) liquid (3) gas (4) plasma

26. How much heat will be liberated between points C and D?
(1) 2.9×10^5 J (2) 4.2×10^5 J (3) 6.8×10^5 J (4) 9.0×10^3 J

27. Between points D and E, the ammonia is changing phase as it
    (1) becomes a liquid   (3) becomes a gas
    (2) condenses   (4) freezes

28. What mass of iron can be changed from solid to liquid if 134,000 joules
is added at its melting point?
    (1) 0.5 kg   (2) 50. kg   (3) 500. kg   (4) 5000 kg

Base your answers to questions 29 through 33 on the graph below, which represents the temperature as heat energy is added to 2.0 kilograms of a substance originally in the solid state

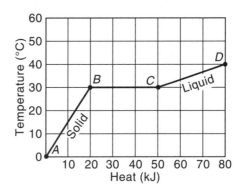

29. The melting point of the substance is
    (1) 10°C   (2) 20°C   (3) 30°C   (4) 50°C

30. How much heat energy is necessary to raise the temperature of the substance in the liquid phase from 30°C to 40°C?
    (1) 15 kJ   (2) 30 kJ   (3) 50 kJ   (4) 60 kJ

31. The total amount of heat energy necessary to raise the temperature of the substance from a solid at 30°C to a liquid at 40°C is
    (1) 20 kJ   (2) 30 kJ   (3) 60 kJ   (4) 80 kJ

32. During which interval is the average kinetic energy of the molecules of the substance unchanged?
    (1) AB (2) BC (3) CD (4) AC

33. The specific heat of the substance in the solid phase is approximately
    (1) 10 kJ/kg · C°   (3) 0.33 kJ/kg · C°
    (2) 0.25 kJ/kg · C°   (4) 1.5 kJ/kg · C°

34. Which graph best represents the relationship between the heat absorbed (Q) by a solid and the temperature (T)? [Assume the specific heat of the solid to be a constant.]

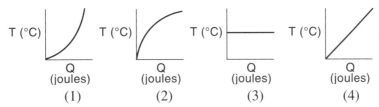

(1)	(2)	(3)	(4)

35. An unknown liquid with a mass of 0.010 kilogram absorbs 0.032 kilojoule of heat. Its temperature rises 8.0C°, with no change in phase. What is the specific heat of the unknown liquid?
(1) 0.0040 kJ/kg · C°   (3) 26 kJ/kg · C°
(2) 0.40 kJ/kg · C°   (4) 260 kJ/kg · C°

36. Equal masses of copper, iron, lead, and silver are heated from 20°C to 100°C. Which substance absorbs the least amount of heat?
(1) lead   (2) iron   (3) copper   (4) silver

Base your answers to questions 37 through 41 on the graph below, which shows the change in temperature of 1.0 kilogram of a substance as it gains heat at a constant rate of 5.0 kilojoules per minute.

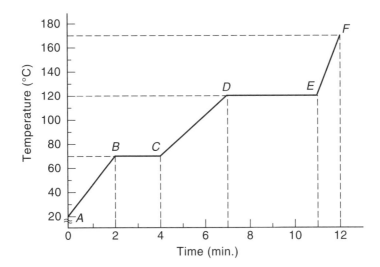

37. How much of the substance will have melted at the end of the first 3 minutes?
(1) 1 kg   (2) 0.5 kg   (3) 0.33 kg   (4) 0 kg

173

38. The specific heat of the substance is
    (1) greatest for the solid phase
    (2) greatest for the liquid phase
    (3) greatest for the gas phase
    (4) the same for all phases

39. Compared to the heat of fusion, the heat of vaporization is
    (1) less   (2) greater   (3) the same

40. During the time interval *BC*, the average kinetic energy of the molecules
    (1) is decreasing
    (2) is increasing
    (3) remains the same

41. During the time interval *DE*, the average potential energy of the molecules
    (1) is decreasing
    (2) is increasing
    (3) remains the same

Base your answers to questions 42 through 46 on the information below and on the Physics Reference Tables.

The temperature of 1 kilogram of mercury is changed from –73°C to 727°C by the addition of heat energy at the rate of 4.19 kilojoule per minute. [Assume atmospheric pressure and no heat loss to the surroundings.]

42. At which temperature can the mercury exist both as a liquid and as a gas?
    (1) 357°C   (2) 396°C   (3) 457°C   (4) 1000°C

43. What is the total range of temperature in which mercury can exist as a liquid?
    (1) 39 C°   (2) 318 C°   (3) 357 C°   (4) 396 C°

44. What amount of heat is necessary to completely melt the mercury at its melting point?
    (1) 11.7 kJ   (2) 35.2 kJ   (3) 163.4 kJ   (4) 297.5 kJ

45. What amount of heat is necessary to raise the temperature of the mercury from –1°C to 1°C?
    (1) 343.6 kJ   (2) 8.4 kJ   (3) 0.13 kJ   (4) 0.28 kJ

46. How many minutes will the mercury take to change completely into a gas after it reaches its boiling point of 357°C.?
    (1) 2.8   (2) 39   (3) 70   (4) 357

47. Which graph best represents the change of phase of a substance?

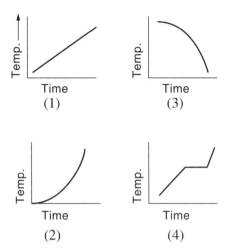

Time
(1)

Time
(3)

Time
(2)

Time
(4)

Base your answers to questions 48 through 50 on the graph below, which represents the relationship between the heat energy added and the temperature of substance X. Substance X has a mass of 1.0 kilogram and is being heated from 20.°C to 750.°C.

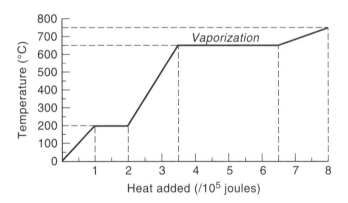

48. In which phase is substance X when its temperature is 250°C?
(1) solid   (2) liquid   (3) gas   (4) plasma

49. What is the heat of fusion of substance X?
(1) 1.0×10^5 J/kg   (3) 3.0×10^5 J/kg
(2) 2.0×10^5 J/kg   (4) 1.5×10^5 J/kg

50. What is the approximate specific heat of the solid phase of substance X?
(1) 1.0×10^5 J/kg · C°   (3) 3.3×10^2 J/kg · C°
(2) 5.0×10^2 J/kg · C°   (4) 1.0×10^3 J/kg · C°

51. Which graph best represents the relationship between temperature and time as a given quantity of a solid melts at its melting point?

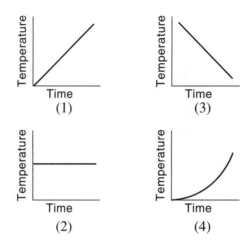

Base your answers to questions 52 through 54 on the graph below, which represents the variation in temperature as 1.0 kilogram of a gas, originally at 200°C, loses heat at a constant rate of 2.0 kilojoules per minute and eventually becomes a solid at room temperature.

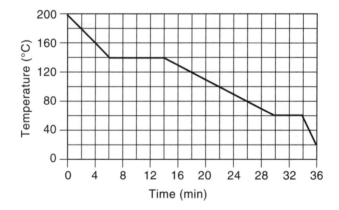

52. What is the heat of fusion of this substance?
(1) 18 kJ/kg   (2) 9 kJ/kg   (3) 8 kJ/kg   (4) 4 kJ/kg

53. What is the boiling temperature of this substance?
(1) 200°C   (2) 140°C   (3) 90°C   (4) 60°C

54. Which phase has the highest specific heat?
(1) solid   (2) liquid   (3) gas

55. Equal masses of aluminum, iron, silver, and copper are heated from 0°C to 10°C. Which absorbs the most heat?
(1) aluminum   (2) iron   (3) silver   (4) copper

56. Equal amounts of heat energy are added to 1 kilogram of lead and 1 kilogram of metal X. If the increase in temperature for metal X is one-half the temperature increase for the lead, then metal X could be
(1) iron   (2) tungsten   (3) zinc   (4) silver

57. Which substance has the highest specific heat?
(1) copper (sol.)   (3) water (liq.)
(2) ammonia (liq.)   (4) steam (gas)

58. The absolute temperature of a fixed mass of ideal gas is tripled while its volume remains constant. The ratio of the final pressure of the gas to its initial pressure is
(1) 1 : 1   (2) 1.5 : 1   (3) 3 : 1   (4) 9 : 1

59. As the temperature of a constant volume of an ideal gas is increased, the pressure of the gas will
(1) decrease   (2) increase   (3) remain the same

60. If the pressure of a fixed mass of an ideal gas is doubled at a constant temperature, the volume of this gas will be
(1) the same   (2) doubled   (3) halved   (4) quartered

61. An ideal gas occupies 50.0 cubic meters at a temperature of 600. K. If the temperature is lowered to 300. K at constant pressure, the volume occupied by the gas will then be
(1) 25.0 m^3 (2) $100. \text{ m}^3$ (3) $200. \text{ m}^3$ (4) $400. \text{ m}^3$

62. If the pressure on a gas is doubled and its absolute temperature is halved, the volume of the gas will be
(1) quartered   (2) doubled   (3) unchanged   (4) halved

63. A sample of an ideal gas at a temperature of 200 K has an average molecular kinetic energy of E. If the temperature of the gas were lowered to 50 K, the average molecular kinetic energy would change to
(1) $\dfrac{E}{2}$ (2) $2E$ (3) $\dfrac{E}{4}$ (4) $4E$

64. Which mathematical expression best represents the relationship between absolute temperature (*T*) and average molecular kinetic energy (*KE*)? [*k* is a constant of proportionality.]

(1) $T = \dfrac{k}{KE^2}$ (2) $T = \dfrac{k}{KE}$ (3) $T = k(KE)^2$ (4) $T = k(KE)$

65. For object *A* to have a higher absolute temperature than object *B*, object *A* must have a
(1) higher average internal potential energy
(2) higher average internal kinetic energy
(3) greater mass
(4) greater specific heat

66. As the temperature of a substance increases, the average kinetic energy of its molecules
(1) decreases   (2) increases   (3) remains the same

67. The graph below represents the relationship between the average kinetic energy of ideal gas molecules and Celsius temperature.

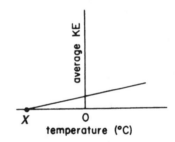

Point *x* represents a temperature of
(1) 273°C   (2) 0°C   (3) –100°C   (4) –273°C

68. According to the kinetic theory of gases, an ideal gas of low density has relatively large
(1) molecules
(2) energy loss in molecular collisions
(3) forces between molecules
(4) distances between molecules

69. A given mass of gas is enclosed in a rigid container. If the velocity of the gas molecules colliding with the sides of the container increases, the
(1) density of the gas will increase
(2) pressure of the gas will increase
(3) density ot the gas will decrease
(4) pressure of the gas will decrease

70. Which graph best represents the relationship between the absolute temperature (T_K) of an ideal gas and the average kinetic energy $(\bar{E}_k)$ of its molecules?

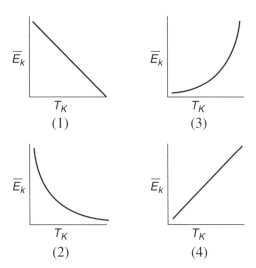

(1)

(3)

(2)

(4)

71. The pressure of a gas exerted on the walls of a balloon is produced by
(1) the collisions of the gas molecules with the walls of the balloon
(2) the repulsion between the gas molecules
(3) collisions between the gas molecules
(4) the expansion of the gas molecules

72. A change in the average kinetic energy of the molecules of an object may best be detected by measuring a change in the object's
(1) mass   (2) speed   (3) weight   (4) temperature

73. Compared to the freezing point of pure water, the freezing point of a salt-water solution is
(1) lower   (2) higher   (3) the same

74. In which diagram below does the water have the highest boiling point?

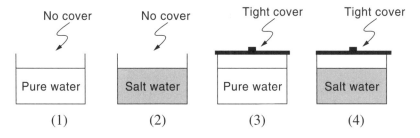

No cover	No cover	Tight cover	Tight cover
Pure water	Salt water	Pure water	Salt water
(1)	(2)	(3)	(4)

75. According to the second law of thermodynamics, as time passes, the total entropy in the universe
    (1) decreases, only   (3) remains the same
    (2) increases, only   (4) cyclically increases and decreases

76. When a student drops a beaker, it shatters, spreading randomly shaped pieces of glass over a large area of the floor. According to the second law of thermodynamics, the measure of the disorder of this system is known as
    (1) vaporization   (3) molecular collision
    (2) absolute order   (4) entropy

77. In an ideal gas, entropy is a measure of the
    (1) volume of the molecules
    (2) mass of the molecules
    (3) forces of attraction between the molecules
    (4) disorder of the molecules

STATIC ELECTRICITY

KEY IDEAS

Electricity is a fundamental property of all matter. There are two types of electric charges: positive and negative. The proton and the electron are, respectively, the fundamental positive and negative charges. If an object has the same number of protons and electrons, it is electrically neutral. Generally, charge is transferred by the gain or loss of electrons.

Electrically charged objects attract or repel each other with a force that is directly proportional to the magnitude of the charges and inversely proportional to the square of the distance between them. The relationship is known as Coulomb's law. Substances that allow charges to move freely through them are called conductors; substances that severely restrict this movement are insulators.

The region of space in which an electric charge is subject to an electric force is known as an electric field. The electric-field concept is an alternative way of explaining how charged objects attract or repel each other. Constructions called field lines are an aid to visualizing the electric fields around various charge configurations.

The work done in moving a unit charge between two points in an electric field is known as the potential difference between these points. Potential difference describes the electric field in terms of energy and work.

The charge that a conductor acquires is proportional to the potential difference across the conductor. The ratio of charge to potential difference is called capacitance. Devices that make use of this property are known as capacitors; they are used to store electric charge for a variety of applications.

KEY OBJECTIVES
At the conclusion of this chapter you will be able to:
- Define the term *electric charge*, and state the SI unit for charge.
- Relate neutral and charged objects to protons and electrons.
- Explain how neutral objects may become charged by contact.
- Solve problems involving elementary charges.
- Define the terms *conductor, insulator,* and *grounding.*

- Describe the difference between charging by induction and charging by conduction.
- Explain how an electroscope operates.
- State the equation for Coulomb's law, and solve problems using the equation.
- Define the term *electric field*, and describe how an electric field is represented by field lines.
- Draw simple field configurations.
- State the equation for measuring electric field intensity, and solve problems using the equation.
- Define the terms *potential difference* and *electric potential*, and state the SI units for measuring these quantities.
- Define the term *electron-volt* and relate it to the joule.
- Relate the electric field strength between oppositely charged parallel plates to the potential difference across them, and use this relationship to solve problems.
- Describe Millikan's oil drop experiment and its contribution to the understanding of electric charge.
- Define the term *capacitance*, and state its SI unit.
- Solve simple capacitance problems.

9.1   WHAT IS ELECTRICITY?

We have all experienced the effects of static electricity: A balloon sticks to a wall after being rubbed on a shirt or blouse or a person receives a shock after walking on a carpet and then touching an electrical appliance.

In Chapter 5, we learned that gravitation is a universal attraction between two masses and obeys an *inverse square* law, known as Newton's law of universal gravitation:

$$\mathbf{F}_G = \frac{Gm_1m_2}{d^2}$$

Imagine a force like gravitation that also obeys an inverse square law but is a *billion–billion–billion–billion* times stronger! There are other differences as well. The electric force between two objects depends on their *charges* rather than on their masses. Also, there are two types of **electric charges**, which we call *positive* and *negative*. Like charges (positive–positive and negative–negative) repel each other, and unlike charges (positive–negative) attract.

The phenomenon of electricity was recognized in ancient Greece nearly 5,000 years ago, but it was not understood completely until the twentieth century when the electrical model of the atom was developed.

9.2 ELECTRIC CHARGES

The fundamental positive charge is the *proton,* which is found in the nucleus of the atom along with uncharged neutrons. The fundamental negative charge is the *electron,* which is located outside the nucleus. The properties of these three particles are compared in the table below:

Particle	Relative Charge	Charge (C)	Mass (kg)
Proton	+1	$+1.60 \times 10^{-19}$	1.66×10^{-27}
Electron	−1	-1.60×10^{-19}	9.11×10^{-31}
Neutron	0	0.00	1.67×10^{-27}

As we can see, the proton and the electron have equal, but opposite, charges. Therefore, a *neutral* object has the same number of protons and electrons. The proton is nearly 2,000 times more massive than the electron and is tightly bound in the nucleus (along with the neutrons). As a result, ordinary objects, such as balloons, become electrically charged by gaining or losing electrons. If an object gains electrons, the *excess* of electrons gives the object a *negative* charge. If an object loses electrons, the *deficiency* of electrons gives the object a *positive* charge.

It has long been known that a hard rubber rod becomes negatively charged when rubbed with animal fur. How does this occur? Since electrons are transferred in charging, the rubber rod *gains* electrons and so becomes negatively charged. The fur *loses* an equivalent number of electrons and becomes positively charged. This example illustrates the fact that electric charge is *conserved*; it cannot be created or destroyed. The law of conservation of electric charge is a fundamental law of physics, as are the laws of conservation of energy and momentum.

PROBLEM
A glass rod becomes positively charged when it is rubbed with silk. Explain how this occurs.

SOLUTION
The glass rod loses electrons to the silk, which becomes negatively charged.

As indicated in the preceding table, the unit of charge in the SI system of measurement is the *coulomb* (C). By inspecting the table, we see that the magnitude of the charge on a proton or an electron is 1.60×10^{-19} coulomb. This quantity is known as the *elementary charge* and is denoted by the letter *e*.

Since electric charges ultimately come from protons and electrons, charge must be some multiple of the elementary charge. We can write this fact as an equation:

PHYSICS CONCEPTS

$$Q = ne$$

---

where Q is the charge on the object (in coulombs), n is the number of elementary charges, and e is the elementary charge itself.

PROBLEM
A balloon has acquired a charge of -3.20×10^{-17} coulomb. How many excess electrons does this charge represent?

SOLUTION
We use the equation $Q = ne$ to solve the problem. In this case, n represents the number of *excess* electrons on the balloon (since the charge it has acquired is negative), and e represents the charge on one electron:

$$Q = ne$$
$$-3.20 \times 10^{-17} \text{ C} = n(-1.60 \times 10^{-19} \text{ C})$$
$$n = 200. \text{ excess electrons}$$

PROBLEM
How many elementary charges are present in 1.00 coulomb of charge?

SOLUTION
We solve this problem as we did the one above:

$$Q = ne$$
$$1.00 \text{ C} = n(1.60 \times 10^{-19} \text{ C})$$
$$n = 6.25 \times 10^{18} \text{ elementary charges}$$

Certain substances, such as sodium chloride (common table salt), consist of positive and negative *ions*—atoms that have lost or gained electrons. In a water solution of sodium chloride, charge is transferred by ions, rather than by free electrons.

9.3 CONDUCTORS AND INSULATORS

Certain materials, for example, metals and solutions of ionic substances, permit charged particles such as electrons or ions to move freely through them; these materials are known as **conductors**. Copper, silver, and a water solution of sodium chloride are examples of electrical conductors. Conductors cannot hold a charge if they are in contact with other materials since the charged particles move easily through them.

Other materials, such as nonmetals, do not readily permit the free movement of charges; these materials are called **insulators**. Rubber, glass, and air are examples of electrical insulators. When an insulator (e.g., a glass rod) is given an electric charge, the charge remains confined to the area where the charge was placed.

The Earth is an electrical conductor and can accept or donate large numbers of electrons. If a charged object is placed in contact with the Earth, it loses its own charge to the Earth. Because of its large size, the Earth remains essentially neutral.

The process of allowing the flow of charge into the Earth is known as *grounding*. Lightning rods are conductors through which dangerously large buildups of atmospheric charge pass harmlessly into the ground.

Human beings, because of the dissolved salts they contain, are also conductors and can act as "grounds" for electric charge. Touching an exposed electrical wire serves to confirm this shocking point.

9.4 CHARGING OBJECTS

The diagram below represents a negatively charged rod that is brought near a neutral object.

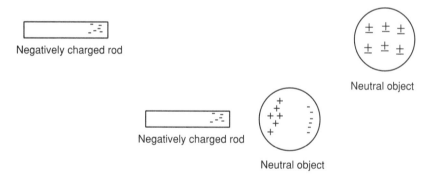

As the rod is brought near the object, the excess electrons in the rod repel the electrons in the object. The result is a *redistribution* of charge within the object. It is still neutral, but some of the charges have been separated, as shown in the lower half of the diagram. This phenomenon is known as *induction*. If the rod were removed, the original distribution of charge would return.

If the rod *touched* the object, some of the excess electrons from the rod would be transferred to the neutral object, giving it a permanent negative charge. This process, known as *charging by conduction,* is shown in the following diagram.

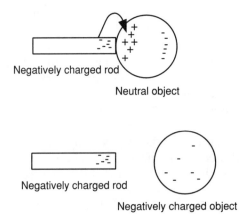

Negatively charged rod

Neutral object

Negatively charged rod

Negatively charged object

Another way in which an object can be charged is illustrated in the following diagram:

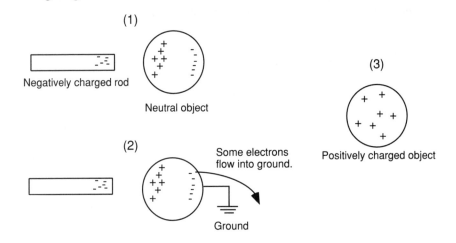

(1)

Negatively charged rod

Neutral object

(3)

Positively charged object

(2)

Some electrons flow into ground.

Ground

In step (1) the negatively charged rod is brought near the object, causing a redistribution of the object's charges. When the object is grounded in step (2), some of the electrons flow into the ground because they are repelled by the rod. When the ground is removed, the object is then *deficient* in electrons and therefore has acquired a positive charge. This process is known as *charging by induction.*

9.5 THE ELECTROSCOPE

The *Braun electroscope*, diagramed below, is a device for detecting the presence of electric charge. It consists of a flat plate, a vertical post, and a "leaf," all of which are conductors. In addition, there is a circular shield, which prevents stray charges from affecting the electroscope. The shield is separated from the rest of the device by an insulating collar placed under the plate.

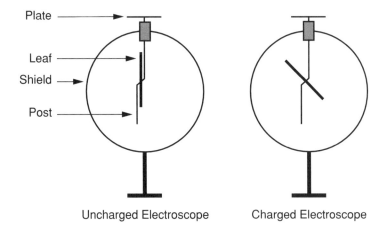

Uncharged Electroscope   Charged Electroscope

When the electroscope is uncharged, the leaf is collapsed; that is, it rests against the vertical post. If, however, the electroscope is brought into contact with a charged object, some of the charge is transferred to the electroscope. Since like charges repel, the leaf moves away from the post as shown in the diagram at the right. The only way to tell whether the charge on a charged electroscope is positive or negative is to test the electroscope by using another object of known charge.

9.6 COULOMB'S LAW

We know that like charges repel and unlike charges attract. Now we will investigate how the force on each charge is calculated. Suppose we have two point charges, Q_1 (+) and Q_2 (−), separated in space by a distance d. (By "point charges" we mean charged objects separated by a distance that is much larger than the size of either object.) This situation is represented in the diagram.

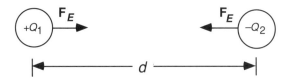

In Section 9.1 we indicated that electric charges obey a law similar to Newton's law of universal gravitation. This law, known as *Coulomb's law*, takes this form:

PHYSICS CONCEPTS

$$F_E = \frac{kQ_1Q_2}{d^2}$$

Here, Q_1 and Q_2 represent the two charges, d represents the distance between them, and k, known as the electrostatic constant, has the value 9.0×10^9 $N \cdot m^2/C^2$. F_E is the magnitude of the force on either charge. In the diagram above, these forces are shown as attractive forces because the charges are unlike.

PROBLEM

(a) Calculate the magnitude of the force between two positive charges, $Q_1 = 3.0 \times 10^{-6}$ coulomb and $Q_2 = 6.0 \times 10^{-5}$ coulomb, separated by a distance of 9.0 meters.

(b) Draw a diagram representing this situation.

SOLUTION

(a) We solve this problem by substituting the known quantities into Coulomb's law:

$$F_E = \frac{kQ_1Q_2}{d^2}$$

$$= \frac{\left(9.0 \times 10^9 \; \frac{N \cdot m^2}{C^2}\right)(3.0 \times 10^{-6} \; C)(6.0 \times 10^{-5} \; C)}{(9.0 \; m)^2}$$

$$= 2.0 \times 10^{-2} \; N$$

(b)

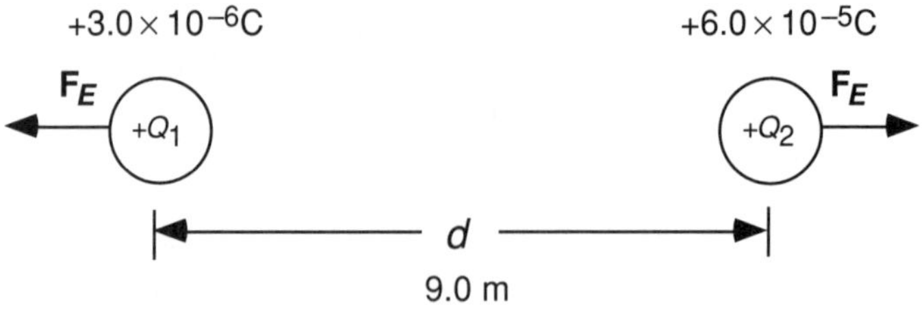

+3.0×10⁻⁶C   +6.0×10⁻⁵C

F_E +Q_1 +Q_2 F_E

|← d →|

9.0 m

9.7 THE ELECTRIC FIELD

An **electric field** is a region of space that influences (attracts or repels) electric charges. The idea of an electric field was first developed by the great English scientist Michael Faraday and was perfected by other physicists during the nineteenth century.

Before the concept of the electric field was developed, it was assumed that charges affected each other directly. This idea, known as *action at a distance*, is illustrated in the diagram below:

Action at a Distance

The problem with action at a distance is that there is no way to explain how one charge "knows" that another charge is near it.

The electric field concept assumes that one charge (Q_1 in the diagram below) somehow changes the space around it. A second charge (Q_2) then interacts with the field with the result that a force is exerted on this charge. One type of interaction is represented in the diagram below.

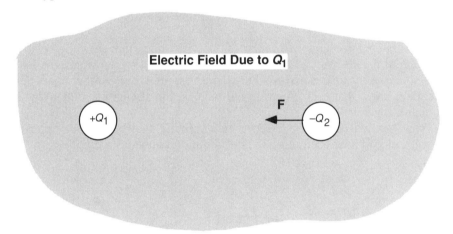

9.8   ELECTRIC FIELD LINES (LINES OF FORCE)

Field lines, also called *lines of force*, are models that we create in order to visualize an electric field. A field line is the path that a very small *positive* charge (known as a *test charge*) takes while in the field; it is drawn as a line (straight or curved) with an arrow to indicate the proper direction. A *negative* charge would move along the same field line but in the opposite direction.

When a sufficient number of field lines have been drawn, the result is a visual representation of the field. The diagrams that follow represent the electric fields in the vicinity of isolated positive and negative charges:

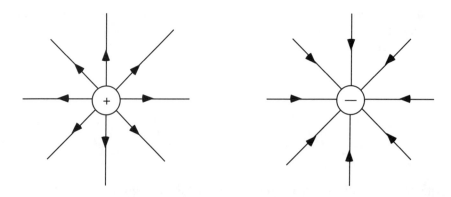

As we can see, the field surrounding each charge has a great deal of *symmetry* associated with it. A positive test charge would move radially *outward* from the isolated positive charge (left diagram) and *inward* toward the isolated negative charge (right diagram). The *number* of lines is an indication of the magnitude of the charge. For example, we might have drawn 16 lines to represent the field of an electric charge 2 times as large as the one shown above.

The *spacing* (or *concentration*) of the field lines indicates the relative *strength* of the electric field. The field is stronger in a region where the lines are more closely spaced and weaker where they are spread farther apart. The diagrams show that each electric field increases in strength as either charge is approached.

Now we will draw the electric field around two equal and opposite charges placed near each other (a configuration known as a *dipole*):

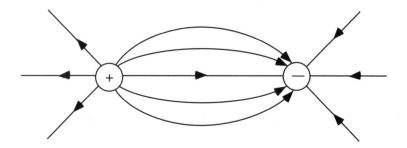

For contrast the electric field in the vicinity of two positive (like) charges is shown in the diagram that follows:

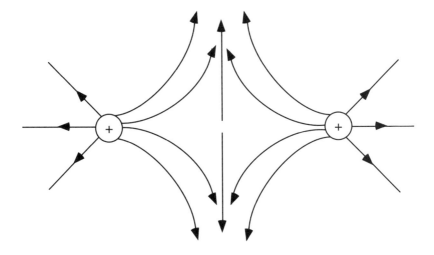

Notice that no field lines pass through the point midway between the charges. If another charge were placed at that point, it would experience no net force; in other words, the field strength is zero at that point. If we had used two *negative* charges, the field lines would have had the same shapes but would have pointed inward toward the charges.

Now let's draw the field lines between two oppositely charged parallel plates. This configuration (see Section 9.14) known as a *parallel plate capacitor*, is shown in the diagram below. We assume that the plates are very large in size. Actually, this situation can be approximated by placing the plates very close together and considering the field near the center of the plates.

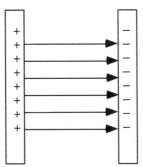

Notice that the field lines are parallel. The result is that the electric field between the plates is uniform; it does not increase or decrease in strength.

Finally, we will consider the electric field in the vicinity of a (positively) charged hollow conductor, as shown in the following diagram:

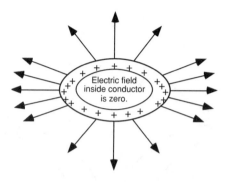

We can see that the field lines are not uniformly distributed: they are more concentrated around the curved parts of the conductor and are less concentrated around the flatter parts. Consequently, pointed objects have a greater buildup of charge. This is the principle upon which the lightning rod operates. The lightning tends to discharge on the pointed rod. Then, since the rod is grounded, any excess charge is passed harmlessly into the earth.

Also note that no electric field exists *inside* the conductor. This statement is true for every hollow conductor, regardless of its shape. As a result, hollow conductors can act as shields against electric charges.

9.9   ELECTRIC FIELD STRENGTH

The electric field is a vector quantity. We can verify this fact by referring to the diagrams in Section 9.8: the arrows point in the direction of the field, and the concentrations of the field lines indicate the magnitude, or strength, of the field.

To *measure* the strength of an electric field, we take a very small positive test charge, place it in the field, and measure the force on it, as the diagram illustrates.

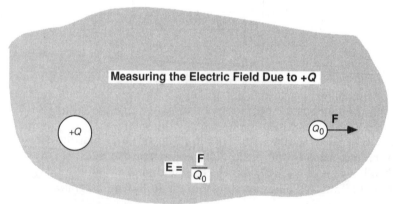

We define the strength of the electric field (E) as the ratio of the force (F) to the magnitude of the test charge (Q_0):

$$E = \frac{F}{Q_0}$$

We *divide* by the test charge so that the result (**E**) depends on the charge (or charges) producing the field, *not* on the test charge itself.

The unit of electric field strength is the *newton per coulomb* (N/C), and the direction of the field is the direction of the force on the (positive) test charge.

PROBLEM

A test charge of $+2.0 \times 10^{-6}$ coulomb experiences a force of 2.4×10^{-3} newton [east] when placed in an electric field. Determine the magnitude and the direction of the electric field.

SOLUTION

We solve the problem by using the definition of electric field:

$$\mathbf{E} = \frac{\mathbf{F}}{Q_0}$$

$$= \frac{2.4 \times 10^{-3} \text{ N [E]}}{2.0 \times 10^{-6} \text{ C}}$$

$$= 1.2 \times 10^3 \text{ N/C [E]}$$

9.10   POTENTIAL DIFFERENCE

We have described the electric field in terms of the force on a charged particle. We can also describe the electric field in terms of work and energy as shown in the diagram.

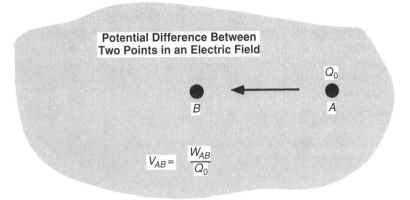

Potential Difference Between Two Points in an Electric Field

Q_0

B   A

$$V_{AB} = \frac{W_{AB}}{Q_0}$$

In the diagram, we move a test charge Q_0 between two points, A and B, in an electric field. If the charge is repelled by the field, we must do work on the charge to move it between the two points. The work we do against the field (W_{AB}) will increase the potential energy of the test charge.

Another way of describing this situation is to say that a **potential difference** exists between points A and B in the electric field. We define this potential difference (V_{AB}) as follows:

PHYSICS CONCEPTS

$$V_{AB} = \frac{W_{AB}}{Q_0}$$

Potential difference is a scalar quantity, as is work. The unit of potential difference is the *joule per coulomb* (J/C) called the *volt* (V) in honor of Alessandro Volta, an Italian scientist.

PROBLEM

When a charge of -4×10^{-3} coulomb is moved between two points in an electric field, 0.8 joule of work is done on the charge. Calculate the potential difference between the two points.

SOLUTION

The *sign* of the charge is not needed to solve the problem. We need know only the magnitude of the charge and the work done on it.

$$V = \frac{W}{Q}$$

$$= \frac{0.8 \text{ J}}{4 \times 10^{-3} \text{ C}}$$

$$= 200 \text{ V}$$

PROBLEM

Calculate the work done on an elementary charge that is moved between two points in an electric field with a potential difference of 1 volt.

SOLUTION

We can rearrange our equation as follows:

$$W = Q V$$

Then we have

$$W = (1.6 \times 10^{-19} \text{ C})(1.0 \text{ V})$$

$$= 1.6 \times 10^{-19} \text{ J}$$

The very small quantity of work in the problem shown above is frequently used as a unit of energy in atomic and nuclear physics. This unit is known as an **electron-volt** (eV). Some multiples of the electron-volt are shown in the table, where the letter M stands for *mega-*; the letter G for *giga-*; and the letter T for *tera.*

Multiple of eV	Abbreviation
10^6	MeV
10^9	GeV
10^{12}	TeV

PROBLEM
A charge, equal to 2×10^7 elementary charges, is moved through a potential difference of 3000 volts. What is the change in the potential energy of the charge?

SOLUTION
The change in the potential energy of the charge is equal to the work done on it. We can calculate this work *directly* in electron-volts by using the unit *elementary charge,* rather than coulomb, for the charge:

$$W = Q\,V$$

$$= (2 \times 10^7 \text{ el. ch.})(3000 \text{ V})$$

$$= 6 \times 10^{10} \text{ eV} = 60 \text{ GeV}$$

9.11   ELECTRIC POTENTIAL

We know that it is possible to measure the potential difference between two points, A and B, by dividing the work done on a charge by the magnitude of the charge. Suppose, however, that we wish to know the **electric potential** at just *one* of the points, A or B. How can we accomplish this?

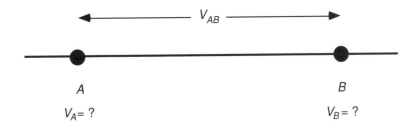

This situation is similar to climbing 525 meters vertically between two points on a hill. It seems only natural to ask how *high* the first and second points are.

The secret is to establish a reference point whose value is zero. With regard to the hill, sea level, with a height of 0 meter, is the reference point, and the height (or altitude) of each point is the distance of that point above sea level. The distance *between* two points is then the *difference* in their altitudes. The diagram below illustrates this concept.

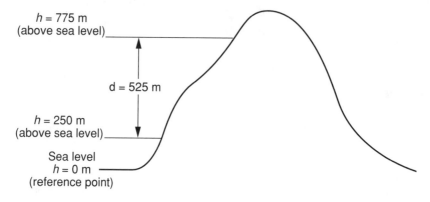

Similarly, to assign electric potentials, we establish a reference point of 0 volt. For an isolated charge, the reference point is taken to be infinitely far from the charge. For other situations, the ground may be taken as a reference point. We then measure the potential difference between the point in question and the reference point, assigning this value as the electric potential of the point. The diagram below, where A represents the point in question, illustrates how this is accomplished.

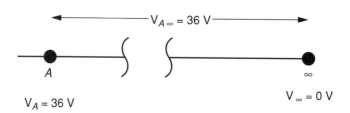

The electric potential at a point is defined as the work needed to move a charge of +1 coulomb from infinity to the point in question.

9.12 FIELD STRENGTH AND POTENTIAL DIFFERENCE

Both field strength and potential difference measure qualities of an electric field, and so it seems only natural that they should be related. We will establish this relationship for a very simple field configuration: oppositely charged parallel plates.

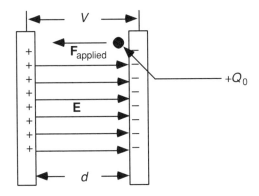

The diagram shows a pair of oppositely charged parallel plates a distance d apart. The potential difference between the plates is V, and a charge of $+Q_0$ is to be moved from the negative to the positive plate against the electric field (E) at constant velocity. To do this, we must supply a force ($\mathbf{F}_{applied}$) that is exactly *equal* to the force that the electric field places on the charge.

The work (W) that we do in moving the charge is $\mathbf{F}_{applied}d$, which is also equal to $V \cdot +Q_0$. The force placed on the charge by the field ($\mathbf{F}$) is $E \cdot +Q_0$. We can combine and simplify all of these relationships as shown below:

$$W = \mathbf{F}_{applied}d = V \cdot +Q_0$$

$$\mathbf{F} = \mathbf{E} \cdot +Q_0$$

$$= \mathbf{F}_{applied}$$

Combining the three relationships gives

$$\frac{V}{d} = \frac{\mathbf{F}}{+Q_0} = \mathbf{E}$$

We have now produced the relationship between field strength and potential difference:

PHYSICS CONCEPTS

$$E = \frac{V}{d}$$

When this equation is used, the unit for electric field is the *volt per meter* (V/m), which is equivalent to the unit *newton per coulomb*.

PROBLEM

Calculate the uniform electric field between two parallel plates if the potential difference between them is 50 volts and they are 2.5 millimeters apart.

SOLUTION

To solve the problem, we need to use the relationship $E = \dfrac{V}{d}$.

First, however, we must convert the given distance to meters:

$$2.5 \text{ mm} \cdot \frac{1\text{m}}{1000 \text{ mm}} = 2.5 \times 10^{-3} \text{ m}$$

Now we can substitute in the equation and do the computation:

$$E = \frac{V}{d}$$

$$= \frac{50 \text{ V}}{2.5 \times 10^{-3} \text{ m}}$$

$$= 2.0 \times 10^4 \text{ V/m}$$

9.13   THE MILLIKAN OIL DROP EXPERIMENT

Robert Millikan, an American physicist, used a modified pair of charged parallel plates to measure the charges on microscopic oil droplets. The diagram shows how the observation was performed.

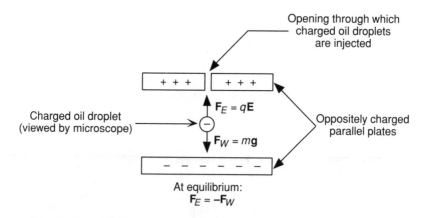

The charged droplet was injected through the opening into the uniform field between the parallel plates, where it was observed by a microscope. By changing the potential difference between the plates, the electric field strength was varied until the upward electric force on the droplet was balanced by the weight of the droplet. Millikan was then able to calculate the electric charge on each oil droplet that he observed. By measuring thousands of such charges, he concluded that the smallest charge, the *elementary charge*, is 1.60×10^{-19} coulomb.

9.14   CAPACITANCE

When a pair of uncharged parallel conducting plates are connected to a source of potential difference, the plates begin to charge and they continue charging until the potential difference between them is equal to the potential difference of the source. These oppositely charged plates constitute a device known as a *parallel plate capacitor*. Capacitors store charge and energy, which may be used in applications such as the flash attachment of a camera. When the flash is turned on, the capacitor builds up charge. As a picture is taken, the flash is triggered by the discharging capacitor.

The charge that the capacitor accumulates depends directly on the potential difference, and we can write this equation:

===================== **PHYSICS CONCEPTS** =====================

$$Q = CV$$

The constant of proportionality, C, is known as the **capacitance**. The more capacitance a capacitor has, the more charge it will store for a given potential difference. The capacitance of a parallel plate capacitor depends on the area of the plates, the distance between the plates, and the material present between the plates.

The SI unit of capacitance is the *coulomb per volt* (C/V) which is known as a *farad* (F) in honor of English chemist and physicist Michael Faraday. The farad is a very large unit, and for most practical purposes the range of capacitance extends from 10^{-12} farad (a *pico*farad, pF) to 10^{-6} farad (a microfarad, μF).

PROBLEM
A pair of parallel plates with a capacitance of 3.0×10^{-6} farad is charged using a 12-volt battery. Calculate the charge on each plate.

SOLUTION

$$Q = CV$$
$$= (3.0 \times 10^{-6} \text{ F})(12 \text{ V})$$
$$= 3.6 \times 10^{-5} \text{ C}$$

QUESTIONS

1. A glass rod becomes positively charged when rubbed with silk. The silk becomes charged because it
   (1) loses protons   (3) gains protons
   (2) loses electrons   (4) gains electrons

2. A charged body may cause the temporary redistribution of charge on another body without coming into contact with it. This process is called
   (1) conduction   (2) potential   (3) permeability   (4) induction

3. A negatively charged object is brought near the knob of a negatively charged electroscope. The leaves of the electroscope will
   (1) move closer together   (3) become positively charged
   (2) move farther apart   (4) become neutral

4. A neutral rubber rod is rubbed with fur and acquires a charge of -2×10^{-6} coulomb. The charge on the fur is
   (1) $+1 \times 10^{-6}$ C   (3) -1×10^{-6} C
   (2) $+2 \times 10^{-6}$ C   (4) -2×10^{-6} C

5. Which diagram shows the leaves of an electroscope charged negatively by induction?

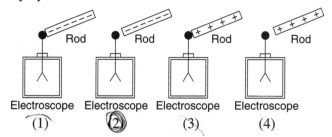

Electroscope   Electroscope   Electroscope   Electroscope
    (1)   (2)   (3)   (4)

6. A small, uncharged metal sphere is placed near a larger, negatively charged sphere. Which diagram best represents the charge distribution on the smaller sphere?

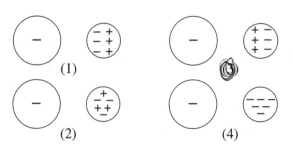

7. The number of excess electrons on a sphere that has a charge of -6.4×10^{-17} coulomb is approximately
   (1) 1024   (2) 400   (3) 100   (4) 4

8. When a rubber rod is rubbed with fur, the rod becomes negatively charged because of the transfer of
   (1) electrons to the fur   (3) electrons to the rod
   (2) protons to the fur   (4) protons to the rod

9. A charge of 10 elementary charges is equivalent to
   (1) 1.60×10^{-19} C   (3) 6.25×10^{18} C
   (2) 1.60×10^{-18} C   (4) 6.25×10^{19} C

10. Sphere *A* has a charge of +2 units and sphere *B*, which is identical to sphere *A*, has a charge of −4 unit. If the two spheres are brought together and then separated, the charge on each sphere will be
    (1) −1 unit   (2) −2 unit   (3) +1 unit   (4) +4 units

11. The unit of charge in the SI system is the
    (1) ohm   (2) ampere   (3) coulomb   (4) volt

12. In the Millikan oil drop experiment, an oil drop is found to have a charge of -4.8×10^{-19} coulomb. How many excess electrons does the oil drop have?
    (1) 1.6×10^{-19} (2) 2   (3) 3   (4) 6.3×10^{18}

13. Two identical metal spheres, charged as shown in the diagram, are brought into contact and then separated.

A
B
$+5 \times 10^{-6}$ coulomb
$+7 \times 10^{-6}$ coulomb

     What will be the charge on sphere *A* after separation?
    (1) -1×10^{-6} C   (3) $+6 \times 10^{-6}$ C
    (2) $+1 \times 10^{-6}$ C   (4) $+12 \times 10^{-6}$ C

14. A body will maintain a constant negative electrostatic charge if the body
    (1) maintains the same excess of electrons
    (2) maintains the same excess of protons
    (3) continuously receives more electrons than it loses
    (4) continuously receives more protons than it loses

15. A positively charged body must have
    (1) an excess of neutrons
    (2) an excess of electrons
    (3) a deficiency of protons
    (4) a deficiency of electrons

16. A neutral object is attracted by a charged object because the neutral object's
    (1) charges are redistributed
    (2) charge is lost to the surroundings
    (3) net charge is changed by induction
    (4) net charge is changed by conduction

17. Negatively charged rod *A* is used to charge rod *B* by induction. Object *C* is then charged by direct contact with rod *B*. The charge on object *C*
    (1) is neutral   (3) is negative
    (2) is positive   (4) cannot be determined

18. A rod is rubbed with wool. Immediately after the rod and wool have been separated, the net charge of the rod-wool system
    (1) decreases   (2) increases   (3) remains the same

19. As shown in the diagram below, a charged rod is held near, but does not touch, a neutral electroscope.

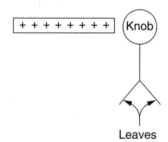

Leaves

The charge on the knob becomes
    (1) positive and the leaves become positive
    (2) positive and the leaves become negative
    (3) negative and the leaves become positive
    (4) negative and the leaves become negative

20. An uncharged metal sphere is placed midway between spheres *A* and *B*, represented in the diagram below.

A   B

Which diagram best represents the arrangement of the charges in the uncharged sphere?

(1)   (2)   (3)   (4)

 $F_e = kq_1q_2$ / r^2

21. The electrostatic force of attraction between two small spheres that are 1.0 meter apart is **F**. If the distance between the spheres is decreased to 0.5 meter, the electrostatic force will then be $F = F$

(1) $\dfrac{F}{2}$ (2) 2F   ~~(3)~~ $\dfrac{F}{4}$ (4) 4F

22. The distance between two point charges is tripled. Compared to the original force, the new electrostatic force between the charges is
(1) decreased to one-ninth   (3) increased by a factor of 3
(2) decreased to one-third   (4) increased by a factor of 9

23. The electrical force of attraction between two point charges is **F**. The charge on one of the objects is quadrupled, and the charge on the other object is doubled. The new force between the objects is

(1) 8F   (2) 2F   (3) $\dfrac{1}{2}$ F   (4) 4F

24. What is the magnitude of the electrostatic force between a charge of $+3.0 \times 10^{-5}$ coulomb and a charge of $+6.0 \times 10^{-6}$ coulomb separated by 0.30 meter? 8.99×10^{9}
(1) 1.8×10^{-3} N   (3) 5.4×10^{0} N
(2) 5.4×10^{-2} N   (4) 1.8×10^{1} N

25. A point charge of -1.0×10^{-9} coulomb is located 5.0×10^{-2} meter from another point charge of $+3.0 \times 10^{-9}$ coulomb. What is the magnitude of the electric force between the two point charges?
(1) 6.0×10^{-17} N   (3) 5.4×10^{-7} N
(2) 1.2×10^{-15} N   (4) 1.1×10^{-5} N

26. Two point charges 1 meter apart repel each other with a force of 9 newtons. What is the force of repulsion when these two charges are 3 meters apart?
    (1) 1 N   (2) 27 N   (3) 3 N   (4) 81 N

27. If the magnitude of the charge on each of two positively charged objects is halved, the electrostatic force between the objects will
    (1) decrease to one-half   (3) decrease to one-sixteenth
    (2) decrease to one-quarter   (4) remain the same

28. Which procedure will double the force between two point charges?
    (1) doubling the distance between the charges
    (2) doubling the magnitude of one charge
    (3) halving the distance between the charges
    (4) halving the magnitude of one charge

29. The force between two fixed, charged spheres is **F**. If the charge on each sphere is halved, the force between them will be
    (1) **F**   (2) $\dfrac{\mathbf{F}}{2}$ (3) $\dfrac{\mathbf{F}}{4}$ (4) 4**F**

30. Which graph best represents the relationship of the electrostatic force between two charges and the distance between them?

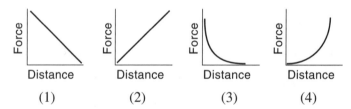

    (1)   (2)   (3)   (4)

31. The diagram below represents two charges with a separation of *d*. Which step would produce the greatest increase in the force between the two charges?

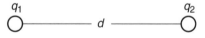

    (1) doubling charge q_1, only   (3) doubling charge q_1, only, and *d*
    (2) doubling *d*, only   (4) doubling both charges and *d*

32. What is the magnitude of the electric field intensity at a point in space where a charge of 100 coulombs experiences a force with a magnitude of 10 newtons?
    (1) 1 N/C   (2) 10 N/C   (3) 0.1 N/C   (4) 100 N/C

33. Which is a vector quantity?
  (1) electric energy
  (2) electric charge
  (3) electric power
  (4) electric field intensity

34. The electric field intensity at a given distance from a point charge is **E**. If the charge is doubled and the distance remains fixed, the electric field intensity will be
  (1) $\dfrac{E}{2}$
  (2) 2E
  (3) $\dfrac{E}{4}$
  (4) 4E

35. The electric field around a point charge is
  (1) radial
  (2) elliptical
  (3) parabolic
  (4) circular

36. Gravitational force is to mass as electrical force is to
  (1) weight
  (2) charge
  (3) gravity
  (4) electricity

37. As the electric field intensity at a point in space decreases, the electrostatic force on a unit charge at this point
  (1) decreases
  (2) increases
  (3) remains the same

38. The diagram below shows some of the lines of electric force around a positive point charge.

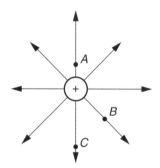

The strength of the electric field is
  (1) greatest at point *A*
  (2) greatest at point *B*
  (3) greatest at point *C*
  (4) equal at points *A*, *B*, and *C*

39. The diagram below represents a uniformly charged rod.

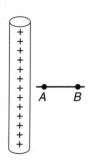

Which graph best represents the relationship between the magnitude of the electric field intensity (E) and the distance from the rod as measured along line AB?

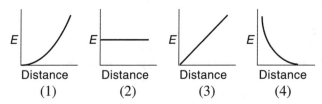

Distance	Distance	Distance	Distance
(1)	(2)	(3)	(4)

40. Millikan determined the charge on an electron by observing the motion of charged oil drops in the presence of
(1) sound waves   (3) standing waves
(2) gold nuclei   (4) an electric field

41. What force will a proton experience in a uniform electric field whose strength is 2.00×10^5 newtons per coulomb?
(1) 8.35×10^{24} N   (3) 3.20×10^{-14} N
(2) 1.67×10^5 N   (4) 1.67×10^{-14} N

42. A proton experiences an electrical force **F** when placed at a point in an electric field. If the proton is replaced by an electron, what force will the electron experience?

(1) **–F**   (2) $-\mathbf{F}\dfrac{m_p}{m_e}$ (3) $\mathbf{F}\dfrac{m_p}{m_e}$ (4) **F**

Base your answers to questions 43 through 46 on the diagram below, which represents two large parallel metal plates with a small charged sphere between them.

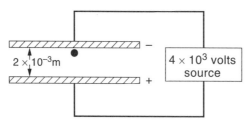

43. The energy gained by the charged sphere as it moves from the negative plate to the positive plate can be measured in
    (1) electron-volts   (3) coulombs/volt
    (2) volt·meters   (4) volts/meter

44. What is the intensity of the electric field between the two charged plates?
    (1) 5.0×10^{-17} m/V   (3) 1.6×10^{-16} c/m
    (2) 2×10^{6} V/m   (4) 8.0 V·m

45. If the distance between the plates were increased, the field intensity would
    (1) decrease   (2) increase   (3) remain the same

46. As the sphere moves from the negative plate to the positive plate, the force on the sphere
    (1) decreases (2) increases   (3) remains the same

Base your answers to questions 47 through 51 on the diagram below, which represents two small charged spheres, A and B, 3 meters apart. Each sphere has a charge of $+2.0 \times 10^{-6}$ coulomb.

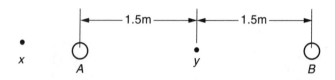

47. Which diagram best illustrates the electric field between charges A and B?

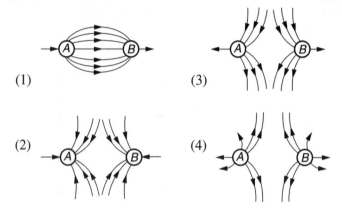

48. The magnitude of the force of sphere A on sphere B is
(1) 1.2×10^{-2} N   (3) 4.4×10^{-13} N
(2) 2.0×10^3 N   (4) 4.0×10^{-3} N

49. If another small sphere with a charge of $+2.0 \times 10^{-6}$ coulomb is placed at point y, the net force on this sphere will be
(1) 0 N   (2) 40. N   (3) 80. N   (4) 240 N

50. If a positive charge is placed at point x, the direction of the net force on the charge will be
(1) into the page   (3) toward the left
(2) out of the page   (4) toward the right

51. If sphere A is moved toward sphere B, the electric field intensity at point x will
(1) decrease   (2) increase   (3) remain the same

52. The diagram below represents a particle of mass *m* and charge +*q* located between two charged parallel plates. The electric field intensity between the plates is *E*.

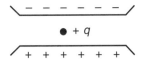

The particle will remain stationary when
(1) *Eq* is less than *mg*
(2) *Eq* is greater than *mg*
(3) *Eq* equals *mg*

53. The diagram shows two parallel metal plates 0.1 meter apart, with a potential difference between them of 10 volts.

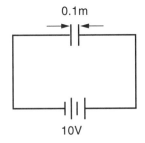

What is the electric field intensity between the plates?
(1) 1 N/C   (2) 100 N/C   (3) 0.001 N/C   (4) 0 N/C

54. The diagram below represents two charged parallel plates.

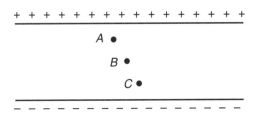

The intensity of the electric field relative to locations *A*, *B*, and *C* is
(1) greater at *A* than at *B*   (3) greater at *B* than at *C*
(2) greater at *C* than at *A*   (4) the same at *A*, *B*, and *C*

55. Two charged spheres are shown in the diagram.

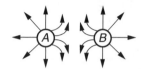

Which polarities will produce the electric field shown?
(1) *A* and *B* both negative   (3) *A* positive and *B* negative
(2) *A* and *B* both positive   (4) *A* negative and *B* positive

Base your answers to questions 56 through 60 on the diagram below, which shows two parallel metal plates, *A* and *B*, connected to a voltage source.

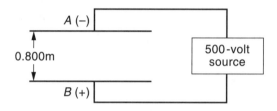

56. What is the electric field intensity between the two plates?
(1) 400. N/C   (2) 500. N/C   (3) 625 N/C   (4) 4000 N/C

57. If an object carrying 2 elementary charges starts from rest at plate *A* and is accelerated to plate *B* by the electric field, the total work done by the electric field on the object is
(1) 200. eV   (2) 250. eV   (3) 500. eV   (4) 1000 eV

58. As a charged object moves from plate *A* to plate *B*, the electric force on the charged object
(1) decreases (2) increases   (3) remains the same

59. If only the original source voltage is increased, the electric field intensity between the plates will
(1) decrease   (2) increase   (3) remain the same

60. If only the distance between the plates is increased, the electric field intensity between the plates will
(1) decrease   (2) increase   (3) remain the same

Base your answers to questions 61 through 64 on the diagram below which shows a positive point charge placed at *A*.

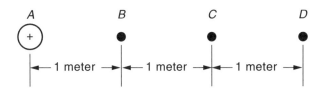

61. The electric field intensity at point *B* is *E*. At point *D* the field intensity will be equal to

(1) $\dfrac{1}{9} E$ (2) $\dfrac{1}{3} E$ (3) $3E$ (4) $9E$

62. If a positive charge is placed at point *B*, the force exerted on this charge by charge *A* will be directed toward
(1) the top of the page   (3) *A*
(2) the bottom of the page   (4) *C*

63. If the charge is moved from point *B* to point *C*, the force between the two charges will
(1) decrease   (2) increase   (3) remain the same

64. The electric field surrounding charge *A* is best represented by which diagram?

(1)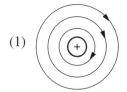

(3)

(2)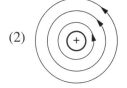

(4)

Base your answers to questions 65 through 69 on the diagram below which shows four point charges, A, B, C, and D, and three points, X, Y, and Z, in the electric field of these point charges.

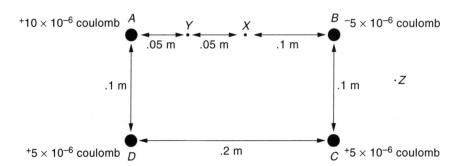

65. Between which pair of charges is the electric force greatest?
(1) A and B   (2) A and D   (3) B and C   (4) B and D

66. The direction of the electric field of charge A at point X is
(1) upward   (3) toward charge A
(2) downward   (4) toward charge B

67. Compared to the intensity of the field of charge A at point X, the intensity of the field of charge A at point Y is
(1) one-fourth as great   (3) twice as great
(2) the same   (4) 4 times as great

68. Compared to the intensity of the field of charge A at point X, the intensity of the field of charge B at point X is
(1) one-fourth as great   (3) the same
(2) one-half as great   (4) twice as great

69. If the electric field intensity at point Z is 1.0×10^5 newtons per coulomb, the force on a 2.0×10^{-6} coulomb charge at point Z will be
(1) 2.0×10^{-1} N   (3) 2.0×10^5 N
(2) 5.0×10^4 N   (4) 5.0×10^{10} N

70. The diagram below represents a positive test charge located near a positively charged sphere.

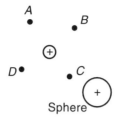

The greatest increase in the electric potential energy of the test charge relative to the sphere would be caused by moving the charge to point
(1) *A*   (2) *B*   (3) *C*   (4) *D*

71. It takes 15 joules of work to bring 3.0 coulombs of positive charge from infinity to a point. What is the electric potential at this point in an electric field?
(1) 45. V   (2) 5.0 V   (3) 0.20 V   (4) 0 V

72. A positive test charge is moving between two oppositely charged plates. As the charge moves toward the negative plate, its potential energy
(1) decreases   (2) increases   (3) remains the same

73. A volt is defined as a
(1) joule/coulomb   (3) coulomb/second
(2) joule/second   (4) joule·second/coulomb

74. The work required to move 2 coulombs of charge through a potential difference of 5 volts is
(1) 10 J   (2) 2 J   (3) 25 J   (4) 50 J

75. If 8.0 joules of work is required to transfer 4.0 coulombs of charge between two points, the potential difference between the two points is
(1) 6.4 V   (2) 2.0 V   (3) 32 V   (4) 40. V

76. The two large metal plates shown in the diagram are charged to a potential difference of 100 volts. How much work is needed to move 1 coulomb of negative charge from point A to point B?

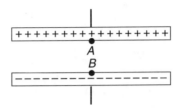

(1) 1 J (2) 100 J (3) 1 eV (4) 100 eV

77. If 1.6×10^{-12} joule of energy is needed to move a charge through a potential difference of 1×10^7 volts then the magnitude of this charge is
(1) 1.6×10^{-19} C (3) 1.6×10^5 C
(2) 1.6×10^{-5} C (4) 1.6×10^{19} C

78. The work done in moving a charge of 2.0×10^{-2} coulomb from one point to another in an electric field is 5.0×10^{-4} joule. What is the potential difference between these two points in the field?
(1) 1.0×10^{-6} V (3) 4.0×10^{-2} V
(2) 2.5×10^{-2} V (4) 2.5×10^2 V

79. An electron-volt is a unit of
(1) potential difference (3) current
(2) charge (4) energy

80. An electron gains 2 electron-volts of energy as it is transferred from point A to point B. The potential difference between points A and B is
(1) 3.2×10^{-19} V (3) 32 V
(2) 2 V (4) 1.25×10^{19} V

81. An energy of 13.6 electron-volts is equivalent to
(1) 1.60×10^{-19} J (3) 6.25×10^{-19} J
(2) 2.18×10^{-18} J (4) 6.63×10^{-18} J

82. Which quantity is equivalent to 3.2×10^{-17} joule?
(1) 8.00×10^{-3} eV (3) 3.20 eV
(2) 3.20×10^{-17} eV (4) 200 eV

83. What is the maximum amount of kinetic energy that may be gained by a proton accelerated through a potential difference of 50 volts?
(1) 1 eV (2) 10 eV (3) 50 eV (4) 100 eV

Base your answers to questions 84 through 88 on the diagram below. The diagram shows a negatively charged pith ball with a mass of 10^{-3} kilogram that is held suspended in the air by the attractive force of a positively charged pith ball. The distance between the centers of the two pith balls is 0.01 meter.

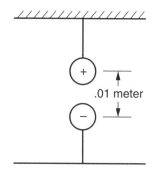

84. The minimum force needed to keep the negatively charged pith ball suspended in the air is approximately
(1) 10^{-5} N   (2) 10^{-4} N   (3) 10^{-3} N   (4) 10^{-2} N

85. The electrostatic field between the two pith balls is best represented by

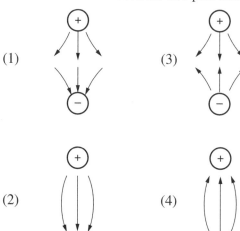

86. If the magnitude of the charge on each pith ball is 1×10^{-7} coulomb, the attractive force between them is
(1) 9×10^{-3} N   (2) 9×10^{-2} N   (3) 9×10^{-1} N   (4) 9.0 N

87. Which graph best represents the attractive force between the two pith balls as a function of the distance between them?

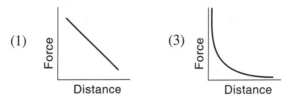

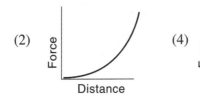

88. If the negatively charged pith ball has a charge of -1.0×10^{-7} coulomb, the number of excess electrons on the pith ball is approximately
    (1) 6×10^4 (2) 6×10^8 (3) 6×10^{11} (4) 6×10^{18}

Base your answers to questions 89 through 91 on the information below.

Charge $+q$ is located a distance r from charge $+Q$. Each charge is 1 coulomb.

89. The magnitude of the electric field due to charge $+Q$ at distance r is equal to
    (1) $\dfrac{kQ}{F}$ (2) $\dfrac{kQq}{r}$ (3) $\dfrac{Q}{r^2}$ (4) $\dfrac{kQ}{r^2}$

90. If 200 joules of work was required to move $+q$ through distance r to $+Q$, the potential difference between the two charges would be
    (1) 100 V   (2) 200 V   (3) 800 V   (4) 50 V

91. If distance r is doubled, the force that $+Q$ exerts on $+q$ is
    (1) quartered   (2) halved   (3) unchanged   (4) doubled

Base your answers to questions 92 through 96 on the diagram below, which shows two vertical plates in a vacuum that are separated by a distance of 4.0 × 10⁻⁴ meter. The plates are connected to a potential difference of 500 volts.

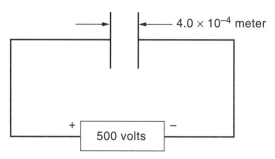

92. What is the field intensity between the plates?
    (1) 2.0×10^{-1} N/C   (3) 5.9×10^{2} N/C
    (2) 8.0×10^{-7} N/C   (4) 1.25×10^{6} N/C

93. How much kinetic energy will an alpha particle (+2 elementary charges) acquire as it moves from one plate to the other?
    (1) 4.0×10^{-3} eV   (3) 5.0×10^{2} eV
    (2) 2.5×10^{2} eV   (4) 1.0×10^{3} eV

94. The work needed to move an electron from the positive plate to the negative plate is
    (1) 1.6×10^{-19} J   (3) 1.7×10^{-16} J
    (2) 8.0×10^{-17} J   (4) 3.4×10^{-16} J

95. As an electron moves from the negative plate to the positive plate, its acceleration
    (1) decreases   (2) increases   (3) remains the same

96. As the plates are moved closer together, the electric field intensity between the plates
    (1) decreases   (2) increases   (3) remains the same

Base your answers to questions 97 through 101 on the diagram below, which shows two identical metal spheres. Sphere A has a charge of +12 coulombs, and sphere B is a neutral sphere.

+ 12 coulombs

$\textbf{A}$ $\textbf{B}$

97. When spheres A and B come into contact, sphere B will
    (1) gain 6 C of protons   (3) gain 6 C of electrons
    (2) lose 6 C of protons   (4) lose 6 C of electrons

98. When spheres *A* and *B* are in contact, the total charge of the system is
(1) neutral   (2) +6 C   (3) +12 C   (4) +24 C

99. When spheres *A* and *B* are separated, the charge on sphere *A* is
(1) +12 C
(2) one-fourth of the original amount
(3) one-half of the original amount
(4) 4 times the original amount

100. After spheres *A* and *B* are separated, which graph best represents the relationship of the force between the spheres and distance between them?

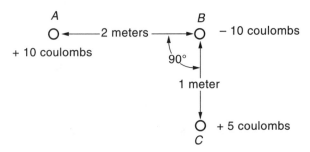

101. After contact, the spheres are moved apart. As the distance between the spheres is increased, the electric potential energy of the system
(1) decreases (2) increases   (3) remains the same

Base your answers to questions 102 through 106 on the diagram below, which shows three small metal spheres with different charges.

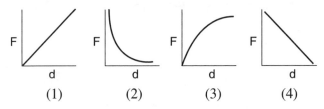

102. Which vector best represents the net force on sphere *B*?

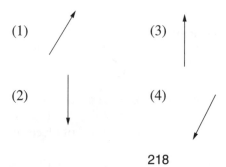

103. Compared to the force between spheres *A* and *B*, the force between spheres *B* and *C* is
(1) one-quarter as great   (3) one-half as great
(2) twice as great   (4) 4 times as great

104. If sphere *A* is moved further to the left, the magnitude of the net force on sphere *B* will
(1) decrease   (2) increase   (3) remain the same

105. If the charge on sphere *B* were decreased, the magnitude of the net force on sphere *B* would
(1) decrease   (2) increase   (3) remain the same

106. If sphere *B* were removed, the force of sphere *C* on sphere *A* would
(1) decrease   (2) increase   (3) remain the same

Base your answers to questions 107 through 112 on the diagram below, which represents two charged spheres, *X* and *Y*.

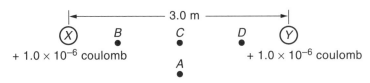

107. At which point is the magnitude of the electric field equal to zero?
(1) *A*   (2) *B*   (3) *C*   (4) *D*

108. Which arrow best represents the direction of the electric field at point *A*?

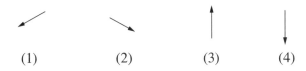

(1)   (2)   (3)   (4)

109. If a unit positive charge moves directly from point *B* to point *D*, the potential energy of the charge will
(1) decrease, only   (3) decrease, then increase
(2) increase, only   (4) increase, then decrease

110. The magnitude of the force between the two spheres is
(1) 1.0×10^{-3} N   (3) 3.0×10^{-3} N
(2) 1.0×10^{3} N   (4) 3.0×10^{3} N

111. Moving the two spheres toward each other would cause their electric potential energy to
(1) decrease   (2) increase   (3) remain the same

112. Compared to the force of the electric field of sphere X on sphere Y, the force of the electric field of sphere Y on sphere X is
(1) less   (2) greater   (3) the same

Base your answers to questions 113 through 117 on the information and diagram below.

This diagram represents two equal negative point charges, A and B, that are a distance d apart.

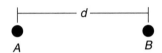

113. Which diagram best represents the electrostatic lines of force between charge A and charge B?

(1)   (3)

(2)   (4)

114. Where would the electric field intensity of the two charges be minimum?
(1) one-quarter of the way between A and B
(2) midway between A and B
(3) on the surface of B
(4) on the surface of A

115. Which graph expresses the relationship of the distance between the charges and the force between them?

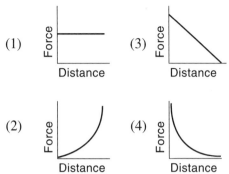

(1)   (3)

(2)   (4)

116. If charge *A* is doubled, the force between *A* and *B* will be
(1) quartered (2) halved   (3) doubled   (4) quadrupled

117. If A has an excess of 2.5×10^{19} electrons, its charge, in coulombs, is
(1) 1.6   (2) 2.5   (3) 6.5   (4) 4.0

Base your answers to questions 118 through 122 on the diagram below, which represents two charged metal spheres.

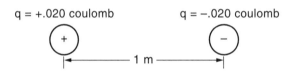

q = +.020 coulomb   q = −.020 coulomb

1 m

118. Which diagram below best represents the electric field around the two spheres?

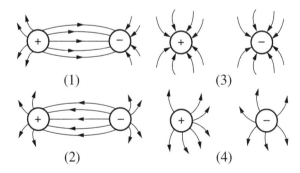

(1)   (3)

(2)   (4)

119. What is the magnitude of the force between the two spheres?
(1) 3.6×10^6 N   (3) 3.6×10^9 N
(2) 1.8×10^8 N   (4) 9.0×10^9 N

120. When a charge of 0.04 coulomb is placed at a point in the electric field, the force on the charge is 100 newtons. What is the magnitude of the electric field at that point?
    (1) 0.00040 N/C
    (2) 2500 N/C
    (3) 100 N/C
    (4) 4.0 N/C

121. The work required to move a charge of 0.04 coulomb from one point to another point in the electric field is 200 joules. What is the potential difference between the two points?
    (1) 0.0002 V   (2) 8 V
    (3) 200 V
    (4) 5,000 V

122. The two spheres are brought into contact with each other and then separated. After separation, what is the charge on each sphere?
    (1) 0.01 C   (2) 0.02 C   (3) 0 C   (4) 0.04 C

Base your answers to questions 123 through 127 on the diagram below, which shows two large parallel metal plates connected to a source of electric potential. The plates are 4.0×10^{-3} meter apart, and the potential difference across the plates is 100 volts.

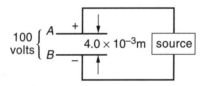

123. The magnitude of the electric field intensity between the plates is
    (1) 2.5×10^{-1} N/C
    (2) 2.5×10^{4} N/C
    (3) 4.0×10^{-1} N/C
    (4) 4.0×10^{4} N/C

124. As an electron moves from plate A to plate B, how much work is done on the electron?
    (1) 1.0×10^{0} eV
    (2) 1.6×10^{-19} eV
    (3) 1.0×10^{2} eV
    (4) 4.0×10^{2} eV

125. If the potential difference across the plates is doubled, the field intensity between the plates will be
    (1) halved   (2) doubled   (3) quartered   (4) quadrupled

126. As an electron moves from plate B toward plate A, the electric force on the electron
    (1) decreases (2) increases   (3) remains the same

127. If the distance between the plates were to decrease, with the potential difference remaining constant, the magnitude of the field intensity would
    (1) decrease   (2) increase   (3) remain the same

Chapter Ten

ELECTRIC CURRENT AND CIRCUITS

KEY IDEAS

Electric current is defined as the time rate at which charge flows through a substance. Currents are established and maintained through a conductor by the application of a potential difference across the conductor.

The resistance of a substance is the ratio of the potential difference across the substance to the current through it. Resistance depends on the nature of the substance, its length, its cross-sectional area, and the temperature. In metallic conductors, the resistance at a given temperature is relatively constant, a fact that is the basis for a relationship known as Ohm's law. Certain substances lose all resistance at low temperatures, a phenomenon known as superconductivity.

Current and potential difference are usually measured in an electric circuit that is a closed path for the flow of charge. A circuit usually contains a source of potential difference and one or more resistances, and may include other devices as well. A circuit that has only one current path is known as a series circuit; a circuit with more than one current path, as a parallel circuit.

Complex circuits may consist of both series and parallel branches or be even more complicated. Such circuits need to be solved by means of Kirchhoff's two laws, which are mathematical statements of the laws of conservation of energy and of electric charge.

KEY OBJECTIVES

At the conclusion of this chapter you will be able to:

- Define the term *electric current*, and state the SI unit for it.
- Solve problems involving current, charge, and time.
- Distinguish between conventional current and electron flow.
- Define the term *resistance*, and state the SI unit for it.
- Solve problems that relate current, potential difference, and resistance.
- Relate the resistance of a material to its length, cross section, resistivity, and temperature.
- Solve problems that relate resistivity, cross section, and length.

- Define the terms *superconductor, series circuit,* and *parallel circuit.*
- State Ohm's law, relate it to metallic conductors, and solve Ohm's law problems.
- Define the term *electric circuit,* and list the components in a simple circuit.
- Draw the circuit symbols for a resistor and a source of potential difference.
- State the equations for determining power and energy output in electric circuits, and solve problems using these equations.
- State the relationships in a series circuit, and solve problems using these relationships.
- State the relationships in a parallel circuit, and solve problems using these relationships.
- Compare and contrast series and parallel circuits.
- State Kirchhoff's rules as they apply to electric circuits.

10.1   INTRODUCTION

In Chapter 9, we learned about static charges. In this chapter we learn about charges in motion and the way they function in electric circuits. Without electric circuits it would be impossible to operate most of the devices that we take for granted, such as televisions, telephones, blenders, and vacuum cleaners.

10.2   ELECTRIC CURRENT

The word *current* means "flow" and **electric current** means "flow of charge." When we use the term *electric current,* we are referring, not to the speed of the charged particles, but to the *quantity* of charge that passes a single point in time.

As an analogy, consider the movement of cars on a highway. We measure the speed of a single car by calculating the distance it travels in a given amount of time. We measure the flow of traffic by counting the number of cars passing a given point in a given amount of time. During rush hour, the flow of traffic will be high, but the speed of any individual car may be quite small. At 4 A.M., however, the reverse is likely to be true.

The SI unit of current is the *ampere* (A) which is equivalent to 1 *coulomb per second*. The symbol used to represent current is $I,$ and we can write

---

PHYSICS CONCEPTS

$$I = \frac{Q}{t}$$

PROBLEM

What is the electric current in a conductor if 240 coulombs of charge pass through it in 1.0 minute?

SOLUTION

We solve the problem by using the relation $I = \dfrac{Q}{t}$ and remembering that time must be measured in seconds.

$$I = \frac{Q}{t}$$

$$= \frac{240 \text{ C}}{60. \text{ s}} = 4.0 \text{ A}$$

In metallic conductors, the current is carried by freely moving electrons; in solutions, the current is carried by ions; in gases, the current is carried by both electrons and ions.

10.3 CURRENT AND POTENTIAL DIFFERENCE

How do we establish a current in a conductor? To initiate the flow of charge, we find that a *potential difference* is necessary:

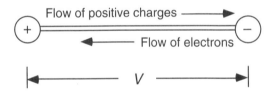

We can see from the diagram that the potential difference is oriented so that positive charges will flow toward the right (the negative terminal) while negative charges (including electrons) will flow toward the left (the positive terminal).

We call the flow of positive charges a *conventional current,* and its direction is always opposite to the direction of electron flow. A conventional current is equivalent to an *electron flow* of the same magnitude in every respect *except direction.* In this book, when we speak of the direction of the current, we are referring to conventional current.

10.4 RESISTANCE

Suppose we removed the source of potential difference in the diagram in Section 14.3. What would happen to the flow of charge in the conductor? We know what would happen: the current would cease. A real-life situation of this type is turning off the switch of a lit lamp.

The reason the current ceases is that, as charges travel through the conductor, they lose energy because of collisions with the atoms of the conductor. This situation is roughly analogous to an object's coming to rest on a surface because of friction. We call the electrical analog of friction *resistance*. An effect of electrical resistance is the almost total conversion of the lost energy into heat. Resistance is the reason that wires become hot as they conduct a current.

We measure the resistance of a material by placing a potential difference across it and then measuring the amount of current that passes through the material. **Resistance** is defined as the ratio of potential difference to current:

PHYSICS CONCEPTS

$$R = \frac{V}{I}$$

The SI unit of resistance is the *volt per ampere* (V/A), which is called the *ohm* (Ω) in honor of German physicist Georg Ohm.

PROBLEM

When a conductor has a potential difference of 110 volts placed across it, the current through it is 0.50 ampere. What is the resistance of the conductor?

SOLUTION

$$R = \frac{V}{I}$$

$$= \frac{110 \text{ V}}{0.50 \text{ A}}$$

$$= 220 \ \Omega$$

The resistance of a material depends on (1) the nature of the material, (2) the geometry of the conductor, and (3) the temperature at which the resistance is measured.

1. Metallic substances are good conductors; that is, they have low resistances. Silver and copper are the best metallic conductors. The quantity that measures how well a substance resists carrying a current is known as the *resistivity* (ρ); its unit is the *ohm·meter* ($\Omega \cdot$ m). The table that follows lists the resistivities of some common materials.

Substance	Resistivity ($\Omega \cdot$ m at 20°C)
Silver	1.59×10^{-8}
Copper	1.68×10^{-8}
Aluminum	2.65×10^{-8}
Iron	9.71×10^{-8}
Silicon (semiconductor)	$0.1 - 60$
Glass (insulator)	$10^9 - 10^{12}$

2. The resistance of a regularly shaped conductor is directly proportional to its length and inversely proportional to its cross-sectional area. This fact makes sense because making a conductor longer increases the likelihood that the electron will collide with the atoms of the conductor, thereby increasing the resistance. Making a conductor wider increases the number of *paths* that the electrons can take and, therefore, decreases the resistance.

3. Generally, the resistance of a metallic conductor increases with rising temperature. This fact also makes sense because increasing the temperature of a conductor increases the vibrational kinetic energy of its atoms, making collisions with electrons more likely. In other materials, such as semiconductors, resistance actually decreases with increasing temperature. (We will study the reasons for this phenomenon in Chapter 14.) In practice, a temperature such as 20°C is chosen as a standard temperature for comparing resistances.

We can combine all of these factors into one relationship:

PHYSICS CONCEPTS

$$R = \rho \cdot \frac{\ell}{A} \text{ (at a specified temperature)}$$

---

where ℓ represents the length of the conductor and A is its cross-sectional area.

PROBLEM
Calculate the resistance at 20°C of an aluminum wire that is 0.200 meter long and has a cross-sectional area of 1.00×10^{-3} square meter.

SOLUTION
We know from the table given above that aluminum has a resistivity of 2.65 $\times 10^{-8}$ $\Omega \cdot$m. Therefore:

$$R = \rho \cdot \frac{\ell}{A} \text{ (at a specified temperature)}$$

$$= 2.65 \times 10^{-8} \ \Omega\text{·m} \cdot \frac{0.200 \ \text{m}}{1.00 \times 10^{-3} \ \text{m}^2}$$

$$= 5.30 \times 10^{-6} \ \Omega$$

Certain substances, called **superconductors**, lose all of their resistance when cooled to very low temperatures. It is also interesting that superconductors do not conduct very well at higher temperatures. If a substance could be made to be superconducting at or near atmospheric temperatures, vast quantities of energy could be conserved because the heat lost by superconducting wires would be minimal.

10.5   ELECTRIC CIRCUITS AND OHM'S LAW

The word *circuit* means "closed path." By an **electric circuit** we mean an arrangement where electric charges can flow in a closed path. The simplest electric circuit consists of a source of potential difference (a battery or a power source), a single resistance, and connecting wires (which are assumed to have negligible resistance).

A device that provides resistance to a circuit is called a *resistor*, and its symbol is as follows:

The symbol for a source of potential difference is

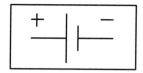

The diagram below represents the completed circuit:

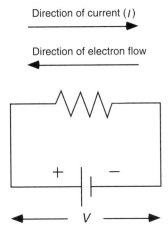

Direction of current (I)

Direction of electron flow

As the electrons begin to flow from the negative terminal, their energy is provided by the source of potential difference. When the electrons reach the positive terminal, all of their energy has been expended (i.e., converted to heat), and the power source must provide them with additional energy for the next trip around the circuit.

Suppose a student performs an experiment using the circuit diagrammed above. The student varies the potential difference across the circuit and measures the current in the circuit with each change. Her results are recorded in the table below.

Potential Difference (V)	Current (A)
0.0	0.0
1.0	1.8
2.0	4.3
3.0	5.7
4.0	8.2
5.0	9.9
6.0	11.8
7.0	14.3
8.0	16.4

To analyze the data, the student finds it useful to plot the data points and then draw a graph, as shown here.

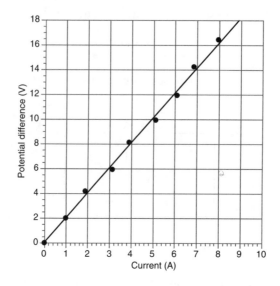

We can see that the graph of the data points, within experimental error, is a straight line that passes through the origin. The *slope* of the graph is the ratio of potential difference to current (which we know as *resistance*). Since the slope is constant, it follows that the resistance of the material is also constant.

PROBLEM
Calculate the resistance of the conductor whose graph is given above.

SOLUTION
To calculate the resistance, we need only calculate the slope of the straight line on the graph.

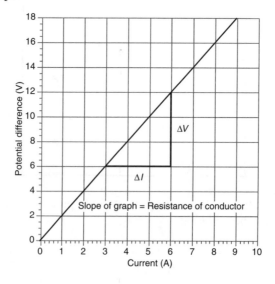

As a result of this calculation ($\Delta V/\Delta I$), we find that the resistance of the conductor is 2.0 ohms.

Materials that have constant resistances are said to obey *Ohm's law*. The mathematical statement of this relationship is as follows:

PHYSICS CONCEPTS

$$V = IR$$

Ohm's law is a fundamental relationship in electric circuits. It describes how much potential difference is required to move charges through a resistance at a given current.

10.6 POWER AND ENERGY IN ELECTRIC CIRCUITS

Suppose we wish to calculate the *rate* at which energy is supplied to the simple circuit drawn in Section 10.5. We know that potential difference is related to work and energy and that current is related to time. If we multiply the potential difference across the circuit by the current in the circuit and examine the *units*, we find that

Potential difference · current = volts · amperes

$$= \frac{\text{joules}}{\text{coulomb}} \cdot \frac{\text{coulombs}}{\text{second}}$$

$$= \frac{\text{joules}}{\text{second}} = \text{watts} \equiv \text{power}$$

Therefore, the *power* supplied to a circuit by the source is given by this relationship:

PHYSICS CONCEPTS

$$P_{\text{source}} = VI$$

PROBLEM

Calculate the rate at which energy is supplied by a 120-volt source to a circuit if the current in the circuit is 5.5 amperes.

SOLUTION

$$P_{source} = VI$$
$$= (120 \text{ V})(5.5 \text{ A}) = 660 \text{ W}$$

Since our "simple" circuit contains only resistance (i.e., no other device, such as a motor, is present), all of the energy is dissipated as heat. We can calculate the rate at which heat energy is produced by using the power relationship ($P = VI$) and Ohm's law to substitute the term IR for V:

PHYSICS CONCEPTS

$$P_{heat} = VI = IR(I) = I^2R$$

This relationship can be used to calculate the rate at which any resistance in a circuit produces heat energy.

PROBLEM

A 150-ohm resistor carries a current of 2.0 amperes. Calculate the rate at which heat energy is produced by the resistor.

SOLUTION

$$P_{heat} = I^2R$$
$$= (2.0 \text{ A})^2(150 \text{ } \Omega)$$
$$= 600 \text{ W}$$

If, however, another device is present in the circuit, the term VI will not be equal to the term I^2R because not all of the electric energy is converted to heat energy.

There is a third relationship for calculating power: $P = \dfrac{V^2}{R}$. See whether you can derive this relationship using Ohm's law and the power equation.

If we know how much power is developed by a circuit, it is an easy matter to calculate the amount of energy produced: we need only to multiply the power (in watts) by the time the circuit operates (in seconds). The result will be the amount of energy (in joules).

PHYSICS CONCEPTS

$$\text{Energy} = \text{Power} \cdot \text{time}$$
$$= VIt = I^2Rt = \frac{V^2}{R} \cdot t$$

PROBLEM

How much energy is produced by 50.-volt source that generates a current of 5.0 amperes for 2.0 minutes?

SOLUTION

$$Energy = VIt$$
$$= (50.\ \text{V})(5.0\ \text{A})(120\ \text{s})$$
$$= 30,000\ \text{J}$$

10.7   SERIES CIRCUITS

At one time, small holiday lights were arranged so that, if one bulb burned out, the entire string of lights remained unlit. We call the type of electric circuit that produced this effect a series circuit. A **series circuit** has only one current path and if that path is interrupted, the *entire* circuit ceases to operate.

The diagram represents a circuit containing three resistors arranged in series. In addition a number of *meters* have been placed in order to measure various characteristics of the circuit.

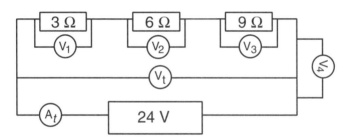

Each symbol ─Ⓥ─ represents a *voltmeter*; this very-high-resistance device measures the potential difference across two points in a circuit. The symbol V_t represents the total potential difference across the circuit. The symbol ─Ⓐ─ represents an *ammeter*; this very-low-resistance device measures the current passing through any part of the circuit. The subscript *t* indicates that the ammeter in the diagram is measuring the total current through the circuit.

Since a series circuit contains only one current path, the current throughout the circuit is constant; therefore, an ammeter placed at any other position in the circuit would record the same value.

The situation is not the same with potential difference: The potential difference across two points depends on the work the source must do in order to move the charge between these two points. In the circuit shown in the diagram, the resistance across any two points will determine how much work needs to be done to transport the charges through the circuit. Our aim is to calculate the readings on all of the meters in the diagram.

233

We can solve this problem by being aware of the following relationships, which hold true for *any series circuit*:

PHYSICS CONCEPTS

$$I_t = I_1 = I_2 = I_3 = \ldots = I_n$$
$$V_t = V_1 + V_2 + V_3 + \ldots + V_n$$
$$V_n = I_n R_n$$

The first relationship states that the current through any resistance in a series circuit is constant throughout the circuit. The second relationship states that the potential difference across the entire circuit (V_t), supplied by the power source, is equal to the sum of the potential differences (V_1, V_2, . . .) across all the resistances. (This is really a statement of the law of conservation of energy and is known, in honor of German physicist Gustan Kirchhoff, as *Kirchhoff's first rule* or is called, more simply, the *loop rule*.) The third relationship states that Ohm's law holds for each resistance.

If we combine the three statements, we can develop a means of finding the resistance of the circuit as a whole:

$$V_t = V_1 + V_2 + V_3 + \ldots + V_n$$
$$I_t R_t = I_1 R_1 + I_2 R_2 + I_3 R_3 + \ldots + I_n R_n$$
$$= I_t R_1 + I_t R_2 + I_t R_3 + \ldots + I_t R_n$$
$$= I_t (R_1 + R_2 + R_3 + \ldots + R_n)$$

PHYSICS CONCEPTS

$$R_t = R_1 + R_2 + R_3 + \ldots + R_n$$

R_t is also known as the *equivalent resistance* of the circuit.

Now let us examine the series circuit diagram again and calculate all of the meter readings.

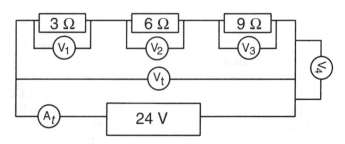

The equivalent resistance (R_t) of the circuit is found from the relationship

$$R_t = R_1 + R_2 + R_3 + \ldots + R_n$$
$$= 3\,\Omega + 6\,\Omega + 9\,\Omega = 18\,\Omega$$

The current through the circuit (I_t) is found from the relationship

$$V_t = I_t R_t$$

We know that V_t equals 24 volts since the source supplies the entire circuit:

$$24\ \text{V} = I_t(18\,\Omega)$$
$$I_t = 1.33\ \text{A}$$

The potential difference across each resistance can be found by using Ohm's law:

$$V_1 = (1.33\ \text{A})(3\,\Omega) = 4\ \text{V}$$
$$V_2 = (1.33\ \text{A})(6\,\Omega) = 8\ \text{V}$$
$$V_3 = (1.33\ \text{A})(9\,\Omega) = 12\ \text{V}$$

What does voltmeter V_4 read? Since we neglect the resistance of the connecting wires, we assume their resistance to be (nearly) 0 ohm, so the potential difference needed to move the charges across that section of the wire is (nearly) 0 volt.

Another fact about series circuits is important: As the *number* of resistances in a series circuit increases, the equivalent resistance of the circuit increases and the current through the circuit *decreases*. This effect is roughly equivalent to that obtained by increasing the length of a conductor. If the resistances were light bulbs, the bulbs would get dimmer as more were added to the circuit. The next problem illustrates this point.

PROBLEM
Suppose a fourth resistance of 18 ohms is added to the series circuit we have been considering. Calculate (a) the equivalent resistance of the circuit and (b) the current through the circuit.

SOLUTION

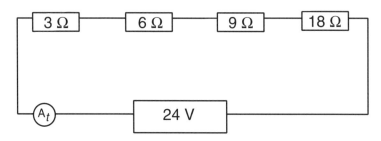

(a) The equivalent resistance (R_t) is now:

$$R_t = 3\ \Omega + 6\ \Omega + 9\ \Omega + 18\ \Omega = 36\ \Omega$$

(b) The current through the circuit is found, once again, from Ohm's law:

$$V_t = I_t R_t$$
$$24\ \text{V} = I_t(36\ \Omega)$$
$$I_t = 0.67\ \text{A}$$

As stated above, the equivalent resistance has increased and the current through the circuit has decreased in comparison to the original circuit.

10.8 PARALLEL CIRCUITS

In contrast to a series circuit, a **parallel circuit** has more than one current path. If a segment of a parallel circuit is interrupted, the result will not necessarily be that the enitire circuit ceases to operate. In a home, for example, the burning out of a single bulb does not usually darken the entire house.

The diagram below represents a parallel circuit containing two resistances and a number of suitably placed meters.

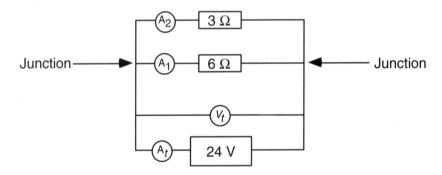

In this type of circuit, the current separates into more than one path. The point (or points) where this separation occurs is known as a *junction*. As a consequence of the law of conservation of electric charge, the sum of the currents *entering* a junction must be equal to the sum of the currents *leaving* the junction. This statement is known as *Kirchhoff's second rule* or, more simply, as the *junction rule*.

PROBLEM
In the diagram below, what are the magnitude and the direction of the current in wire X?

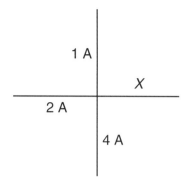

SOLUTION

There is *more than one* answer to this problem! The value of the current in X depends on the directions of the other currents. We know that, in each case, the sum of the currents *entering* the junction must be equal to the sum of the currents *leaving* it.

Two solutions are shown in the diagrams. Can you find other solutions?

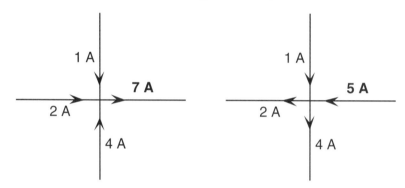

We can analyze the circuit shown on page 236 by using the following relationships, which are valid for *any parallel circuit*:

PHYSICS CONCEPTS

$$V_t = V_1 = V_2 = V_3 = \ldots = V_n$$
$$I_t = I_1 + I_2 + I_3 + \ldots + I_n$$
$$V_n = I_n R_n$$

The first relationship states that the potential difference across a parallel circuit is constant. This relationship follows from the fact that each resistance comprises an independent path for the flowing charges and that, if one resistance ceases to operate, the others can continue to function. The second relationship states that the current through the entire circuit is equal to the sum of the currents through all the resistances. (This is really an application of Kirchhoff's second rule.) Once again, the third relationship states that Ohm's law holds for each resistance.

If we combine the three statements, we can develop a means of finding the resistance of the parallel circuit as a whole:

$$V_n = I_n R_n n \Rightarrow I_n = \frac{V_n}{R_n}$$

$$I_t = I_1 + I_2 + I_3 + \ldots + I_n$$

$$\frac{V_t}{R_t} = \frac{V_1}{R_1} + \frac{V_2}{R_2} + \frac{V_3}{R_3} + \ldots + \frac{V_n}{R_n}$$

$$= \frac{V_t}{R_1} + \frac{V_t}{R_2} + \frac{V_t}{R_3} + \ldots + \frac{V_t}{R_n}$$

$$= V_t \left(\frac{1}{R_t} = \frac{1}{R_1} + \frac{1}{R_2} + \frac{1}{R_3} + \ldots + \frac{1}{R_n} \right)$$

PHYSICS CONCEPTS

$$\frac{1}{R_t} = \frac{1}{R_1} + \frac{1}{R_2} + \frac{1}{R_3} + \ldots + \frac{1}{R_n}$$

Here, R_t is the equivalent resistance of the parallel circuit. We will use these relationships to calculate the meter readings in the diagram of the parallel circuit on page 236.

First, we can use Ohm's law $\left(I = \frac{V}{R} \right)$ to calculate currents I_1 and I_2:

$$I_1 = \frac{V_t}{R_1} = \frac{24 \text{ V}}{3 \, \Omega} = 8 \text{ A}$$

$$I_2 = \frac{V_t}{R_2} = \frac{24 \text{ V}}{6 \, \Omega} = 4 \text{ A}$$

Next, we calculate the total current I_t by adding I_1 and I_2:

$$I_t = 8 \text{ A} + 4 \text{ A} = 12 \text{ A}$$

We find the equivalent resistance R_t from the relationship

$$\frac{1}{R_t} = \frac{1}{R_1} = \frac{1}{R_2}$$

$$= \frac{1}{3\ \Omega} + \frac{1}{6\ \Omega} = \frac{1}{2\ \Omega}$$

$$R_t = 2\ \Omega$$

We could also have used Ohm's law ($V_t = I_t R_t$) to calculate the equivalent resistance of this circuit.

We note that the equivalent resistance is less than any single resistance in the circuit. This is characteristic of parallel circuits in general.

If more resistance is added in parallel, the equivalent resistance *decreases* and the total current *increases* because each new parallel resistance creates another independent path in which charges can flow. The result is roughly equivalent to that obtained by increasing the cross-sectional area of a conductor. For this reason, overloading a household circuit by connecting too many electrical appliances is dangerous. As the current in the house wires increases, the amount of heat energy also increases, a situation that may lead to fires in unprotected circuits. Fortunately, fuses and circuit-breakers are designed to prevent such fires from occurring. The next problem illustrates this effect.

PROBLEM

A 2-ohm resistor is added in parallel to the parallel circuit shown at the beginning of this section. Calculate (a) the equivalent resistance and (b) the total current of the altered circuit.

SOLUTION

The diagram of the modified circuit is as follows:

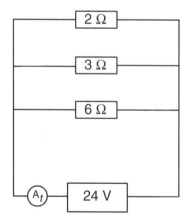

(a)

$$\frac{1}{R_t} = \frac{1}{R_1} + \frac{1}{R_2} + \frac{1}{R_3}$$

$$= \frac{1}{3\,\Omega} + \frac{1}{6\,\Omega} + \frac{1}{2\,\Omega} = \frac{1}{1\,\Omega}$$

$$R_t = 1\,\Omega$$

(b)

$$I_t = \frac{V_t}{R_t} = \frac{24\text{ V}}{1\,\Omega} = 24\text{ A}$$

As we can see, the equivalent resistance has decreased to $1\,\Omega$ and the total current has increased to 24 A.

Most circuits represent more complex combinations of series and parallel arrangements than are shown in this chapter. In addition, they may include additional power sources, current loops, and junctions. These complex circuits will not be analyzed in this book. You should be aware, however, that Kirchhoff's rules and some fancy algebra can be used for these analyses. Ask your physics teacher to show you how to analyze one of these complex circuits. Here is a situation where your teacher earns his or her richly deserved pay!

Questions

1. An ampere can be defined as 1
   (1) C/s   (3) J/C
   (2) Ω/V   (4) N/C

2. Electrical conductivity in liquid solutions depends on the presence of free
   (1) neutrons   (2) protons   (3) molecules   (4) ions

3. An electric current in a metal consists of moving
   (1) nuclei   (2) protons   (3) neutrons   (4) electrons

4. As the temperature of a coil of copper wire increases, its electrical resistance
   (1) decreases   (2) increases   (3) remains the same

5. If the cross-sectional area of a metallic conductor is halved and the length of the conductor is doubled, the resistance of the conductor will be
   (1) halved   (2) doubled   (3) unchanged   (4) quadrupled

6. Which segment of copper wire has the highest resistance at room temperature?
   (1) 1.0 m length, 1.0×10^{-6} m^2 cross-sectional area
   (2) 2.0 m length, 1.0×10^{-6} m^2 cross-sectional area
   (3) 1.0 m length, 3.0×10^{-6} m^2 cross-sectional area
   (4) 2.0 m length, 3.0×10^{-6} m^2 cross-sectional area

7. The ratio of the potential difference across a conductor to the current in the conductor is called $V = I \cdot R$
   (1) conductivity   (3) charge
   (2) resistance   (4) power

8. Most metals are good electrical conductors because
   (1) their molecules are close together
   (2) they have high melting points
   (3) they have many intermolecular spaces through which the current can flow
   (4) they have a large number of free electrons

9. Which graph represents a circuit element at constant temperature that obeys Ohm's law? $R = \dfrac{V}{I}$

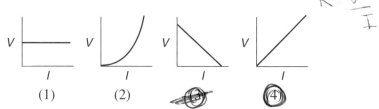

   (1)   (2)   (3)   (4)

10. Which graph best represents the relationship between the resistance (R) of a solid conductor of constant cross section and its length (L)?

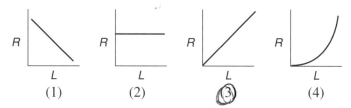

   (1)   (2)   (3)   (4)

11. The resistance of a wire at constant temperature depends on the wire's
    (1) length, only
    (2) type of metal, only
    (3) length and cross-sectional area, only
    (4) length, cross-sectional area, and type of metal

12. The current in a circuit is supplied by a generator. If the resistance in the circuit is increased, the force required to keep the generator turning at the same speed is

   (1) decreased   (2) increased   (3) the same

13. The ratio of the potential difference across a conductor to the current in the conductor is called

   (1) energy   (2) charge   (3) resistance   (4) power

14. The graph below shows how the voltage and current are related in a simple electric circuit.

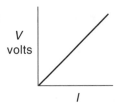

For any point on the line, what does the ratio of V to I represent?

   (1) work, in joules   (3) resistance, in ohms
   (2) power, in watts   (4) charge, in coulombs

15. The graph below represents the relationship between potential difference and current for four different resistors. Whieh resistor has the greatest resistance?

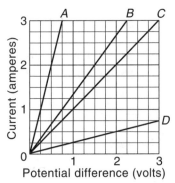

   (1) *A*   (2) *B*   (3) *C*   (4) *D*

16. Three ammeters are located near junction P in a direct current circuit as shown in the diagram below. If ammeter A_1 reads 3 amperes and ammeter A_2 reads 4 amperes, what does ammeter A_3 read?

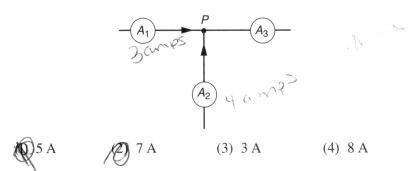

(1) 5 A   (2) 7 A   (3) 3 A   (4) 8 A

17. In the diagram shown, how many amperes is the reading of ammeter A?

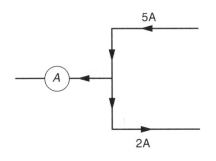

(1) 5 A   (2) 2 A   (3) 3 A   (4) 7 A

18. The diagram below represents currents flowing in branches of an electric circuit. What is the reading on ammeter A?

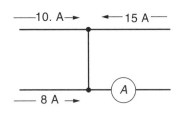

(1) 13 A   (2) 17 A   (3) 3 A   (4) 33 A

19. The diagram below represents currents flowing in branches of an electric circuit. What is the reading on ammeter A?

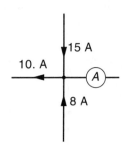

(1) 13 A   (2) 17 A   (3) 3 A   (4) 33 A

20. If the potential difference across a 30.-ohm resistor is 10. volts, what is the current through the resistor?

(1) 0.25 A   (2) 0.33 A   (3) 3.0 A   (4) 0.50 A

21. If the voltage across a 4-ohm resistor is 12 volts, the current through the resistor is

(1) 0.33 A   (2) 48 A   (3) 3.0 A   (4) 4.0 A

22. A charge of 5.0 coulombs moves through a circuit in 0.50 second. How much current is flowing through the circuit?

(1) 2.5 A   (2) 5.0 A   (3) 7.0 A   (4) 10. A

23. A flow rate of 1 coulomb per 0.01 second is measured in a wire. What is the electrical current in the wire?

(1) 1 A   (2) 0.1 A   (3) 10 A   (4) 100 A

24. What is the current in a conductor if 6.25×10^{18} electrons pass a given point each second?

(1) 1 A   (2) 1.6×10^{-19} A   (3) 2.6 A   (4) 6.25×10^{18} A

25. A resistor carries a current of 0.10 ampere when the potential difference across it is 5.0 volts. The resistance of the resistor is

(1) 0.020 Ω   (2) 0.50 Ω   (3) 5.0 Ω   (4) 50. Ω

26. If the potential difference across a 12-ohm resistor is 6 volts, the current through the resistor is

(1) 0.33 A   (2) 0.50 A   (3) 3.0 A   (4) 4.0 A

27. The circuit represented in the diagram below is a series circuit.

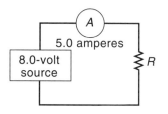

Power =
(volts)
(Ampere)

The electric energy expended in resistor R in 2.0 seconds is
(1) 20. J (2) 40. J (3) 80. J (4) 120 J

28. Which combination of current and electromotive force would use energy at the greatest rate?
(1) 10 A at 110 V (3) 3 A at 220 V
(2) 8 A at 110 V (4) 5 A at 110 V

29. A 10-volt potential difference maintains a 2-ampere current in a resistor. The total energy expended by this resistor in 5 seconds is
(1) 10 J (2) 20 J (3) 50 J (4) 100 J

30. What is the current in a normally operating 60-watt, 120-volt lamp?
(1) 1.0 A (2) 2.0 A (3) 0.50 A (4) 4.0 A

31. How much electric energy will be used by a 115-volt, 60-watt light bulb in 1 minute?
(1) 60 J (2) 115 J (3) 3,600 J (4) 6,900 J

32. If the current and the resistance of an electric circuit are each doubled, the power will
(1) remain the same (3) be 8 times as large
(2) be doubled (4) be quadrupled

33. An ampere-volt is a unit of
(1) work (2) resistance (3) energy (4) power

34. An electric heater raises the temperature of a measured quantity of water. The water absorbs 6000 joules of energy from the heater in 30.0 seconds. What is the minimum power supplied to the heater?
(1) 5.00×10^2 W (3) 1.80×10^5 W
(2) 2.00×10^2 W (4) 2.00×10^3 W

35. A circuit contains a rheostat (variable resistor) connected to a source of constant voltage. As the resistance of the rheostat increases, the power dissipated in the circuit
(1) decreases (2) increases (3) remains the same

36. In the circuit represented below, which switches must be closed to produce a current in conductor *AB*?

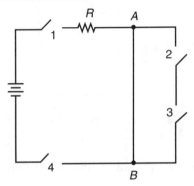

    (1) 1 and 4   (2) 2 and 3   (3) 1, 2, and 3   (4) 2, 3, and 4

Base your answers to questions 37 through 41 on the information and diagram below.

The diagram represents a direct-current motor connected in series with a 12-volt battery, a resistance *R*, and an ammeter *A*. Mass *M* is suspended from a pulley that is attached to the shaft of the motor.

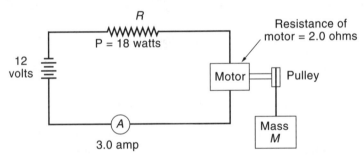

37. What is the resistance of resistor *R*?
    (1) $8.0\ \Omega$ (2) $2.0\ \Omega$ (3) $3.0\ \Omega$ (4) $6.0\ \Omega$

38. What is the potential difference across resistor *R*?
    (1) 0 V   (2) 12 V   (3) 6.0 V   (4) 4.0 V

39. What is the rate at which the motor uses electric energy?
    (1) 0 W   (2) 18 W   (3) 36 W   (4) 4.0 W

40. What is the total charge that will pass through the motor in 5.0 seconds?
    (1) 0.60 C   (2) 15 C   (3) 3.0 C   (4) 4.0 C

41. If the motor is 100% efficient, how much will the gravitational potential energy of mass M change in 10 seconds?
   (1) 9.8 J   (2) 12 J   (3) 18 J   (4) 180 J

42. The algebraic sum of all the potential drops and applied voltages around a complete circuit is equal to zero. This is an application of the law of conservation of
   (1) mass   (2) energy   (3) charge   (4) momentum

Base your answers to questions 43 through 47 on the information below.

An electric heater rated at 4800 watts is operated on 120 volts.

43. What is the resistance of the heater?
   (1) 576,000 Ω   (2) 120 Ω   (3) 3.0 Ω   (4) 40. Ω

44. How much energy is used by this heater in 10.0 seconds?
   (1) 1.15 J   (2) 40. J   (3) 4.8×10^3 J   (4) 4.8×10^4 J

45. If the heater were replaced by one having a greater resistance, the amount of heat produced each second would
   (1) decrease   (2) increase   (3) remain the same

46. If another heater is connected in parallel with the first one and both operate at 120 volts, the current in the first heater will
   (1) decrease   (2) increase   (3) remain the same

47. If the original heater were operated at fewer than 120 volts, the amount of heat produced would
   (1) decrease   (2) increase   (3) remain the same

Base your answers to questions 48 through 53 on the diagram below, which represents an electric circuit. Charge is transferred from point A to point B at the rate of 5 coulombs per second for 120 seconds. Six joules of work are done in transferring each coulomb of charge.

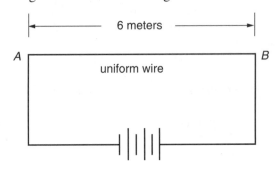

48. What is the current in the wire?
   (1) 1.2 A   (2) 5 A   (3) 6 A   (4) 24 A

49. What is the potential difference between points A and B?
   (1) 6 V   (2) 30 V   (3) 100 V   (4) 600 V

50. What is the rate at which work is done in this circuit?
   (1) 6 W   (2) 30 W   (3) 144 W   (4) 720 W

51. What is the total work done in this circuit?
   (1) 6 J   (2) 30 J   (3) 720 J   (4) 3600 J

52. What is the rate at which electrons are transferred in this circuit?
   (1) 8×10^{-19} electron/s   (3) 1.1×10^{18} electrons/s
   (2) 5 electrons/s   (4) 3.1×10^{19} electrons/s

53. What force moves 1 coulomb of charge from point A to point B?
   (1) 1 N   (2) 5 N   (3) 30 C   (4) 120 C

54. If V_2 in the diagram reads 24 volts, V_1 will read

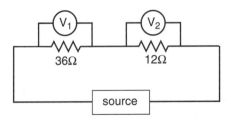

   (1) 8 V   (2) 24 V   (3) 48 V   (4) 72 V

Base your answers to questions 55 through 59 on the diagram below, which shows three resistors connected to a 15-volt source.

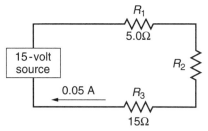

55. The equivalent resistance of the circuit is
    (1) 10 Ω   (2) 20 Ω   (3) 30 Ω   (4) 40 Ω

56. The potential difference across resistor R_2 is
    (1) 2.5 V   (2) 5.0 V   (3) 7.5 V   (4) 10 V

57. The total power developed in the circuit is
    (1) 2.5 W   (2) 5.0 W   (3) 7.5 W   (4) 10 W

58. Compared to the heat developed in resistor R_1, the heat developed in resistor R_3 is
    (1) one-third as great   (3) 3 times as great
    (2) two times as great   (4) one-fourth as great

59. If resistor R_3 is removed and replaced by a resistor of lower value, the resistance of the circuit will
    (1) decrease   (2) increase   (3) remain the same

60. If three resistors of 9 ohms each are connected in series, their total resistance will be
    (1) 4.5 Ω   (2) 27 Ω   (3) 3 Ω   (4) 9 Ω

Base your answers to questions 61 through 64 on the diagram below.

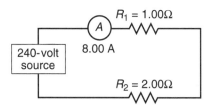

61. The potential difference across R_1 is
    (1) 0 V   (2) 8.00 V   (3) 12.0 V   (4) 24.0 V

62. What is the total resistance of the circuit?
(1) 0.500 Ω   (2) 2.00 Ω   (3) 3.00 Ω   (4) 4.00 Ω

63. What power is supplied by the source?
(1) 24.0 W   (2) 90.0 W   (3) 3.00 W   (4) 192 W

64. What is the current in resistor R_2?
(1) 8.00 A   (2) 2.00 A   (3) 16.0 A   (4) 4.00 A

65. The diagram below represents an electric circuit.

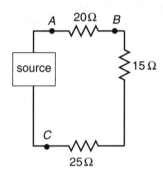

If the voltage between *A* and *B* is 10 volts, the voltage between *B* and *C* is
(1) 5 V   (2) 10 V   (3) 15 V   (4) 20 V

66. Compared to the potential drop across the 10-ohm resistor shown in the diagram, the potential difference across the 5-ohm resistor is

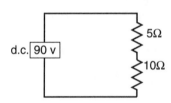

(1) the same   (3) one-half as great
(2) twice as great   (4) 4 times as great

67. Ammeter *A* in the diagram below will read 2 amperes when switch *B* makes contact with which terminal?

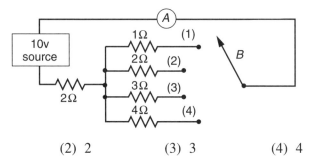

    (1) 1   (2) 2   (3) 3   (4) 4

68. As more resistors are added in series across a battery, the potential drop across each resistor
(1) decreases   (2) increases   (3) remains the same

69. A 5-ohm resistor and a 10-ohm resistor are connected in series. If the current in the 5-ohm resistor is doubled, the current in the 10-ohm resistor will
(1) be halved   (3) be doubled
(2) remain the same   (4) be quadrupled

Base your answers to questions 70 through 73 on the diagram below. The reading of voltmeter V is 6.0 volts [Neglect the resistance of the connecting wires.]

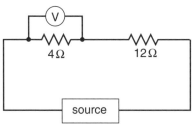

70. The current in the 4.0-ohm resistor is
   (1) 0.38 A   (2) 0.67 A   (3) 1.5 A   (4) 24 A

71. The total resistance of the 4.0-ohm and 12-ohm resistors is
   (1) 48 Ω   (2) 22 Ω   (3) 3 Ω   (4) 16 Ω

72. Heat is developed in the 4.0-ohm resistor at the rate of
   (1) 96 W   (2) 24 W   (3) 10 W   (4) 9 W

73. The potential difference of the source is
   (1) 16 V   (2) 24 V   (3) 3.0 V   (4) 32 V

Base your answers to questions 74 through 77 on the circuit diagram below. The reading of ammeter A_1 is 2.0 amperes. Neglect the resistance of the connecting wires and the battery.

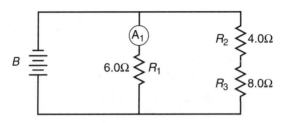

74. The potential difference supplied by battery B is
    (1) 1.0 V   (2) 2.0 V   (3) 6.0 V   (4) 12 V

75. How much heat energy will be produced by resistor R_1 in 3.0 seconds?
    (1) 72 J   (2) 2.0 J   (3) 12 J   (4) 24 J

76. How much charge will pass through resistor R_1 in 3.0 seconds?
    (1) 1.0 C   (2) 2.0 C   (3) 6.0 C   (4) 12 C

77. Compared to the potential difference across resistor R_3, the potential difference across resistor R_2 is
    (1) one-half as much   (3) twice as much
    (2) the same   (4) 3 times as much

78. If the current in the 10-ohm resistor in the diagram is 1 ampere, then the current in the 40-ohm resistor is

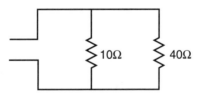

    (1) 1 A   (2) 0.25 A   (3) 5 A   (4) 4 A

79. As additional resistors are connected in parallel to a source of constant voltage, the current in the circuit
    (1) decreases   (2) increases   (3) remains the same

Base your answers to questions 80 through 84 on the diagram below, which shows three resistors connected in parallel to a 300-volt direct-current source.

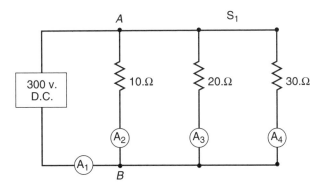

80. A voltmeter connected between points *A* and *B* will have a reading of
(1) 20 V   (2) 50 V   (3) 100 V   (4) 300 V

81. What is the reading of ammeter A_3?
(1) 5.0 A   (2) 15 A   (3) 30. A   (4) 55 A

82. The greatest amount of current is through ammeter
(1) A_1 (2) A_2 (3) A_3 (4) A_4

83. With switch S_1 open, the equivalent resistance of the circuit is
(1) 5.5 Ω   (2) 6.7 Ω   (3) 10. Ω   (4) 30. Ω

84. Heat is produced by the 10-ohm resistor at a rate of
(1) 30. W   (2) 3.0×10^2 W   (3) 9.0×10^3 W (4) 3.0×10^3 W

85. Which circuit below would have the lowest voltmeter reading?

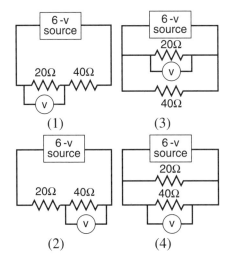

253

86. Which circuit segment has an equivalent resistance of 6 ohms?

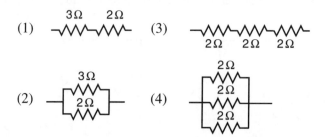

Base your answers to questions 87 through 90 on the diagram below.

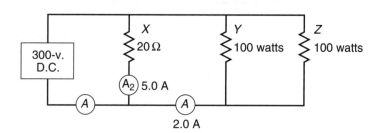

87. The reading of ammeter A_3 is
    (1) 1 A   (2) 5 A   (3) 7 A   (4) 15 A

88. The current in resistor Y is
    (1) 1.0 A   (2) 2.0 A   (3) 0.5 A   (4) 5.0 A

89. The power dissipated by resistor X is
    (1) 100 W   (2) 200 W   (3) 220 W   (4) 500 W

90. The energy used by resistor Y in 10 minutes is
    (1) 10 J   (2) 100 J   (3) 1000 J   (4) 60,000 J

Base your answers to questions 91 through 96 on the diagram below, which represents two metal-wire resistors connected to a source of voltage. The ammeter reads 5.0 amperes.

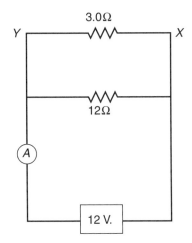

91. The rate at which the source of voltage supplies energy to these resistors can be expressed in units of
(1) amperes   (2) joules   (3) volts   (4) watts

92. What is the current in the 12-ohm resistor?
(1) 1.0 A   (2) 5.0 A   (3) 3.0 A   (4) 4.0 A

93. The work done in moving an electron through the 3.0-ohm resistor is
(1) 1.6×10^{-19} J   (3) 12. J
(2) 1.9×10^{-18} J   (4) 4.0 J

94. Compared to the heat produced by the 12-ohm resistor in one second, the heat produced by the 3-ohm resistor in one second is
(1) one-fourth as great   (3) 4 times as great
(2) one-half as great   (4) 16 times as great

95. If the 3.0-ohm resistor were removed from the circuit, the equivalent resistance of the circuit would
(1) decrease   (2) increase   (3) remain the same

96. If the 12-ohm resistor were heated, its resistance would
(1) decrease   (2) increase   (3) remain the same

Base your answers to questions 97 through 101 on the diagram below, which shows a 40.-watt light bulb and a 120-watt light bulb connected in parallel to a 120-volt direct-current source.

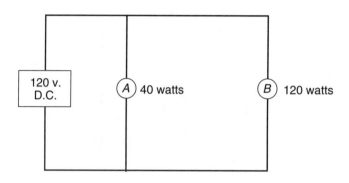

97. What is the current in bulb *A*?
    (1) 4800 A   (2) 40. A   (3) 3.0 A   (4) 0.33 A

98. Energy is supplied by the source at a rate of
    (1) 40 W   (2) 80 W   (3) 120 W   (4) 160 W

99. What is the potential difference across bulb *B*?
    (1) 60. V   (2) 120 V   (3) 240 V   (4) 480 V

100. How much energy is consumed by bulb *B* in 1 minute?
    (1) 60 J   (2) 2 J   (3) 120 J   (4) 7200 J

101. Compared to the resistance of bulb *A*, the resistance of bulb *B* is
    (1) one-third as great   (3) twice as great
    (2) one-half as great   (4) 3 times as great

MAGNETISM; ELECTROMAGNETISM AND ITS APPLICATIONS

Chapter Eleven

KEY IDEAS

One of the phenomena associated with all substances is magnetism. Magnets have north and south poles, named for the way they orient themselves in the Earth's magnetic field. Unlike electric charges, single magnetic poles have not been discovered: A north pole is found in conjunction with a south pole. Like poles repel each other and unlike poles attract, a property similar to that observed in electric charges. The magnetic field around a magnetic configuration may be mapped using field lines in the same way an electric field is mapped.

Magnetic fields exist whenever an electric current is present. The direction of the magnetic field is always perpendicular to the electric field. Current-carrying wires and charges moving perpendicularly through a magnetic field experience forces that are perpendicular to both the direction of the magnetic field and the motion of the charges. If the conductor is a loop, a torque will result; this fact is the basis for electric motors and meters.

If a conductor is moved perpendicularly through a magnetic field, a potential difference is established across the conductor. If the conductor is rotated through the field, as occurs in a generator, the potential difference and the current will alternate in direction. Whenever potential difference is induced in a conductor, a secondary magnetic field is established that always opposes the original motion. This statement, known as Lenz's law, is a consequence of the law of conservation of energy.

An application of alternating current is the transformer. This device uses a changing magnetic field to increase or decrease the potential difference in a secondary circuit.

If a charge is accelerated, the changing electric and magnetic fields will give rise to electromagnetic waves that are carriers of energy.

KEY OBJECTIVES

At the conclusion of this chapter you will be able to:
- Define the terms *magnet, north pole, south pole, temporary magnet, permanent magnet.*

- Define the term *domain*, and describe how domains contribute to the magnetic properties of a metal such as iron.
- State the conventions for drawing magnetic field lines, and draw simple magnetic field configurations.
- Define the term *magnetic induction*, and state the SI unit for magnetic induction (field strength).
- Use an appropriate hand rule to describe the magnetic field around a current-carrying wire.
- Use an appropriate hand rule to determine the magnetic polarity of a current-carrying coil (solenoid).
- State the factors that influence the magnetic induction in a straight wire and in a solenoid.
- Use an appropriate hand rule to determine the force on a current-carrying wire in an external magnetic field.
- State the equation that determines the magnitude of the force on a current-carrying wire in an external magnetic field, and use this equation to solve related problems.
- Describe the mutual effect of two parallel current-carrying wires.
- Define the term *torque*, and describe how a torque arises as a result of a current-carrying loop in a magnetic field.
- Describe the principle upon which a galvanometer operates.
- Describe how a galvanometer may be converted into an ammeter or a voltmeter.
- Describe how a direct-current motor is constructed.
- Use an appropriate hand rule to determine the force on a charged particle moving in an external magnetic field.
- State the equation that determines the magnitude of the force on a charged particle moving in an external magnetic field, and use this equation to solve related problems.
- Describe the principle upon which mass spectrometry operates.
- Describe how a potential difference may be induced across a conductor moving in a magnetic field.
- State the equations that govern electromagnetic induction, and use them to solve related problems.
- Describe the construction and operation of a simple alternating-current generator.
- State Lenz's law, and demonstrate how it applies to electromagnetic induction.
- Define the term *electromotive force*, state Lenz's law, and define the term *back emf.*
- Define the term *transformer*, and describe the principle upon which a transformer operates.
- State the equations relevant to transformer operation, and apply them to the solution of problems.
- Describe how electromagnetic waves may be produced from accelerating charges.

11.1 INTRODUCTION

The phenomenon of magnetism was known in ancient times, when it was observed that certain rocks (called lodestones) attracted iron. It was also observed that, when pieces of iron were rubbed with lodestones, the iron became magnetized and that, if a very thin magnet was floated on water, one end of the magnet always pointed in the northern direction. As a result of these discoveries, the Chinese used magnets to create compasses with which to navigate their waters.

11.2 GENERAL PROPERTIES OF MAGNETS

A **magnet**, then, is any substance that possesses the properties discussed in Section 11.1. One common shape of a magnet is a rectangular bar, as shown in the diagram:

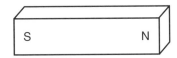

We know that magnets have "polarity." The end that points northward is the **north pole** of the magnet, and the end that points southward is the **south pole**. Also, when two magnets are brought near one another, it is observed that *like poles repel* and *unlike poles attract*. For these reasons we can conclude that the Earth itself behaves as a giant magnet. If a piece of metal is placed in the vicinity of a magnet, the metal itself will become magnetized. If the metal retains its magnetism after the original magnet is removed, it is called a **permanent magnet**; otherwise it is a **temporary magnet**. Alloys such as ALNICO make good permanent magnets, while soft iron produces excellent temporary magnets.

Why do certain substances, such as iron, have magnetic properties? It has been discovered that groups of atoms align their unpaired electrons so that they spin in the same direction, giving rise to microscopic magnets called **domains**. An external magnetic field causes all the domains to align in the same direction, as shown in the diagram on the right.

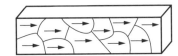

11.3 MAGNETIC FIELDS

In the same way that electrostatic and gravitational forces can be explained by electric and gravitational fields, the existence of magnetic forces can be explained by the presence of magnetic fields. Also, just as field lines are used to visualize electric and gravitational fields, magnetic field lines (called *flux lines*) are used to visualize a magnetic field.

The magnetic fields between various poles of two adjacent magnets are shown in the diagrams, as well as the field around a horseshoe magnet, which is simply a bar magnet that has been bent so that the north and south poles are near each other. By agreement the field lines point away from the north and toward the south.

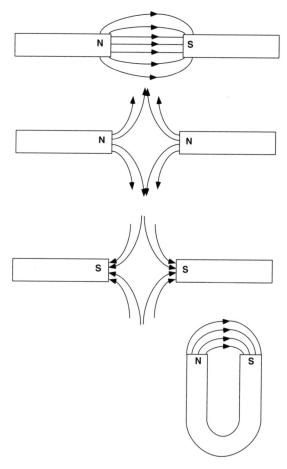

We can also draw the magnetic field around a single bar magnet, as shown below.

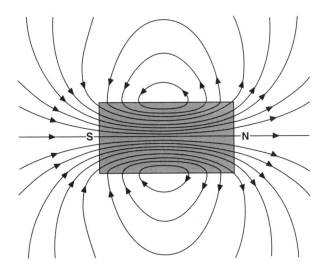

Note that the field lines form closed loops inside the magnet. As a consequence of this fact, magnetic "monopoles" are not believed to exist: a north pole of a magnet is always accompanied by a corresponding south pole.

Each magnetic flux line has been standardized and has the SI unit of 1 weber (Wb). The strength of the magnetic field, known as the **magnetic induction**, is given by the concentration of these flux lines, that is, the number of flux lines per unit area. We represent magnetic induction by the letter **B,** and its unit is the *weber per square meter* (Wb/m^2), also known as the tesla (T). Magnetic induction is a vector quantity because it has both magnitude and direction. A weak field, such as the Earth's magnetic field, has a magnetic induction of approximately 5×10^{-5} tesla. A field of 1 tesla is extremely strong and is used in applications such as magnetic resonance imaging (MRI) and nuclear particle accelerators.

11.4 ELECTROMAGNETISM

In 1820, the Danish physicist Hans Oersted discovered that a wire carrying a current produced a magnetic field as shown in the diagram.

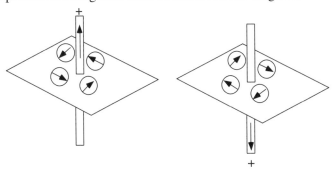

We note that the magnetic field is circular and that its plane is perpendicular to the direction of the wire carrying the current. We can determine the direction of the magnetic field by using what we call *right hand rule 1:*

━━━━━━━━━━━━━━ **PHYSICS CONCEPTS** ━━━━━━━━━━━━━━

The thumb of the right hand is pointed in the direction of the conventional current. The fingers of the right hand (from wrist to fingertips) will curl in the direction of the magnetic field. (*Note:* If you use electron flow instead of conventional current, use your *left* hand instead of your right hand.)

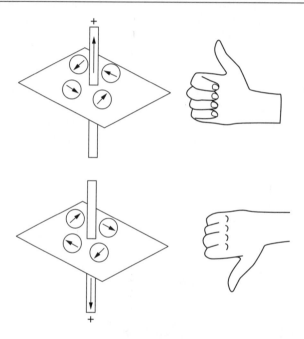

To represent the direction of a magnetic field in two dimensions, we use dots (•) to indicate that the direction is out of the plane of the paper and X's (X) to indicate that the direction is into the plane of the paper, as shown in the diagram.

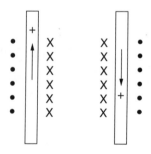

The magnetic induction in the vicinity of a straight wire is directly proportional to the current in the wire and is inversely proportional to the distance from the wire. A uniform magnetic field into or out of the plane of the paper can be represented as shown below:

Uniform field out of the page   Uniform field into the page

Magnetic Field Around a Coil

If we bend our wire into a single loop, the magnetic field will appear as illustrated.

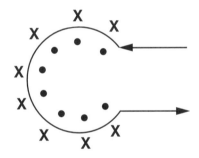

We note below that the magnetic field around the loop has the appearance of a very thin bar magnet. Each face of the single loop is a magnetic pole.

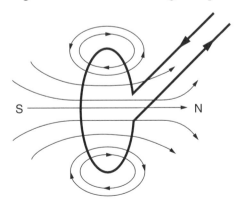

If we link a number of loops together, we produce a coil (or a *solenoid*) whose magnetic field is the result of the fields of the individual loops. The magnetic field around a solenoid is shown below.

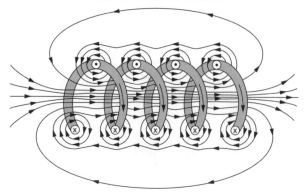

We note that the magnetic field of the solenoid is nearly identical to the magnetic field of a bar magnet. We can determine the north pole of our coil by using what we call *right hand rule 2:*

PHYSICS CONCEPTS

The fingers of the right hand are wrapped in the direction of the conventional current (electron-flow users, be sure to use the left hand). The thumb will point to the end of the coil, which is the north pole.

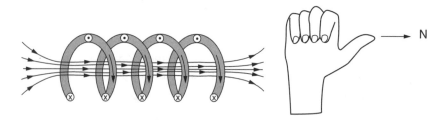

We note that the magnetic field is strongest inside the coil and is also uniform because the lines are closely spaced and are parallel. The magnetic induction depends on the current in the coil, the number of turns per unit length of the coil, and the nature of the core. If a *ferromagnetic* core such as soft iron is placed inside the coil, the magnetic induction can be increased thousands of times—a fact that is applied in making commercial electromagnets.

Forces on a Current Carrier in a Magnetic Field

If a wire carrying a current is placed in a magnetic field, so that the direction of the current is perpendicular to the direction of the magnetic field, the magnetic field of the wire will interact with the magnetic field of the magnet to produce a force on the wire. This is illustrated in the diagram.

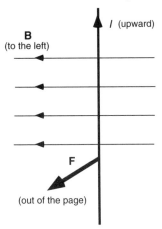

If, however, the current is parallel to the magnetic field, no force will be present on the wire.

When the direction of the current is perpendicular to the magnetic field, the magnitude of the force on the wire can be found from this relationship:

Physics Concepts

$$F = I\ell B$$

This equation tells us that the magnitude of the force on the wire is proportional to the current in the wire, the length of the wire in the magnetic field, and the magnetic induction. If we solve the equation for B $\left(B = \dfrac{F}{I\ell} \right)$, we find that the unit of magnetic induction, the tesla (T), is equivalent to the newton per ampere·meter N/A·m.

PROBLEM
Calculate the force on the wire shown in the diagram above if the current in the wire is 100. amperes, and the length of the wire in a magnetic field whose induction is 3.0×10^{-3} teslas is 2.0 meters.

SOLUTION
$$F = I\ell B$$
$$= (100.\ \text{A})(2.0\ \text{m}) \left(3.0 \times 10^{-3}\ \frac{\text{N}}{\text{A·m}} \right)$$
$$= 6.0 \times 10^{-1}\ \text{N}$$

To determine the direction of the force on the wire, we use what we call *right hand rule 3,* illustrated in the diagram below.

The thumb points in the direction of conventional current, the fingers point in the direction of the magnetic field, and the direction of the force points away from the palm (electron-flow users, left hand, please!).

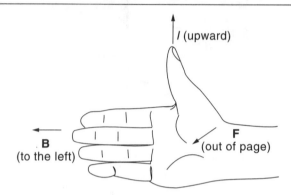

Suppose the wire was neither perpendicular nor parallel to the magnetic field. In that case, there would be a force on the wire but it would be less than that expressed by the relationship $F = I\ell B$. Here, the force is given by the equation $F = I\ell B \sin \theta$, where θ is the angle between the magnetic field and the current-carrying wire.

Forces Between Current-Carrying Parallel Wires

If each of two parallel wires carries a current, the magnetic fields will interact. If the currents are oriented in the same direction, they will attract each other; if oriented in opposite directions, they will repel each other. This situation is illustrated in the diagram.

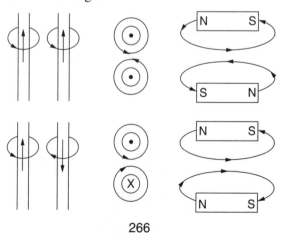

The diagram shows that the orientation of the magnetic field between the two parallel wires is similar to the orientation of the magnetic field between the bar magnets. Because the bar magnets attract each other, we can infer that the wires will attract each other.

The force on either wire is directly proportional to the currents in the two wires and the lengths of the wires, and is inversely proportional to the distance between the wires. This parallel-wire arrangement is used to define the ampere as a unit of current: Two parallel wires will each carry a current of 1 ampere if both wires are 1 meter in length and 1 meter apart, and a force of 2×10^{-7} newton (exactly) exists between them.

Torque and Electromagnetism

The diagram represents a rectangular loop placed in a uniform magnetic field.

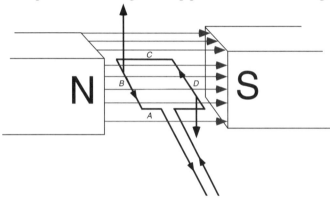

The portions of the loop that are parallel to the magnetic field (*A* and *C*) will experience no force. If we apply right hand rule 3 to section *B* of the loop, we find that it experiences a force directly upward out of the page. Similarly, section *D* will experience a force directly downward into the page. The net result of these two forces is to rotate the loop.

Forces that produce rotational motion are called **torques**. The motion produced by the rotation is the basis for the electric motor. The torque on the loop is proportional to the current in the loop. The loop will not rotate continuously, but will stop after a 90° rotation because the net torque on the loop will then be zero.

Electric Meters

A practical application of the effect described above is the construction of meters for measuring current and potential difference. The simplest device, the *galvanometer*, is used to measure small electric currents. A galvanometer is constructed by placing a coil of wire in a permanent magnetic field. The coil is attached to a spring and a calibrated dial. Since the torque is proportional to the current in the coil, the deflection of the galvanometer needle will depend on this current.

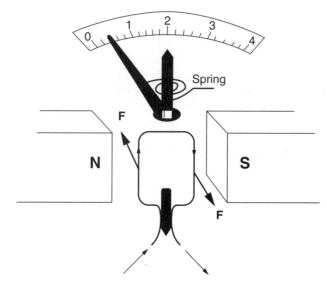

An ammeter is a galvanometer that has been modified in order to measure larger currents. A low-resistance device, known as a shunt, is placed in parallel with the wires of the galvanometer as shown in the diagram.

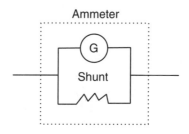

Most of the current in the circuit passes through the shunt, not through the galvanometer wires. Nevertheless, the dial on the ammeter is calibrated to read the current passing through both the shunt and the galvanometer wires. An ammeter is a low-resistance device connected in series with the circuit it is measuring. The low resistance of the ammeter causes a minimal effect on the circuit.

A voltmeter is a galvanometer that has been modified to measure the potential difference across two points in a circuit. Therefore, the voltmeter must be connected in parallel with the points of the circuit it is measuring, as shown in the diagram.

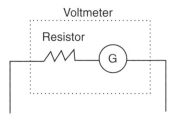

For the voltmeter to have a minimal effect on the circuit, it must be a high-resistance device so that most of the current will flow through the circuit, not through the voltmeter. This high resistance is gained by placing a large resistor in series with the galvanometer coil.

Electric Motors

Another application of a current-carrying loop in a magnetic field is the electric motor. The electrical energy supplied by the circuit is converted by means of the magnetic field into the mechanical energy of rotation. In a motor a coil is wound around a soft-iron core that concentrates the magnetic field and increases the magnitude of the torque.

To continue the rotation past 90°, the current in the coil must be reversed after each half rotation. In a direct current (dc) motor, this reversal is accomplished by means of a device called a *split-ring commutator*. The diagram represents a typical motor.

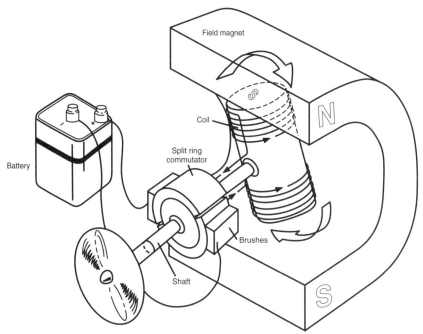

Free Charges in a Magnetic Field

When a charged particle moves freely, the direction of its velocity is analogous to the direction of the current in a wire. If the charged particle is introduced into a magnetic field, it will experience a force that is always perpendicular to the direction of its velocity. We can derive an expression for the magnitude of the force by using these relationships:

$$F = I\ell B \quad \text{and} \quad I = \frac{q}{t}$$

$$\therefore F = \frac{q\ell B}{t} = q\left(\frac{1}{t}\right)B$$

PHYSICS CONCEPTS

$$F = qvB$$

PROBLEM

Calculate the force on a proton moving at a speed of 2.0×10^6 in a magnetic field of 0.50 tesla.

SOLUTION

$$F = qvB$$
$$= (1.6 \times 10^{-19} \text{ C})(2.0 \times 10^6 \text{ m/s})\left(0.50 \frac{N}{A \cdot m}\right)$$
$$= 1.6 \times 10^{-13} \text{ N}$$

We can modify right hand rule 3 as follows:

PHYSICS CONCEPTS

The thumb of the right hand now represents the direction of the velocity of the positively charged particle. The fingers in the direction of the magnetic field and the force point away from the palm (for negative particles, use the left hand).

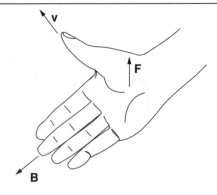

Since the particle is free to move, its velocity continually changes direction and the force always remains perpendicular to the direction of the velocity. This force is a centripetal force; consequently the object moves in a circular path, as shown in the diagram.

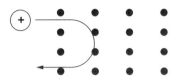

Mass Spectrometry

A practical application of the effect described above is a device known as a *mass spectrometer*, which is used to measure the *charge-to-mass ratio* of ions. A beam of charged particles is introduced into a magnetic field. By measuring the radius of the circular path and the speed of the particles, we can calculate the charge-to-mass $\left(\dfrac{q}{m}\right)$ ratio:

$$F = qvB = \frac{mv^2}{r} = F_c$$

PHYSICS CONCEPTS

$$\frac{q}{m} = \frac{v}{Br}$$

If the charges of the particles are known, then the particle masses can be determined.

The mass of a particle is directly proportional to its radius of curvature in the magnetic field.

The use of mass spectrometry has helped to demonstrate the existence of the electron and the existence of atomic isotopes.

The diagram below represents a mass spectrometer.

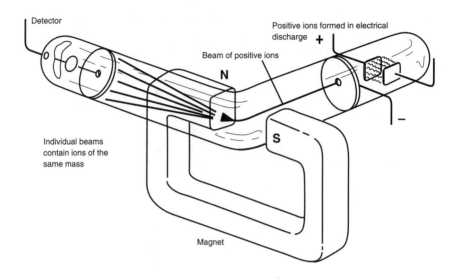

Electron Beams

A device such as a computer monitor or a television tube depends on the production and control of beams of electrons. An electron beam is generated when a filament is heated until it ejects electrons. This process, called *thermionic emission*, is much like the evaporation of water molecules from the surface of liquid water.

Once the electrons are ejected, they are controlled by both electric and magnetic fields. The diagram shows that the direction of a beam of electrons is dependent on the electrical charges of the parallel plates above and below the beam.

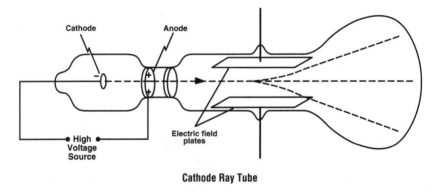

Cathode Ray Tube

We have already seen how magnetic fields can be used to change the direction of a charged particle. Originally, a device known as a cathode ray tube was used to demonstrate the existence and control of electron beams. The cathode ray tube is an evacuated glass tube that contains a source of electrons at one end, a fluorescent screen inside the surface at the other end, and two pairs of deflecting plates in between. When an electron beam strikes the screen, it produces a fluorescent spot of light. The brightness of the spot is directly related to the intensity of the electron beam striking the screen, and the location of the spot is controlled by the electric fields.

11.5 ELECTROMAGNETIC INDUCTION

A motor uses a magnetic field to convert electrical energy into mechanical energy. It is also possible to accomplish the reverse process, that is, to use a magnetic field to convert mechanical energy into electrical energy. Devices that accomplish this purpose are known as *generators*.

Let's begin by moving a wire perpendicularly through a magnetic field.

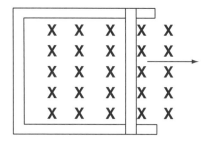

If we focus on one electron in the wire, indicated in the diagram below:

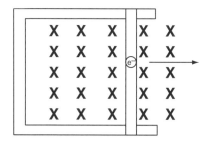

we can apply *right hand rule 3,* but we must use the *left* hand (why?), to show that there is a downward force on the electron. This is true for all electrons passing through the moving segment of the wire. These moving electrons constitute an electric current.

Since work has been done in moving the electrons through the wire, a potential difference has been induced across the ends of the wire. This potential difference depends on the strength of the magnetic field, the length of the wire in the magnetic field, and the speed with which the wire is moved. The induced potential difference is represented by this relationship:

PHYSICS CONCEPTS

$$V = B\ell v$$

PROBLEM

Calculate the potential difference induced across the ends of a conductor that is 0.20 meter long and is moved perpendicularly through a magnetic field of 4.0×10^{-1} tesla, with a speed of 8.0 meters per second.

SOLUTION

$$V = B\ell v$$
$$= (4.0 \times 10^{-1} \text{ T})(0.20 \text{ m})(8.0 \text{ m/s})$$
$$= 6.4 \times 10^{-1} \text{ V}$$

Induced potential difference is also known as **electromotive force** (emf), symbolized as ε.

To induce a potential difference across a conductor, the only requirement is that the magnetic field be interrupted. This interruption is accomplished when the conductor "cuts through" the magnetic field lines. However, physical motion need not be present, only a change in the magnetic field is required.

Our equation $V = B\ell v$ is completely equivalent to the following equation:

PHYSICS CONCEPTS

$$V = \frac{\Delta \phi}{\Delta t}$$

This equation shows that it is the change in the magnetic flux with time that produces the potential difference. The Greek letter ϕ (phi) represents the magnetic flux, which is measured in webers.

Generators

A generator is a commercial device that converts mechanical energy into electrical energy. The source of mechanical energy can be falling water, steam expansion, or something as simple as a hand crank.

In a very simple electric generator, such as the one diagramed below, a coil of wire is wrapped around an iron core. This arrangement is placed in a magnetic field and rotated in the field.

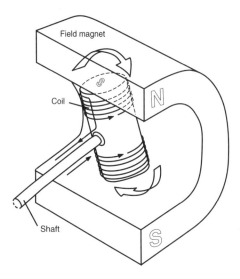

In this arrangement, it is the ends of the coil that interrupt the field and produce the potential difference across the ends of the coil. Let's rotate the coil clockwise through 360° and show the rotation at 90° intervals. If we apply right hand rule 3 to the rotating coil, we can show the direction of the current in the coil at each position.

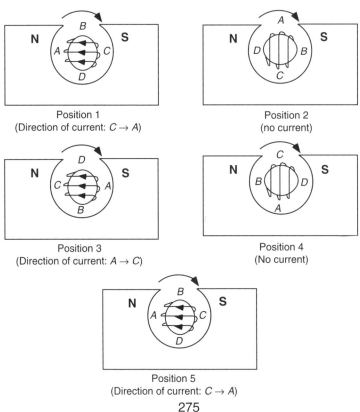

Position 1
(Direction of current: $C \rightarrow A$)

Position 2
(no current)

Position 3
(Direction of current: $A \rightarrow C$)

Position 4
(No current)

Position 5
(Direction of current: $C \rightarrow A$)

275

Points *A*, *B*, *C*, and *D*, are shown in order to orient the coil during each 90° rotation. At positions 2 and 4 in the diagram the current in the coil is zero because the ends of the coil do not interrupt the magnetic field. We also note that the direction of the current in position 3 is opposite to the direction of the current in positions 1 and 5. This type of generator is known as an alternating-current (ac) generator because it produces a current that reverses its direction regularly. Until this point, we considered currents that have only one direction and are known as direct currents (dc).

This alternating current is a direct result of the fact that the potential difference across the ends of the coil has reversed direction. If we included all the intermediate positions of the coil, we would find that the potential difference (or current) varies as shown in the diagram.

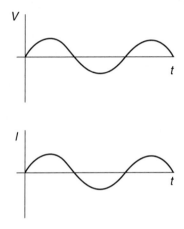

This variation traces out a *cosine* or *sine* wave. The speed with which the coil is rotated determines the "frequency" of the ac potential difference. In the United States alternating current is produced at 60 hertz (Hz), indicating that the current reverses its direction 60 times each second.

11.6 LENZ'S LAW

Suppose a bar magnet is brought toward a coil of wire as shown below.

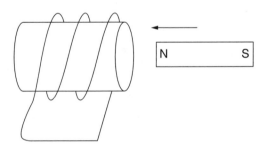

Since the coil will interrupt the magnetic field, a potential difference will be induced across the ends of the coil and a current will be established in it. What is the direction of the current in the coil?

Obviously, there are only two choices. In either case the coil will behave as an electromagnet with a north and a south pole.

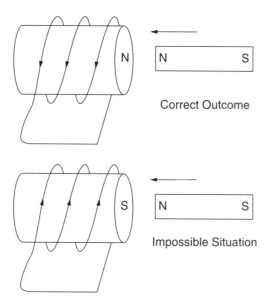

Correct Outcome

Impossible Situation

In the upper part of the diagram above, the magnets repel each other; in the lower part, they attract each other. The lower diagram cannot be correct because the attraction would eliminate the need to use mechanical energy to induce electrical energy—a clear violation of the law of conservation of energy. In the upper, however, the repulsion ensures that outside work will be required in order to induce electrical energy. In 1834, the German physicist H.F.E. Lenz recognized this principle and stated it as follows:

> When a potential difference is induced across a conductor, its direction must oppose the motion that induced it.

Whenever a motor is operated, a secondary potential difference is established that reduces the effective potential difference of the circuit used to operate the motor. This secondary potential difference, which is a result of Lenz's law, is known as **back emf**.

Transformers

When electricity is transmitted across long distances, high voltages are used to reduce losses due to heat (I^2R). At the power plant the electricity is first generated at a lower potential difference, and it must be *stepped up* for transmission.

When electricity enters a building or home, however, its potential difference must be reduced or *stepped down*.

A **transformer** is a device that allows the potential difference to be increased or decreased. A diagram of a transformer is shown below.

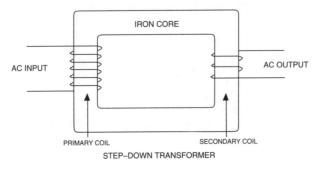

STEP–DOWN TRANSFORMER

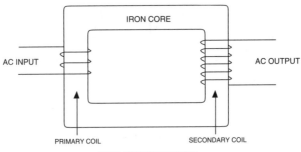

STEP–UP TRANSFORMER

One of the coils, known as the *primary* coil, is attached to an alternating-current source, and the other, the *secondary* coil, to the circuit that requires the stepped-up or stepped-down voltage. The alternating current in the primary coil produces a changing electric field that, in turn, produces a changing magnetic field. The changing magnetic field is carried by the iron core and induces a changing electric field in the secondary coil.

The ratio of the voltages in the primary and secondary coils depends on the relative numbers of turns of wire in these coils, as indicated in this equation:

PHYSICS CONCEPTS

$$\frac{V_p}{V_s} = \frac{N_p}{N_s}$$

If the voltage is increased, that is, if the number of turns is greater in the secondary coil, the transformer is termed a step-up transformer. If the voltage is decreased (the number of turns in the secondary coil is less), the transformer is a step-down transformer.

PROBLEM

The primary coil on a transformer has 1200 turns. If it is desired to step down the voltage from 220 volts to 110 volts, how many turns should the secondary coil have?

SOLUTION

$$\frac{V_p}{V_s} = \frac{N_p}{N_s}$$

$$N_s = \frac{N_p V_s}{V_p}$$

$$= \frac{(1200 \text{ turns})(110 \text{ V})}{(220 \text{ V})}$$

$$= 600 \text{ turns}$$

Before we conclude that a transformer enables us to get something for nothing, we should remember that we cannot violate the law of conservation of energy. Specifically, the power output at the secondary coil ($P_s = V_s I_s$) cannot exceed the power input at the primary coil ($P_p = V_p I_p$). Therefore, a step-up transformer will produce *decreased* current at the secondary coil, and a step-down transformer will produce *increased* current at the secondary coil. The ratio of the power output to the power input is known as the *efficiency* of the transformer.

PHYSICS CONCEPTS

$$Percent\ efficiency = \frac{V_s I_s}{V_p I_p} \times 100$$

Generally, heat losses produce transformers with efficiencies somewhat less than 100%.

PROBLEM

A transformer steps up potential difference from 300. volts to 600. volts. The current in the primary coil is 6.0 amperes, while the current in the secondary coil is 2.0 amperes. Calculate the efficiency of the transformer.

SOLUTION

$$Percent\ efficiency = \frac{V_s I_s}{V_p I_p} \times 100$$

$$= \frac{(600.\ \text{V})(2.0\ \text{A})}{(300.\ \text{V})(6.0\ \text{A})} \times 100$$

$$= 67\%$$

Electromagnetic Waves

We have seen that a changing electric field produces a changing magnetic field and vice versa. If the changing electric field is produced by an *accelerating* charge, energy will be radiated away from the charge in the form of *electromagnetic waves*. In an electromagnetic wave, both the electric field and the magnetic field vary as a sine wave and the two fields are perpendicular to each other and to their direction of motion. All electromagnetic waves (collectively known as the *electromagnetic spectrum*) travel in space at the speed of light.

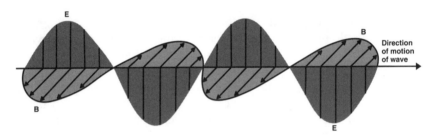

Questions

1. Which is the unit of magnetic flux in the SI system?
   (1) weber
   (2) joule
   (3) coulomb
   (4) newton per ampere-meter

2. The presence of a uniform magnetic field may be detected by using a
   (1) stationary charge
   (2) small mass
   (3) beam of neutrons
   (4) magnetic compass

3. In the diagram below, what is the direction of the magnetic field at point *P*?

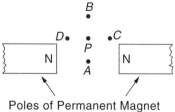

Poles of Permanent Magnet

(1) toward *A*   (2) toward *B*   (3) toward *C*   (4) toward *D*

4. Which diagram best illustrates the direction of the magnetic field between the unlike poles of two bar magnets?

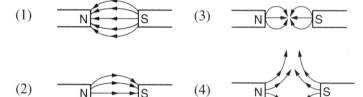

5. Which diagram best represents the magnetic field near the poles of a horseshoe magnet?

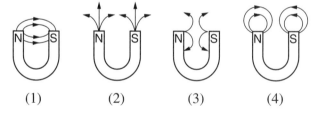

(1)   (2)   (3)   (4)

6. Which vector best represents the direction of the magnetic field at point A near the two north magnetic poles shown in the diagram?

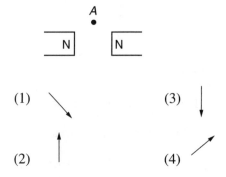

    (1)   (3)

    (2)   (4)

7. Which arrow in the diagram below represents the direction of the flux *inside* the bar magnet?

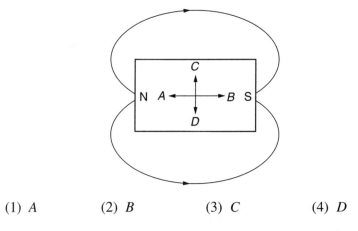

(1) *A*   (2) *B*   (3) *C*   (4) *D*

8. As the distance between two opposite magnetic poles increases, the flux density midway between them
(1) decreases   (2) increases   (3) remains the same

9. Which diagram best represents the magnetic field around an iron bar placed between unlike magnetic poles?

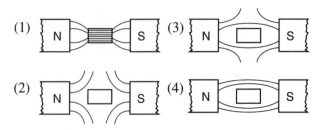

Base your answers to questions 10 through 14 on the diagram below, which represents a cross section of an operating solenoid. A compass is located at point C.

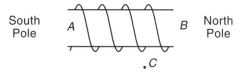

10. Which diagram best represents the shape of the magnetic field around the solenoid?

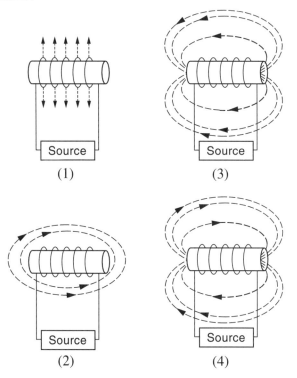

11. Which shows the direction of the compass needle at point C?

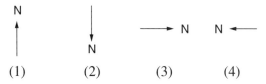

12. If *B* is the north pole of the solenoid, which diagram best represents the direction of *electron flow* in one of the wire loops?

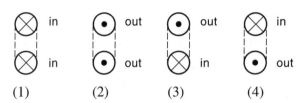

(1)　　　　　(2)　　　　　(3)　　　　　(4)

13. If the current in the solenoid is doubled and the number of turns halved, the magnetic field strength of the solenoid will
(1) decrease　　(2) increase　　(3) remain the same

14. If an iron rod were inserted into the solenoid, the strength of the magnetic field inside the solenoid would
(1) decrease　　(2) increase　　(3) remain the same

15. Each diagram below represents a cross section of a long, straight, current-carrying wire with the electron flow into the page. Which diagram best represents the magnetic field near the wire?

(1) 　　　　(3)

(2) 　　　　(4)

16. A magnetic field will be produced by
(1) moving electrons　　　　(3) stationary protons
(2) moving neutrons　　　　(4) stationary ions

17. What is the direction of the magnetic field near the center of the current-carrying wire loop shown in the diagram below?

(1) into the page   (3) to the left
(2) out of the page   (4) to the right

18. The diagram represents a wire with electrons moving in the direction shown. At point *A*, the magnetic field is directed

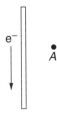

(1) out of the page   (3) from left to right
(2) into the page   (4) from right to left

19. Which diagram best represents the direction of the magnetic field around a wire conductor in which the electrons are moving as indicated? [The X's indicate that the field is directed into the paper, and the dots indicate that the field is directed out of the page.]

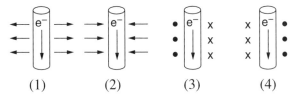

(1)   (2)   (3)   (4)

20. As the permeability of a substance in a magnetic field decreases, the flux density within the substance
(1) decreases   (2) increases   (3) remains the same

21. In the diagram below, electron current is passed through a solenoid. The north pole of the solenoid is nearest to point

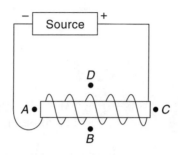

(1) *A*   (2) *B*   (3) *C*   (4) *D*

22. If the current through a solenoid increases, the magnetic field strength of the solenoid
(1) decreases   (2) increases   (3) remains the same

23. As materials of increasing permeability are placed at position *X* in the diagram, the flux density at position *X*

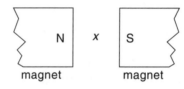

(1) decreases   (2) increases   (3) remains the same

24. As the current increases in a wire placed perpendicular to a magnetic field, the force on the wire
(1) decreases   (2) increases   (3) remains the same

25. The diagram below shows a loop of wire between the poles of a magnet. The plane of the loop is parallel to the magnetic field. If an electron flow is established in the direction shown in the loop, in which direction will a magnetic force be exerted on segment *AB*?

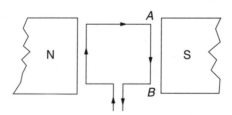

(1) toward the top of the page   (3) into the page
(2) toward the bottom of the page (4) out of the page

26. The diagram below shows an end view of a current-carrying wire between the poles of a magnet. The wire is perpendicular to the magnetic field.

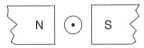

If the direction of the electron flow is out of the page, which arrow correctly shows the direction of the magnetic force **F** acting on the wire?

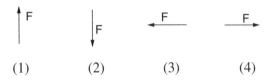

(1)   (2)   (3)   (4)

27. Wires *x* and *y* experience a force when electrons pass through them as shown in the diagram below.

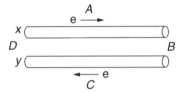

The force on wire *y* will be toward
(1) *A*   (2) *B*   (3) *C*   (4) *D*

28. The currents in two straight, parallel, fixed wires are in the same direction. If the current direction in one is reversed, the magnitude of the magnetic force between the two conductors
(1) decreases   (2) increases   (3) remains the same

29. When electrons flow from point *A* to point *B* in the wire shown in the diagram, a force will be produced on the wire

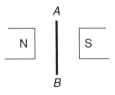

(1) toward N   (3) into the page
(2) toward S   (4) out of the page

30. Two long, straight, parallel conductors carry equal currents and are spaced 1.0 meter apart. If the current in each conductor is doubled, the magnitude of the magnetic force acting between the conductors will be
(1) unchanged   (2) doubled   (3) halved   (4) quadrupled

31. In the diagram below, a solenoid that is free to rotate around an axis at its center, *C*, is placed between the poles of a permanent magnet.

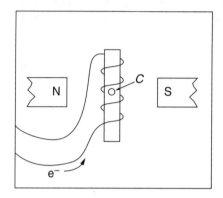

As an electron current starts through the solenoid in the direction shown, the solenoid will
(1) remain motionless   (3) start turning clockwise
(2) vibrate back and forth   (4) start turning counterclockwise

32. As the angle between a current-carrying wire and an external magnetic field changes from 90° to 0°, the magnetic force on the wire
(1) decreases   (2) increases   (3) remains the same

33. The magnitude of the magnetic force between two straight, parallel conductors a given distance apart depends on
(1) the magnitudes of the currents, only
(2) the directions of the currents, only
(3) the magnitudes and the directions of the currents
(4) neither the magnitudes nor the directions of the currents

34. When an electron moves across a magnetic field, the angle between the direction of the magnetic force on the electron and the direction of the magnetic field is always
(1) 0°   (2) 45°   (3) 90°   (4) 180°

35. A beam of electrons is moving through a uniform magnetic field perpendicularly to the field. Which graph shows how the magnetic force varies as the speed of the electrons is increased?

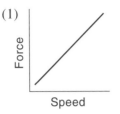

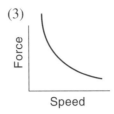

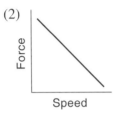

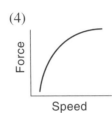

36. A magnetic force is experienced by an electron moving through a magnetic field. If the electron were replaced by a proton traveling at the same velocity, the magnitude of the magnetic force experienced by the proton would be
(1) the same
(2) twice as great
(3) half as great
(4) zero

37. A charge moves perpendicularly to a uniform magnetic field and experiences a force of magnitude F. If the speed of the charge is doubled, the magnitude of the force will be
(1) $\dfrac{1}{4}$ F
(2) $\dfrac{1}{2}$ F
(3) 2F
(4) 4F

38. An electron moves at 1.0 meter per second perpendicularly across a magnetic field whose intensity is 1.0 newton per ampere-meter. The magnetic force exerted on the electron is
(1) 1.0 N
(2) 1.6×10^{-19} N
(3) 9.8 N
(4) 6.25×10^{18} N

39. A particle is being accelerated by a magnetic field. This particle must be
(1) neutral and stationary
(2) neutral and in motion
(3) charged and stationary
(4) charged and in motion

40. A wire conductor is moved with constant speed at right angles to a magnetic field. If the strength of the magnetic field is increased, the induced potential difference across the ends of the conductor
(1) decreases   (2) increases   (3) remains the same

41. What is the potential difference induced in a wire 0.10 meter long as it moves with a speed of 50. meters per second perpendicularly to a magnetic field that has a magnetic flux density of 0.050 tesla?
(1) 0.25 V   (2) 25 V   (3) 250 V   (4) 2500 V

42. The magnitude of the electric potential difference induced across the ends of a conductor moving in a magnetic field may be increased by
(1) increasing the diameter of the conductor
(2) increasing the speed of the conductor
(3) decreasing the resistance of the conductor
(4) decreasing the length of the conductor

Base your answers to questions 43 through 45 on the diagram below, which represents a bar magnet falling vertically through a stationary horizontal loop of conducting wire.

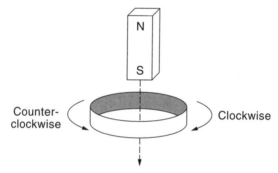

43. As the south pole of the magnet approaches the loop, the current in the loop
(1) decreases   (2) increases   (3) remains the same

44. As the south pole of the magnet approaches the loop, the direction of the electron flow in the loop is
(1) counterclockwise   (2) clockwise   (3) alternating

45. If the magnet were inverted and dropped from the same height as before, the maximum electron flow in the loop would change in
(1) rate, only
(2) direction, only
(3) rate and direction

46. The diagram below represents a rectangular loop of wire that is rotating about the axis shown. Side *AB* is 0.50 meter long and is moving at a speed of 2.0 meters per second. The strength of the uniform magnetic field is 4.0 webers per square meter. What is the maximum electric potential difference induced across side *AB*?

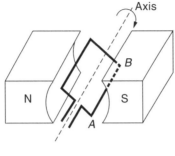

(1) 16 V   (2) 12 V   (3) 10. V   (4) 4.0 V

47. A conducting loop is rotated one full turn (360°) in a uniform magnetic field. Which graph best represents the induced potential difference across the ends of the loop as a function of the angular rotation?

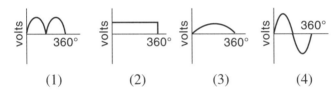

(1)   (2)   (3)   (4)

48. As the speed of a conducting loop rotating in a magnetic field decreases, the magnitude of the induced current in the loop
(1) decreases   (2) increases   (3) remains the same

49. The diagram below shows conductor *C* between two opposite magnetic poles. Which procedure will produce the greatest induced potential difference in the conductor?

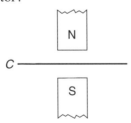

(1) holding the conductor stationary between the poles
(2) moving the conductor out of the page
(3) moving the conductor toward the right side of the page
(4) moving the conductor toward the north pole

50. A galvanometer with a low-resistance shunt in parallel with its moving coil is
    (1) a motor   (2) a generator   (3) a voltmeter   (4) an ammeter

51. As the armature of a motor turns, an emf is induced that is opposite to the applied voltage. The existence of this back emf can best be accounted for by
    (1) the conservation of momentum
    (2) the conservation of energy
    (3) Coulomb's laws
    (4) Newton's laws of motion

52. The only difference between two motors is the material of their armature cores. Motor A has its coil wrapped around a piece of soft iron, and motor B has its coil wrapped around a piece of wood. Compared to the force exerted on the armature of motor A, the force exerted on the armature of motor B is
    (1) less   (2) greater   (3) the same

53. In the diagram below of a DC motor, the arrows represent the direction of electron current. In what direction will end A of the armature spin?

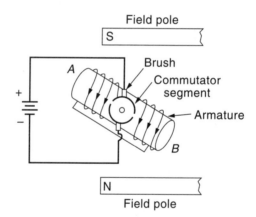

    (1) into the page   (3) clockwise
    (2) out of the page   (4) counterclockwise

54. Which device can be used to separate isotopes of an element?
    (1) a mass spectrometer   (3) an induction coil
    (2) an electroscope   (4) two closely spaced double slits

Base your answers to questions 55 through 57 on the diagram below, which shows an apparatus for demonstrating the effect of a uniform magnetic field on a beam of electrons moving in the direction shown.

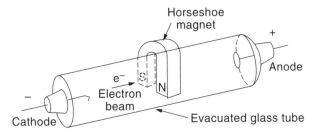

55. As the electron beam enters the magnetic field of the horseshoe magnet, the beam will be deflected
    (1) toward the south pole of the magnet
    (2) toward the north pole of the magnet
    (3) downward, toward the bottom of the tube
    (4) upward, toward the top of the tube

56. The velocity of the electron beam is 3.0×10^6 meters per second, perpendicular to the 5.0×10^{-3} tesla magnetic field. What is the magnitude of the force acting on each electron in the beam?
    (1) 8.0×10^{-22} N   (3) 1.7×10^{-9} N
    (2) 2.4×10^{-15} N   (4) 1.5×10^4 N

57. If the speed of the electrons traveling through the magnetic field increases, the magnetic force on the electrons will
    (1) decrease   (2) increase   (3) remain the same

58. The charge-to-mass ratio of an electron is approximately 1.76×10^{11} coulombs per kilogram. This value indicates that the
    (1) electron's charge is about equal to its mass
    (2) charge on an electron is extremely small compared to its mass
    (3) mass of an electron is the same as that of an atom
    (4) mass of an electron is extremely small compared to its charge

59. Which event would produce electromagnetic radiation?
    (1) an accelerating neutron
    (2) an accelerating electron
    (3) a neutron moving with constant velocity
    (4) an electron moving with constant velocity

60. Some fluorescent ceiling lights operate at higher voltage than that supplied by household circuits. Which device is used to increase the voltage for these lights?
    (1) laser   (2) transformer   (3) motor   (4) generator

61. When a 12-volt potential difference is applied to the primary coil of a transformer, an 8.0-volt potential difference is induced in the secondary coil. If the primary coil has 24 turns, how many turns does the secondary coil have? [Assume 100% efficiency.]
(1) 36   (2) 16   (3) 3   (4) 4

62. Compared to the power developed in the primary coil of a 100% efficient transformer, the power developed in the secondary coil is
(1) less   (2) greater   (3) the same

63. A potential difference of 50. volts is required to operate an electrical device. The potential difference of the source is 120 volts. The table shows the primary and secondary windings for four available transformers. Which transformer is suitable for this application?

Transformer	Primary	Secondary
A	250	600
B	600	250
C	240	150
D	150	240

(1) A   (2) B   (3) C   (4) D

64. A transformer has 150 turns of wire in the primary coil and 1200 turns of wire in the secondary coil. The potential difference across the primary is 110 volts. What is the potential difference induced across the secondary coil?
(1) 14 V   (2) 110 V   (3) 150 V   (4) 880 V

Base your answers to questions 65 and 66 on the diagram and the information below. The diagram represents a 100% efficient transformer connected to an AC source. This transformer has two turns in the secondary coil for each turn in the primary coil.

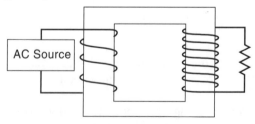

65. The voltage across the secondary coil is
   (1) the same as that in the primary
   (2) twice that in the primary
   (3) half that in the primary
   (4) zero

66. Compared to the power in the primary coil, the power in the secondary coil is
   (1) less   (2) greater   (3) the same

67. An ideal transformer has a current of 2.0 amperes and a potential difference of 120 volts across its primary coil. If the current in the secondary coil is 0.50 ampere, the potential difference across the secondary coil is
   (1) 480 V   (2) 120 V   (3) 60. V   (4) 30. V

Base your answers to questions 68 through 72 on the diagram below, which represents a beam of electrons entering a magnetic field between the poles of a magnet. The magnetic flux density between the poles of the magnet is 5.0×10^{-5} weber per meter squared.

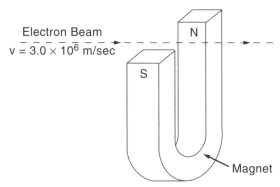

68. As the electrons enter the magnetic field between the poles of the magnet, they will be deflected
   (1) toward the south pole   (3) downward
   (2) toward the north pole   (4) upward

69. The magnitude of the force acting on each electron as it travels through the magnetic field is
   (1) 1.5×10^{-13} N   (3) 2.4×10^{-17} N
   (2) 8.0×10^{-17} N   (4) 4.8×10^{-24} N

70. The electrons are replaced by helium ions with a charge of +2 elementary charges. The ions enter the magnetic field between the poles of the magnet with the same velocity as the electrons. Compared to the force on each electron, the force on each helium ion will be
(1) the same   (3) 16 times as great
(2) twice as great   (4) 4 times as great

71. If the velocity of the helium ions entering the magnetic field between the poles is increased, the force on each helium ion will
(1) decrease   (2) increase   (3) remain the same

72. Both helium ions and helium atoms enter the magnetic field with the same velocity. Compared to the force exerted by the magnetic field on the helium ions, the force exerted on the helium atoms will be
(1) less   (2) greater   (3) the same

Base your answers to questions 73 through 77 on the diagram below, which represents a U-shaped wire conductor positioned perpendicular to a uniform magnetic field that acts into the page. *AB* represents a second wire, which is free to slide along the U-shaped wire. The length of wire *AB* is 1 meter, and the magnitude of the magnetic field is 8.0 webers per meter^2.

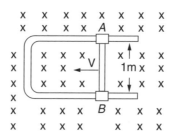

73. If wire *AB* is moved to the left at a constant speed, the direction of the induced electron motion in wire *AB* will be
(1) toward *A*, only   (3) first toward *A* and then toward *B*
(2) toward *B*, only   (4) first toward *B* and then toward *A*

74. If wire *AB* is moved to the left with a constant speed of 10. meters per second, the potential difference induced across wire *AB* will be
(1) 0.8 V   (2) 8.0 V   (3) 10 V   (4) 80 V

75. Wire *AB* is moved at a constant speed to the left. The current induced in the conducting loop will produce a force on wire *AB* that acts
(1) to the right   (3) into the page
(2) to the left   (4) out of the page

76. The resistance of wire *AB* is increased, and the wire is moved to the left at a constant speed of 10 meters per second. Compared to the induced potential difference before the resistance was increased, the new potential difference will be
(1) less　　　　(2) greater　　　　(3) the same

77. If wire *AB* is accelerating to the left, the potential difference induced across *AB*
(1) decreases　　(2) increases　　(3) remains the same

Base your answers to questions 78 through 82 on the diagram below, which represents an electron beam entering the space between two parallel, oppositely charged plates. A uniform magnetic field, acting out of the page, exists between the plates.

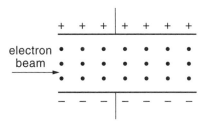

78. If the magnitude of the electric force on each electron and the magnetic force on each electron are the same, which diagram best represents the direction of the vector sum of the forces acting on one of the electrons?

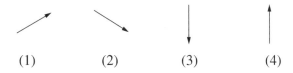

79. In which direction would the magnetic field have to point in order for the magnetic force on the electrons to be opposite in direction from the electric force on the electrons?
(1) toward the bottom of the page　(3) out of the page
(2) toward the top of the page　　(4) into the page

80. The potential difference across the plates is *V*, the distance between the plates is *d*, and the charge on each electron is *q*. Which expression best represents the magnitude of the electric field intensity between the plates?
(1) $\dfrac{V}{d}q$　　　(2) Vqd　　　(3) $\dfrac{V}{d}$　　　(4) $\dfrac{q}{V}d$

81. If the electric force were equal and opposite to the magnetic force on the electrons, which diagram would best represent the path of the electrons as they travel in the space between the plates?

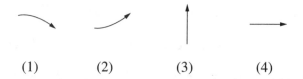

(1)　　　　　　(2)　　　　　　(3)　　　　　　(4)

82. If only the potential difference between the plates is increased, the force on the electron beam will
(1) decrease　　　(2) increase　　　(3) remain the same

Base your answers to questions 83 through 86 on the diagram below, which represents wires *A* and *B* both carrying a flow of electrons out of the paper. The wires are 1 meter apart.

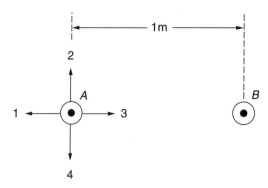

83. As a result of the magnetic fields associated with the wires, wire *A* will experience a force directed toward
(1) 1　　　　　(2) 2　　　　　(3) 3　　　　　(4) 4

84. The current in each of the wires is doubled. Compared to the original magnetic force on wire *A*, the new magnetic force on wire *A* will be
(1) the same　　　　　(3) 3 times as much
(2) twice as much　　　(4) 4 times as much

85. If the distance between the two wires is increased, the magnetic force on wire *A* will then be
(1) less　　　(2) more　　　(3) the same

86. If only the direction of the electron flow in wire *B* were reversed, the magnitude of the force on wire *A* would
(1) decrease　　　(2) increase　　　(3) remain the same

Base your answers to questions 87 through 90 on the diagram below, which shows a cathode ray tube. The electrons in the tube are emitted from a heated cathode and travel in a beam to the face of the tube, causing a bright spot to appear where they hit.

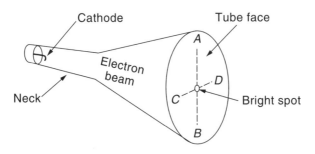

87. The process by which the electrons are emitted from the cathode as it becomes hot is called
    (1) thermionic emission
    (2) alpha particle emission
    (3) photoelectric emission
    (4) photon emission

88. If an upward magnetic field is applied to the neck of the tube, the bright spot will move toward point
    (1) *A* (2) *B* (3) *C* (4) *D*

89. If an upward electric field is applied to the neck of the tube, the bright spot will move toward point
    (1) *A* (2) *B* (3) *C* (4) *D*

90. A 5.0×10^{-4} tesla magnetic field is placed so that its direction is perpendicular to the path of the electrons. The electrons move toward the tube face at a speed of 4.0×10^6 meters per second. What is the magnitude of the force exerted on each electron by this magnetic field?
    (1) 1.6×10^{-9} N
    (2) 2.0×10^3 N
    (3) 3.2×10^{-16} N
    (4) 8.0×10^{-23} N

91. When a galvanometer is used, the deflection of the needle occurs as a result of
    (1) electrostatic force
    (2) magnetic force
    (3) gravitational force
    (4) photoelectric effect

92. To convert a galvanometer into a voltmeter, it is necessary to connect a
    (1) resistor in series with the coil
    (2) resistor in parallel with the coil
    (3) shunt wire in series with the coil
    (4) shunt wire in parallel with the coil

93. The purpose of the shunt in an ammeter is to provide
    (1) electrostatic deflection of the coil
    (2) magnetic deflection of the coil
    (3) resistance to current flow
    (4) a path for some current to bypass the coil

94. The current in the armature of an electric motor switches direction with each rotation. Which motor part produces this phenomenon?
    (1) magnet   (3) armature
    (2) split-ring commutator   (4) stator

95. If the amount of current flowing in a current-carrying loop is doubled, the strength of the twisting force is
    (1) halved   (2) doubled   (3) the same   (4) quadrupled

96. Compared to the voltage in the coil of a transformer with more turns of wire, the voltage in the coil with fewer turns is
    (1) smaller   (2) greater   (3) the same

97. As a twisting torque causes the current-carrying loop in an electric motor to begin rotating, the current in that loop
    (1) decreases   (2) increases   (3) remains the same

Base your answers to questions 98 through 102 on the information and diagram below.

A stream of electrons from heated filament F is accelerated by a potential difference of 80.0 volts toward plate P. Some of the electrons in the beam pass through the hole in plate P and follow the path shown in the diagram. The magnetic and electric fields are uniform.

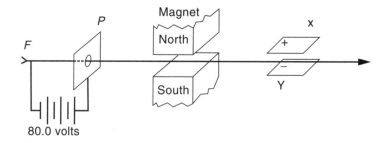

98. What is the kinetic energy of an electron as it reaches plate P?
    (1) 1.28×10^{-17} J   (3) 9.80 J
    (2) 2.40×10^{-10} J   (4) 80.0 J

99. With only the magnetic field present, the beam would be deflected
   (1) into the page (3) upward
   (2) out of the page (4) downward

100. With the magnetic field removed and the electric field turned on, the beam would be deflected
   (1) upward (3) into the page
   (2) downward (4) out of the page

In answering questions 101 and 102, assume that the magnetic field produced by the magnet and the electric field produced by plates *X* and *Y* remain constant.

101. If the distance between the filament and the plate is decreased, the kinetic energy gained by an electron as it moves from the filament to the plate will
   (1) decrease (2) increase (3) remain the same

102. If the potential difference between the filament and the plate is increased, the force of the magnetic field on the electrons will
   (1) decrease (2) increase (3) remain the same

Chapter Twelve

WAVES AND SOUND

Waves transfer energy without the transfer of mass. Mechanical waves, such as sound, require a medium for transmission, whereas electromagnetic waves, such as visible light, do not.

A wave may be longitudinal, transverse, or a combination of both, depending on the direction in which the medium vibrates in relation to the movement of the wave's energy. Longitudinal waves exhibit parallel vibrations; in transverse waves the vibrations are perpendicular.

The characteristics of a periodic wave include speed, wavelength, frequency and period, and amplitude. Among the properties of periodic waves are reflection, refraction (the change in the direction of a wave that enters a medium at an angle), interference (the combination of two or more waves simultaneously in a medium), diffraction (the apparent "bending" of a wave around an obstacle), and the Doppler effect (the apparent change in the frequency of a wave as perceived by an observer because of the relative motion between the wave source and the observer).

KEY OBJECTIVES

At the conclusion of this chapter you will be able to:

- Define the terms *periodic wave, wave motion, transverse wave, longitudinal wave,* and *surface wave,* and provide examples of each.
- Compare and contrast mechanical waves with electromagnetic waves.
- Define the terms *period, frequency, amplitude,* and *wavelength,* and solve problems that relate these quantities to wave speed.
- Use a diagram of a periodic wave to identify the following: crest, trough, amplitude, phase, and wavelength.
- Define the term *reflection*, and apply the law of reflection.
- Define the term *ray*, and apply it to various types of periodic waves.
- Define the term *refraction*, and apply Snell's law.
- Define the terms *constructive interference, destructive interference, resonance,* and *diffraction.*
- Explain how interference can produce standing waves and beats.
- Define the term *Doppler effect*, and explain this phenomenon.

12.1 DEFINITION OF WAVE MOTION

The diagram represents a coiled spring held between two people; a handkerchief is tied to the spring. If one person quickly jerks the end of the spring up and down, there will be a disturbance in the spring. When the disturbance reaches the other person's hand, it will cause the hand to jerk. Therefore, the disturbance, or *wave pulse,* transfers energy. A moving particle can also transfer energy, but its mass is transferred as well.

If we look at the handkerchief tied to the spring in the diagrams below, however, we can see that it has vibrated about the spring's rest position, that is, it has moved up and down, but it has not moved along with the energy. A wave pulse, or a series of identical, repeating, evenly spaced pulses (called a **periodic wave**), transfers energy, but not mass.

12.2 TYPES OF WAVES

A wave is a vibratory disturbance that is transmitted through a material or through space. Water waves, sound waves, and waves that travel along a spring are examples of *mechanical* waves. Mechanical waves require a material medium for transmission. The energy disturbance is propagated by the molecules of water, air, or, as in the example in Section 12.1, the metal atoms of a spring.

Light waves, microwaves, and radio waves are examples of *electromagnetic* waves. Electromagnetic waves do not need a material medium; they are the result of changes in the field strengths of electric and magnetic fields and can travel in space (a vacuum). Since electromagnetic waves cannot be observed directly, we will use mechanical wave models to explore wave properties and behavior. Light will be studied in greater detail in Chapter 13.

Mechanical waves can be divided into three different types: transverse, longitudinal, and surface waves. In a **transverse wave**, the particles of the medium vibrate or exhibit simple harmonic motion (SHM) about a rest position *perpendicular* to the direction of motion of the wave. Waves on a string, as shown below, are an example of a transverse wave.

In a **longitudinal wave,** diagramed below, a disturbance causes the particles of the material to vibrate in SHM in a direction *parallel* to the direction of motion of the wave.

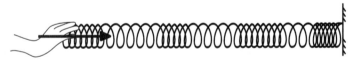

Sound is an example of a longitudinal wave. Fluids usually transmit only longitudinal waves.

On the surface of water, the motions of particles are both parallel and perpendicular to the direction of motion of the wave. In the diagram, the combination of transverse and longitudinal waves produces what is known as a **surface wave**.

Wave motion

12.3 CHARACTERISTICS OF PERIODIC WAVES

A periodic disturbance (e.g., water dripping from a faucet into a sink of water) will produce a traveling wave. As a result each particle in the medium will vibrate in simple harmonic motion in response to the disturbance.

If we look at a single particle in the medium (e.g., the handkerchief on the spring in Section 12.1), we see that it moves up and down as we send a traveling wave through the spring. The time it takes for its motion to repeat itself is called the **period** (T), and it is measured in seconds. The number of times the motion repeats itself in a time interval of one unit of time is known as the **frequency** (f) of the wave. Frequency is measured in hertz (Hz), which is equivalent to cycles per second or reciprocal seconds (s^{-1}).

Frequency and period are inversely proportional to each other and are related by this equation:

PHYSICS CONCEPTS

$$T = \frac{1}{f}$$

In other words, frequency and period are reciprocals of each other.

The maximum displacement of any particle in the medium relative to its rest position is called the **amplitude** of the wave. In a transverse wave the maximum upward displacement is known as a *crest*; the maximum downward displacement, as a *trough*. In a longitudinal wave, the particles in the medium produce areas of maximum compression called *condensations* and areas of maximum separation called *rarefactions*. Condensations are analogous to crests, and rarefactions to troughs. Areas of condensation and rarefaction are shown in the diagram below.

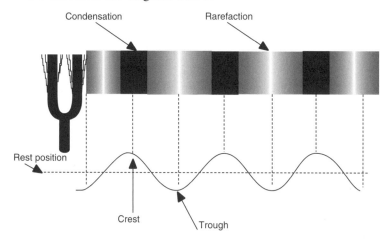

305

Amplitude is related to the energy carried by the wave. In sound waves, amplitude corresponds to loudness; in light, to brightness.

Points on a periodic wave that are at equal displacements from their rest position and are experiencing identical movements, that is, are moving in *the same direction* toward or away from the rest position, are said to be *in phase*. Points on a periodic wave that are at equal displacements from their rest position but are experiencing motion in *opposite directions* from each other are described as being *180° out of phase* or completely out of phase. In the diagram, points 1 and 5, 2 and 6, 3 and 7, 4 and 8 are in phase. Points 1 and 3, 2 and 4, 3 and 5, 4 and 6, 5 and 7, 6 and 8 are completely out of phase.

Wave motion

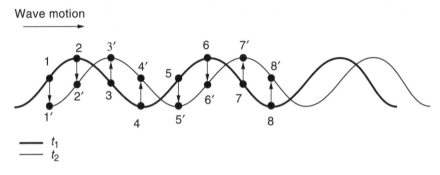

The distance between two successive points on a periodic wave that are in phase is called the **wavelength** (λ). Wavelength is measured in meters. Successive points that are 180° out of phase are therefore separated by a distance of one-half wavelength.

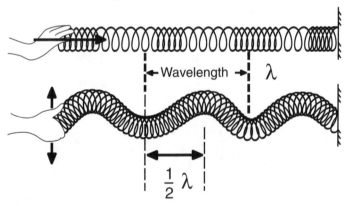

12.4 SPEED OF A WAVE

How fast does a periodic wave move? We know that velocity is the change in displacement per unit time. If the time was equal to T, the period of the wave motion, the wave would move a distance of one wavelength (λ). Therefore, $v = \dfrac{\lambda}{T}$. Since period and frequency are reciprocals of each other, we can rewrite the equation as follows:

PHYSICS CONCEPTS

$$v = f\lambda$$

PROBLEM

What is the wavelength of a sound wave whose speed is 330 meters per second and whose frequency is 990 hertz?

SOLUTION

$$v = f\lambda$$

$$\lambda = \frac{v}{f}$$

$$= \frac{330 \text{ m/s}}{990 \text{ s}^{-1}}$$

$$= 0.33 \text{ m}$$

All electromagnetic waves travel in space at the speed of light, which is denoted by the letter c and approximately equal to 3.0×10^8 meters per second. Generally, the speed of a mechanical wave depends only on the properties of the medium, not on the amplitude or frequency of the wave. A wave with large amplitude transmits more energy than a wave with low amplitude, but both travel at the same speed through a given medium. If two waves have the same speed, the wave with the higher frequency will have a shorter wavelength than the wave with the lower frequency. This is a direct result of the equation $v = f\lambda$.

12.5 REFLECTION

When a wave travels from one medium to another, part of the energy of the wave is transmitted into the new medium with the same frequency, part is absorbed, and part moves back into the original medium, that is, it is *reflected*, with the same frequency.

If the difference between the media is small, most of the wave's energy or amplitude will be transmitted and very little will be reflected. If, however, the two media are very different, very little energy will be transmitted and most will be reflected.

If the wave travels from a less dense to a more dense medium, the reflected wave will be inverted, or 180° out of phase.

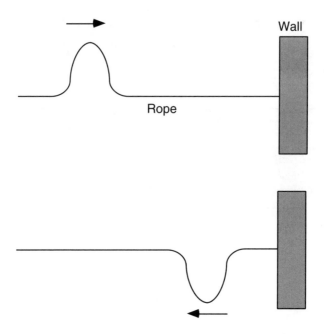

If, on the other hand, the wave travels from a more dense to a less dense medium, the reflected wave will *not* undergo a phase change.

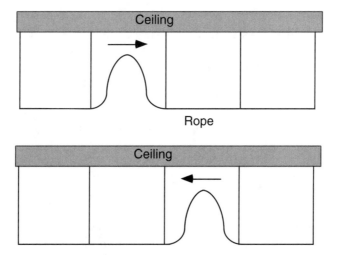

Although transverse waves are used in the two diagrams above to illustrate this property, longitudinal waves behave in the same manner.

Wave Shape

A wave can have various shapes, depending on the source that produces it. If a person drops pebbles into a pond and then views the result from above, the diagram below shows what is seen.

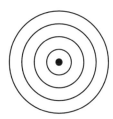

The pebbles are known as a *point source,* and the wave is circular, that is, it spreads out evenly in all directions. The circles in the diagram represent the crests of the wave and are called *wave fronts.*

A line that indicates the direction of motion is called a **ray**. The rays of circular waves are radial lines.

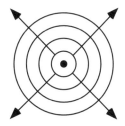

If the source of the waves is broad, such as a wooden plate bobbing in the water, the result will be a series of plane waves, as shown in the diagram below.

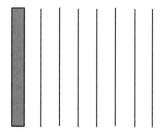

In the next diagram the wave fronts are straight lines. The rays all point in the same direction, that is, they are parallel to one another. At distances very far from a point source, waves become nearly plane in shape.

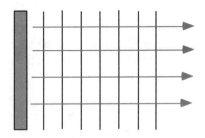

The diagrams show that in all waves the rays are perpendicular to the wave fronts.

Incident and Reflected Waves

When a plane wave strikes a reflecting surface at an angle, it is reflected at the same angle, as shown in the diagram below.

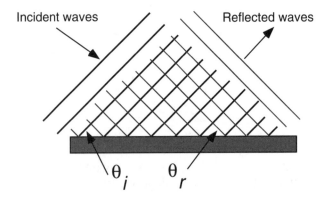

Incident waves   Reflected waves

The wave that strikes the surface is called the *incident* wave. The angle it makes with the surface is the angle of incidence (represented by θ_i in the diagram). Similarly, the wave that leaves the surface is termed the *reflected* wave, and the angle it makes with the surface is the reflected angle (θ_r in the diagram). In every case, and for all waves, the angle of incidence is equal to the angle of reflection. This relationship is known as the law of reflection.

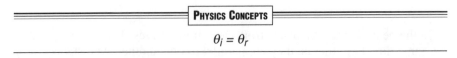

PHYSICS CONCEPTS

$$\theta_i = \theta_r$$

It is not always convenient to refer to the wave fronts themselves in measuring angles of incidence and reflection. Often it is easier to refer to the *rays* associated with the waves and to measure the angles of incidence and reflection with respect to a *normal* (a perpendicular line drawn to the surface). This situation is diagramed below.

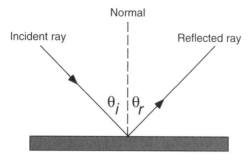

12.6 REFRACTION

If a wave passes from one medium to another at an angle to the boundary, and its speed changes, its direction in the second medium will also change, as shown in the diagram below.

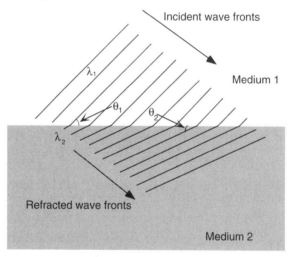

The wave that strikes the boundary in medium 1 is the *incident* wave, and the angle it makes with the boundary surface (θ_1) is the angle of incidence. The wave in medium 2 is called the *refracted* wave, and the angle it makes with the boundary surface (θ_2) is the angle of refraction.

As the wave enters the second medium, its frequency does not change; therefore, its change in velocity is accompanied by a change in its wavelength. In the diagram, the wavelength in medium 2 is less than the wavelength in medium 1 (λ_1); consequently, the speed of the wave in medium 2 is also less than the speed of the wave in medium 1.

We could prove this statement using the relationship $v = f\lambda$. Using simple trigonometry, it can be shown that the ratio of the speeds in the two media is related to the ratio of the sines of the angles in these media. This relationship is known as Snell's law, after the Dutch astronomer and mathematician Willebrord Snell.

PHYSICS CONCEPTS

$$\frac{\sin \theta_2}{\sin \theta_1} = \frac{v_2}{v_1}$$

If we refer to rays rather than wave fronts, the picture is as follows:

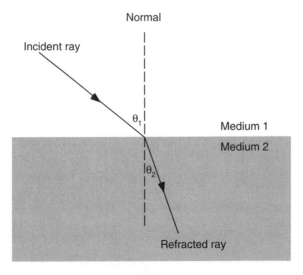

Once again, we measure the angles of incidence and refraction with respect to a normal.

By examining the sizes of the angles, we can draw conclusions about the relative speeds of the waves in the two media. The ray representing a slower wave is positioned closer to the normal than is the ray representing a faster wave. We will explore this phenomenon further in Chapter 13.

12.7 INTERFERENCE

Two or more waves passing simultaneously through the same area of a medium affect the medium independently but do not affect each other. The resultant displacement of any point in the medium is the algebraic sum of the displacements of all the individual waves; this is known as the principle of *superposition*, and the result is called *interference*.

There exist two kinds of interference, constructive and destructive. **Constructive interference** occurs when the individual wave displacements, *A* and *B* in the diagram below, are in the same direction. In this case, the resulting amplitude, *A+ B*, is greater than any individual wave amplitude.

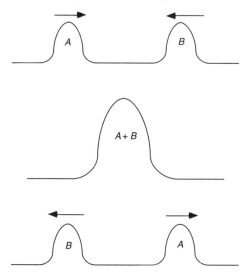

Destructive interference occurs when displacements *A* and *B* are in opposite directions, as illustrated below. In this case, the resulting amplitude is less than any individual wave amplitude. If the displacements are equal in magnitude, complete or maximum destructive interference occurs. If the displacements are not equal in magnitude, the result is partial destructive interference.

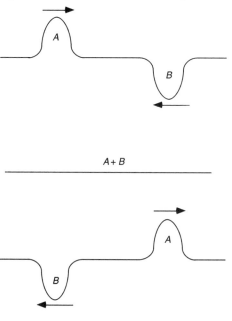

12.8 STANDING WAVES

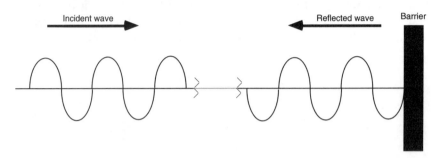

The diagram above illustrates a wave on a string traveling toward a barrier and the reflected wave emerging from it. The incident and reflected waves have the same frequencies and amplitudes, but they are traveling in opposite directions. When the waves pass one another, they will interfere regularly, both constructively and destructively.

This interference will produce a wave that appears to "stand still" in the horizontal direction. Adjacent crests and troughs will move vertically in opposite directions about points that have no motion; the result is known as a *standing wave*. The points that do not move are called *nodes,* and the crest-trough combinations are *antinodes*. The diagram below illustrates this phenomenon.

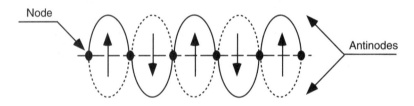

12.9 RESONANCE

If a person were to pluck a stretched guitar string that was not connected to a guitar, the sound would not be heard. What exactly does the guitar *body* contribute to the production of audible sounds?

When a string is plucked, a standing wave pattern is established in the string. The guitar box is capable of vibrating at the frequencies produced by the strings, and a standing-wave pattern is established in the guitar itself. This phenomenon is known as **resonance**. The *amplitudes* of the standing waves in the guitar are much larger than those in the string, however, and therefore we hear the sound. In general, musical instruments act as resonance devices.

Sometimes, resonance can be an unwanted phenomenon. Years ago, a gale-force wind caused a bridge in Tacoma, Washington, to resonate at its *natural frequency of vibration*. The energy produced by the standing-wave pattern was great enough to cause the bridge to collapse. When bridges and like structures are built today, devices are incorporated to prevent the production of these destructive standing-wave patterns.

12.10 BEATS

If two waves of the same frequency and amplitude interfere constructively, they will produce a single wave with the same frequency and twice the amplitude. If the waves interefere destructively, they will cancel each other. Suppose two sound waves with *nearly* the same frequencies (e.g., 256 Hz and 258 Hz) interact with one another. What will happen then?

The result of this interaction will be the production of a "warbling" sound, that is, a sound that is alternately louder and softer. In this case, the alternation will occur 2 times per second, and the frequency of this sound will be 257 Hz.

This phenomenon, known as **beats** and illustrated below, is the result of a regularly alternating pattern of constructive and destructive interference. The number of beats per second (the beat frequency) is found by subtracting the smaller frequency from the larger one; the frequency of the resulting wave (i.e., the *pitch* of the sound wave) is the *average* of the two frequencies.

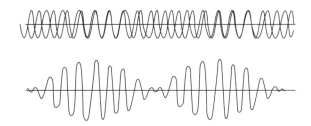

PROBLEM
A person hears tones of 440 hertz and 444 hertz simultaneously. Calculate (a) the number of beats heard each second and (b) the pitch of the resultant sound.

SOLUTION
(a) The number of beats per second (the beat frequency) is found by subtracting the smaller frequency from the larger one:

$$444 \text{ Hz} - 440 \text{ Hz} = 4 \text{ beats/s}$$

(b) The pitch of the resulting wave is the average of the two frequencies:

$$\frac{444 \text{ Hz} + 440 \text{ Hz}}{2} = 442 \text{ Hz}$$

12.11 DIFFRACTION

Diffraction is the bending of a wave around an obstacle. If a person stands beside an open door, he or she can usually hear conversation taking place in the room. As the sound waves emerge through the door, they are able to "bend around" the doorway. Similarly, water waves seem to be able to pass through pier barriers as though no obstruction were present.

The requirement for diffraction is that the size of the opening be on the order of the length of the wave being diffracted. For this reason, light will *not* diffract through a doorway because the opening is far too large in comparison to the wavelength of light. We will examine how wave diffraction occurs in Chapter 13.

12.12 DOPPLER EFFECT

All of us are familiar with the sound of a siren on a moving vehicle—a fire engine, for example. As the vehicle approaches, the *apparent* pitch of the siren is increased; as the vehicle passes us and then recedes, the apparent pitch is decreased.

This phenomenon, known as the **Doppler effect**, occurs with all types of waves, including light. It is the result of relative motion between a source of waves and an observer. As the distance between the source and the observer decreases, the frequency of the source, as perceived by the observer, is increased; as the distance increases, the apparent frequency is decreased.

Effect on Mechanical and Electromagnetic Waves

For mechanical waves, such as sound and water, the effect produced by a source in motion is different from the effect experienced by an observer in motion, even though the general outcome for both is similar. For example, if an observer is moving toward a stationary source of sound, his or her ear drum receives more waves than if the observer were at rest, and the apparent frequency of the sound is increased. If, however, the source is moving toward a stationary observer, the result is a series of sound waves that are crowded together on the side nearest the observer. The result is that the observer's ear drum receives more waves than if the source were at rest, and the frequency of the sound appears to be increased. The following diagram illustrates the situation in which a source of sound is in motion and the observers are stationary.

For electromagnetic waves, such as visible light, the frequency change is recorded as a *color shift*, a phenomenon important in astronomy and astrophysics.

Bow Waves, Shock Waves, and Sonic Booms

If you have ever seen a duck swimming on a lake or pond, you may have observed a V-shaped wave produced by the duck. This phenomenon, known as a *bow wave,* is also produced by a boat in motion on a body of water. A bow wave is a special case of the Doppler effect. As the duck (or boat) travels on the water, it produces water waves. If the speed of the traveler is *greater* than the speed of the water waves, a bow wave results.

When planes exceed the speed of sound, they produce *shock waves,* which are exactly analogous to bow waves. The diagram below illustrates how shock waves are formed. A shock wave is accompanied by an explosionlike sound known as a *sonic boom.*

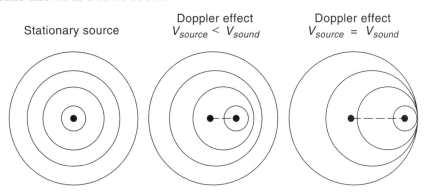

| Stationary source | Doppler effect $V_{source} < V_{sound}$ | Doppler effect $V_{source} = V_{sound}$ |

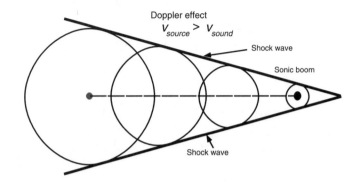

QUESTIONS

1. A series of pulses generated at regular time intervals in an elastic medium will produce
   (1) nodes
   (2) antinodes
   (3) a polarized wave
   (4) a periodic wave

2. Waves transfer energy between two points with no transfer of
   (1) work
   (2) momentum
   (3) mass
   (4) force

3. Compression waves in a spring are an example of
   (1) longitudinal waves
   (2) transverse waves
   (3) polarized waves
   (4) torsional waves

4. A wave in which the vibration is at right angles to the wave's direction of motion is called a
   (1) longitudinal wave
   (2) compressional wave
   (3) transverse wave
   (4) torsional wave

5. Which is an example of a longitudinal wave?
   (1) gamma ray   (2) X ray   (3) sound wave (4) water wave

6. Which wave requires a medium for transmission?
   (1) light   (2) infrared   (3) radio   (4) sound

7. A single pulse in a uniform material medium transfers
   (1) standing waves
   (2) energy
   (3) mass
   (4) wavelength

8. If a disturbance is parallel to the direction of travel of a wave, the wave is classifed as
   (1) longitudinal   (3) transverse
   (2) electromagnetic   (4) torsional

9. As the energy imparted to a mechanical wave increases, the maximum displacement of the particles in the medium
   (1) decreases   (2) increases   (3) remains the same

10. The water wave that will transfer the greatest amount of energy is the water wave that has the
   (1) highest frequency   (3) greatest amplitude
   (2) lowest frequency   (4) longest wavelength

11. A wave is generated in a rope, which is represented by the solid line in the diagram below. As the wave moves to the right, point P on the rope is moving toward which position?

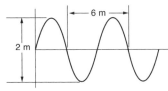

   (1) A (2) B (3) C (4) D

12. What is the amplitude of the wave represented in the diagram?

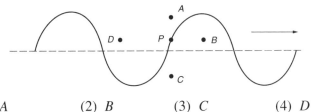

   (1) 1 m   (2) 2 m   (3) 3 m   (4) 6 m

13. Which distance represents the wavelength of the wave shown below?

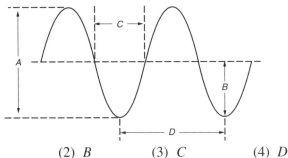

   (1) A (2) B (3) C (4) D

14. Which two wave representations in the diagram below have the same amplitude?

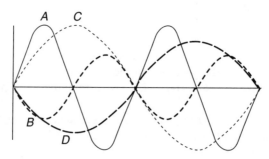

(1) *A* and *C*   (2) *A* and *B*   (3) *B* and *C*   (4) *B* and *D*

15. The graph below represents the displacement of a point in a medium as a function of time as a wave passes through the medium. What is the frequency of the wave?

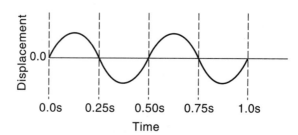

(1) 1 Hz   (2) 2 Hz   (3) $\frac{1}{4}$ Hz   (4) 4 Hz

16. In the diagram below, a train of waves is moving along a string. What is the wavelength?

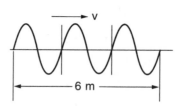

(1) 1 m   (2) 2 m   (3) 3 m   (4) 6 m

17. Which distance on the diagram below identifies the amplitude of the given wave?

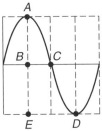

(1) *AE*   (2) *AB*   (3) *AC*   (4) *AD*

18. The number of water waves passing a given point each second is the wave's
(1) frequency   (2) amplitude   (3) wavelength   (4) velocity

19. The wavelength of the periodic wave shown in the diagram below is 4.0 meters. What is the distance from point *B* to point *C*?

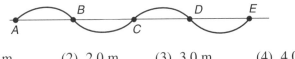

(1) 1.0 m   (2) 2.0 m   (3) 3.0 m   (4) 4.0 m

20. The diagram below represents the series of wave fronts produced by a wave generator that operated for 2 seconds.

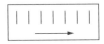

What was the frequency of the wave produced?
(1) 6 Hz   (2) 2 Hz   (3) 3 Hz   (4) 12 Hz

21. As the period of a wave decreases, the wave's frequency
(1) decreases   (2) increases   (3) remains the same

22. Waves are traveling with a speed of 3 meters per second toward *P*, as shown in the diagram.

If three crests pass *P* in 1 second, the wavelength is
(1) 1 m   (2) 6 m   (3) 3 m   (4) 9 m

23. A source produces periodic waves with a frequency of f and a speed of v. The distance traveled by the wave in a time interval equal to one period of the wave is equal to

(1) $\dfrac{v}{f}$ (2) $\dfrac{f}{v}$ (3) $\dfrac{v}{2}$ (4) fv

24. A sound wave takes 1 second to travel from a source to observer A. How long does the same sound wave take to travel in the same medium to observer B, who is located twice as far from the source as observer A?

(1) $\dfrac{1}{4}$ s   (2) 2s   (3) $\dfrac{1}{2}$ s   (4) 4 s

25. What is the period of a wave with a frequency of 250 hertz?
(1) 1.2×10^{-3} s   (3) 9.0×10^{-3} s
(2) 2.5×10^{-3} s   (4) 4.0×10^{-3} s

26. A wave has a frequency of 2.0 hertz and a velocity of 3.0 meters per second. The distance covered by the wave in 5.0 seconds is
(1) 30. m   (2) 15 m   (3) 7.5 m   (4) 6.0 m

27. As the amplitude of a periodic wave increases, its wavelength
(1) decreases   (2) increases   (3) remains the same

28. The number of water waves passing a given point each second is a measure of the wave's
(1) wavelength   (2) amplitude   (3) frequency   (4) velocity

29. If the period of a wave is doubled, its wavelength will be
(1) halved   (2) doubled   (3) unchanged   (4) quartered

30. The speed of a transverse wave in a string is 10. meters per second. If the frequency of the source producing this wave is 2.5 hertz, what is its wavelength?
(1) 0.25 m   (2) 2.0 m   (3) 25 m   (4) 4.0 m

31. The frequency of a water wave is 6.0 hertz. If its wavelength is 2.0 meters, the speed of the wave is
(1) 0.33 m/s   (2) 2.0 m/s   (3) 6.0 m/s   (4) 12 m/s

32. A radio station transmits waves with a wavelength of 30 meters. The frequency of the transmitted waves is
(1) 1×10^{7} Hz   (3) 3×10^{9} Hz
(2) 1×10^{9} Hz   (4) 9×10^{9} Hz

33. If the frequency of a light wave in a vacuum is 5.1×10^{14} hertz, what is its wavelength?
(1) 5.9×10^{-7} m (3) 1.5×10^{-7} m
(2) 1.7×10^{-7} m (4) 8.1×10^{-7} m

34. What is the distance between two consecutive points in phase on a wave called?
(1) frequency (2) period (3) amplitude (4) wavelength

35. The frequency of a sound wave is 100. hertz. If the wave's speed is 330 meters per second, its wavelength is
(1) 50. m (2) 33 m (3) 3.3 m (4) 0.30 m

36. As a wave travels into a medium in which its speed increases, its wavelength
(1) decreases (2) increases (3) remains the same

37. The speed of a transverse wave in a string is 12 meters per second. If the frequency of the source producing this wave is 3.0 hertz, what is its wavelength?
(1) 0.25 m (2) 2.0 m (3) 3.6 m (4) 4.0 m

38. If a wave has a frequency of 110. hertz, its period is
(1) 9.09×10^{-4} s (3) 1.00×10^{-1} s
(2) 9.09×10^{-3} s (4) 1.00×10^{1} s

Base your answers to questions 39 and 40 on the information below.

The frequency of a wave is 2.0 hertz, and its speed is 0.04 meter per second.

39. The period of the wave is
(1) 0.005 s (2) 2.0 s (3) 0.50 s (4) 0.02 s

40. The wavelength of the wave is
(1) 1.0 m (2) 0.02 m (3) 0.08 m (4) 4.0 m

41. Sound waves with a constant frequency of 250 hertz are traveling through air at STP. What is the wavelength of the sound waves?
(1) 0.76 m (2) 1.3 m (3) 250 m (4) 83,000 m

42. The diagram illustrates the wave pattern formed when a stone is dropped into still water. What does the collection of points on the outermost circle represent?

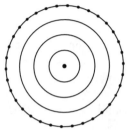

(1) a wave front   (3) the frequency
(2) a wavelength   (4) the period

43. Maximum constructive interference will occur at points where the phase difference between two waves is
(1) 0°   (2) 90°   (3) 180°   (4) 270°

44. Two pulses approach each other as shown.

Which diagram best represents the wave formed when the two pulses meet?

(1)    (3)

(2)    (4) _____

45. The diagram below represents two waves traveling simultaneously in the same medium.

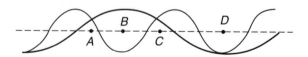

At which point will maximum constructive interference occur?
(1) *A*   (2) *B*   (3) *C*   (4) *D*

46. Two pulses are traveling along a string toward each other as represented in the diagram below.

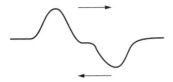

Which phenomenon will occur as the pulses meet?
(1) reflection   (3) interference
(2) polarization   (4) refraction

47. Which phenomenon must occur when two or more waves pass simultaneously through the same region in a medium?
(1) refraction   (3) dispersion
(2) interference   (4) reflection

48. As the phase difference between two superposed waves changes from 180° to 90°, the amount of destructive interference
(1) decreases   (2) increases   (3) remains the same

49. Two pulses in a stretched spring approach P as shown in the diagram.

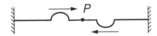

Which diagram best illustrates the appearance of the spring when the two pulses meet at P?

50. Maximum constructive interference between two waves of the same frequency could occur when their phase difference is
(1) λ (2) $\dfrac{\lambda}{2}$ (3) $\dfrac{3\lambda}{2}$ (4) $\dfrac{\lambda}{4}$

51. In the diagram below which point is in phase with point *X?*

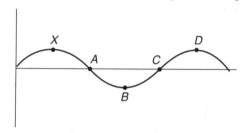

(1) *A*   (2) *B*   (3) *C*   (4) *D*

52. The diagram below shows a rope with two waves moving along it in the directions shown.

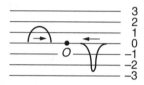

What will be the resultant wave pattern at the instant when the maximum displacement of both pulses is at point *O* on the rope?

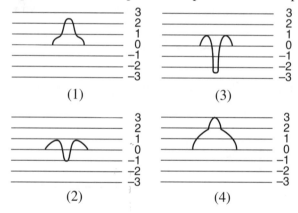

53. Maximum destructive interference will occur between two waves where their phase difference is
(1) 0°   (2) 90°   (3) 180°   (4) 270°

54. Two points on a periodic wave in a medium are said to be in phase if they
(1) have the same amplitude, only
(2) are moving in the same direction, only
(3) have the same period
(4) have the same amplitude and are moving in the same direction

55. Standing waves are produced by the interference of two waves with the same
(1) frequency and amplitude, but opposite directions
(2) frequency and direction, but different amplitudes
(3) amplitude and direction, but different frequencies
(4) frequency, amplitude, and direction

56. If two identical sound waves arriving at the same point are in phase, the resulting wave, compared to the original waves, will have
(1) an increase in speed   (3) a larger amplitude
(2) an increase in frequency   (4) a longer period

57. When the stretched string of the apparatus represented below is made to vibrate, point P does not move.

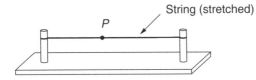

Point P is most probably at the location of
(1) a node   (3) maximum amplitude
(2) an antinode   (4) maximum pulse

58. Two waves of the same wavelength (λ) interfere to form a standing wave pattern, as shown in the diagram.

What is the straight-line distance between consecutive nodes?
(1) 1λ (2) 2λ (3) $\frac{1}{2}\lambda$ (4) $\frac{1}{4}\lambda$

59. The distance between two nodes of standing waves is N meters. The wavelength of the wave producing these standing waves is
(1) $\frac{N}{2}$ m   (2) N m   (3) $\frac{3N}{2}$ m   (4) $2N$ m

60. Which characteristic of a wave is always changed whenever a wave is reflected, refracted, or diffracted?
(1) wavelength   (3) speed
(2) period   (4) direction of travel

61. Which characteristic of a wave changes as the wave travels across a boundary between two different media?
(1) frequency   (2) period   (3) phase   (4) speed

62. When a pulse traveling in a medium strikes the boundary of a different medium, the energy of the pulse will be
(1) completely absorbed by the boundary
(2) entirely transmitted into the new medium
(3) entirely reflected back into the original medium
(4) partly reflected back into the original medium and partly transmitted or absorbed into the new medium

63. Standing waves can be produced in a vibrating rope because of the phenomenon of
(1) reflection   (2) refraction   (3) dispersion   (4) diffraction

64. Compared to the frequency of a source wave, the frequency of the echo as the wave reflects from a stationary object is
(1) smaller   (2) larger   (3) the same

65. As a wave enters a different medium with no change in velocity, the wave will be
(1) reflected but not refracted   (3) both reflected and refracted
(2) refracted but not reflected   (4) neither reflected nor refracted

66. A pulse traveling along a stretched spring is reflected from the fixed end. Compared to the pulse's speed before reflection, its speed after reflection is
(1) less   (2) greater   (3) the same

67. Which diagram best illustrates wave refraction?

68. In the diagram below, ray *AB* is incident on surface *XY* at point *B*. If the corresponding wave is traveling more rapidly in medium 2 than in medium 1, through which point will the ray most likely pass?

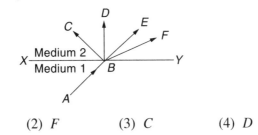

(1) *E* (2) *F* (3) *C* (4) *D*

69. As a wave travels from one medium to another, its speed decreases. The ratio of the angle of incidence to the angle of refraction is
(1) equal to 0 (3) equal to 1
(2) less than 1, but greater than 0 (4) greater than 1

Base your answers to questions 70 through 73 on the information and diagram below.

The diagram shows wave fronts passing from medium 1 into medium 2 at boundary *BB'*.

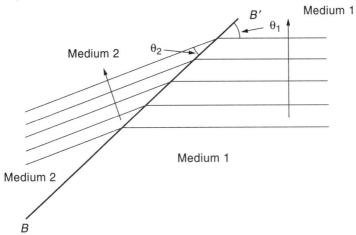

The distance between the wave fronts in medium 1 is 0.04 meter and in medium 2 is 0.02 meter. The frequency of both waves is 10 hertz.

70. The change in the direction of the wave fronts is called
(1) refraction (2) dispersion (3) diffraction (4) reflection

71. Compared to the speed of the waves in medium 2, the speed of the waves in medium 1 is
   (1) one-half as great   (3) twice as great
   (2) the same   (4) 4 times as great

72. If angle θ_1 were increased, angle θ_2 would
   (1) decrease   (2) increase   (3) remain the same

73. Compared to the period of the waves in medium 2, the period of the waves in medium 1 is
   (1) one-half as great   (3) twice as great
   (2) the same   (4) 4 times as great

74. If a ray is bent away from the normal when entering a new medium, the speed of its wave has
   (1) decreased   (2) increased   (3) remained the same

75. Refraction of a wave is caused by a change in the wave's
   (1) amplitude   (2) frequency (3) phase   (4) speed

76. The diagram below represents straight wave fronts approaching an opening in a barrier.

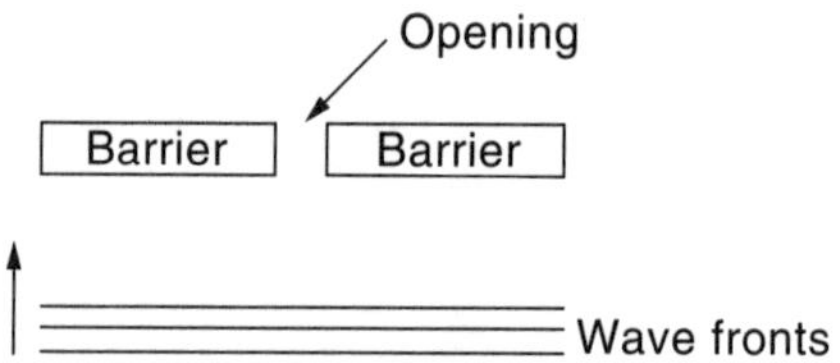

Which diagram best represents the shape of the waves after passing through the opening?

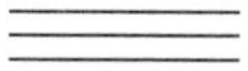

77. A wave spreads into the region behind a barrier. This phenomenon is called
   (1) diffraction   (2) reflection (3) refraction   (4) interference

78. Which wave phenomenon is represented in the diagram below?

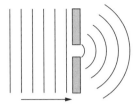

(1) refraction   (2) diffraction (3) reflection   (4) interference

79. An observer detects an apparent change in the frequency of sound waves produced by an airplane passing overhead. This phenomenon illustrates
(1) the Doppler effect   (3) an increase in wave amplitude
(2) the refraction of sound waves (4) an increase in wave intensity

80. A girl moves away from a source of sound at a constant speed. Compared to the frequency of the sound wave produced by the source, the frequency of the sound wave heard by the girl is
(1) lower   (2) higher   (3) the same

81. An Earth satellite in orbit emits a radio signal of constant frequency. Compared to the emitted frequency, the frequency of the signal received by a stationary observer will appear to be
(1) higher as the satellite approaches
(2) higher as the satellite moves away
(3) lower as the satellite approaches
(4) unaffected by the satellite's motion

Base your answers to questions 82 through 85 on the diagram below which represents the wave pattern produced by a vibrating source moving linearly in a shallow tank of water. The pattern is viewed from above, and the lines represent wave crests.

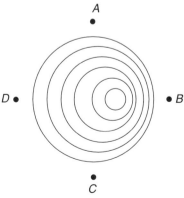

82. The source is moving toward point
    (1) *A*   (2) *B*   (3) *C*   (4) *D*

83. The wave pattern is an illustration of
    (1) diffraction   (3) dispersion
    (2) interference   (4) the Doppler effect

84. Compared to the frequency of the waves observed at point *D*, the frequency of the waves observed at point *B* is
    (1) lower   (2) higher   (3) the same

85. The velocity of the source is increased. The wavelength of the waves observed at point *D* will
    (1) decrease   (2) increase   (3) remain the same

The answers to questions 86 through 90 are to be chosen from the four sets of wave diagrams below. Each diagram represents a graph of the amplitude of a periodic wave as a function of time. The amplitude and time scales for all graphs are identical.

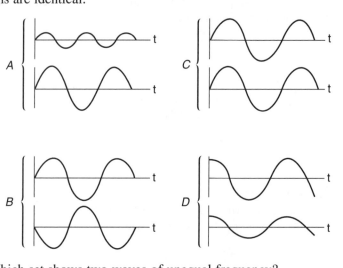

86. Which set shows two waves of unequal frequency?
    (1) *A*   (2) *B*   (3) *C*   (4) *D*

87. Which set of waves would produce the greatest constructive interference?
    (1) *A*   (2) *B*   (3) *C*   (4) *D*

88. Which set of waves shows equal wavelength but different phase?
    (1) *A*   (2) *B*   (3) *C*   (4) *D*

89. Which set shows two waves of equal amplitude that are in phase?
    (1) *A*   (2) *B*   (3) *C*   (4) *D*

90. Which set of waves shows unequal amplitude but the same period?
    (1) *A*   (2) *B*   (3) *C*   (4) *D*

Base your answers to questions 91 through 94 on your knowledge of physics and on the diagram below, which shows the water waves produced by two sources, *X* and *Y*. Three waves are produced by each source every second, and the semicircles represent wave crests.

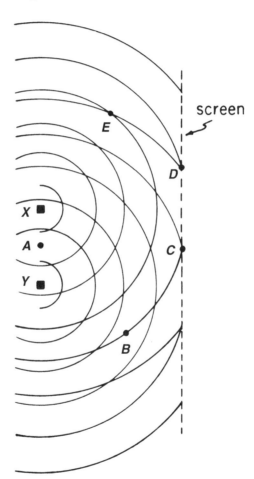

91. A point where destructive interference occurs is
    (1) *E*   (2) *B*   (3) *C*   (4) *D*

92. The distance between points X and D differs from the distance between points Y and D by how many wavelengths?

    (1) 1   (2) 2   (3) $\frac{1}{2}$ (4) $1\frac{1}{2}$

93. The period of the waves is

    (1) 0.33 s   (2) 2.5 s   (3) 3.0 s   (4) 4.0 s

94. If the wave travels into an area of different water depth, there will be a change in the wave's

    (1) velocity, only   (3) frequency and velocity
    (2) wavelength, only   (4) wavelength and velocity

Base your answers to questions 95 through 99 on the information and diagram below.

The diagram represents a wave traveling from left to right along a horizontal elastic medium. The horizontal distance from b to f is 0.08 meter. The vertical distance from x to y is 0.06 meter.

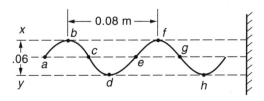

95. If the crest at b takes 2.0 seconds to move from b to f, what is the speed of the wave?

    (1) 0.03 m/s   (2) 0.04 m/s   (3) 0.05 m/s   (4) 0.06 m/s

96. If the period of the wave is 2.0 seconds, what is the frequency?

    (1) 0.5 Hz   (2) 2.0 Hz   (3) 5.0 Hz   (4) 4.0 Hz

97. What is the amplitude of the wave?

    (1) 0.03 m   (2) 0.04 m   (3) 0.05 m   (4) 0.06 m

98. As the wave moves to the right from its present position, in which direction will the medium at point e first move?

    (1) down   (2) up   (3) to the right   (4) to the left

99. The frequency of the wave is now doubled. If the velocity remains constant, its wavelength is

    (1) quartered   (2) halved   (3) unchanged   (4) doubled

Base your answers to questions 100 through 102 on the diagram below, which represents periodic water waves in a ripple tank. The speed of a wave decreases as it moves from the deep to the shallow portion of the tank.

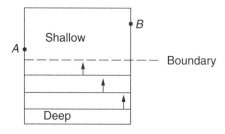

100. Which drawing best represents the waves after they enter the shallow section?

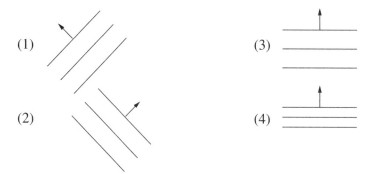

(1)

(3)

(2)

(4)

101. Which diagram best represents the pattern produced when the waves are reflected from the boundary between the deep and shallow sections of the ripple tank?

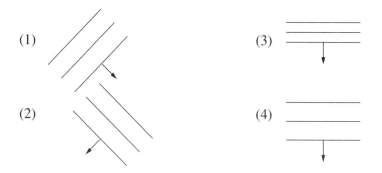

(1)

(3)

(2)

(4)

102. If a barrier is placed in the ripple tank connecting points A and B, which drawing best represents the waves after reflection from the barrier?

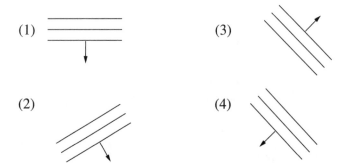

(1)

(3)

(2)

(4)

Base your answers to questions 103 through 107 on the diagram and information below.

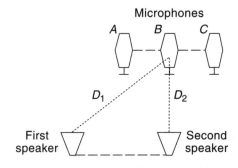

Two speakers are arranged as shown so that initially they will emit tones that are in phase, equal in volume, and equal in frequency. A microphone is placed at position A, which is equidistant from both speakers, and then is moved along a line parallel to the line joining the speakers until no sound is heard. [This is position B.] The microphone is then moved to position C, where sound is again picked up by the microphone.

103. Which phenomenon caused the sound to be louder at position C than at position B?
(1) reflection   (2) dispersion (3) polarization (4) interference

104. Distance D_2 is shorter than distance D_1 by an amount equal to
(1) the wavelength of the emitted sound
(2) one-half the wavelength of the emitted sound
(3) twice the wavelength of the emitted sound
(4) the distance between the two speakers

336

105. If the sound waves emitted by D_1 and D_2 have a frequency of 660 hertz and a speed of 330 meters per second, their wavelength is
(1) 1.0 m   (2) 2.0 m   (3) 0.25 m   (4) .50 m

106. As the first speaker is adjusted so that the sound that it emits is 180° out of phase with the sound emitted by the second speaker, the loudness of the sound received at A is
(1) greater   (2) less   (3) the same

107. If speaker D_1 were removed and speaker D_2 were accelerated toward microphone B, the frequency of the waves detected at B would
(1) decrease   (2) increase   (3) remain the same

Base your answers to questions 108 through 113 on the information and diagram below. The diagram represents two sound waves that are produced in air by two tuning forks. The frequency of wave A is 400 hertz.

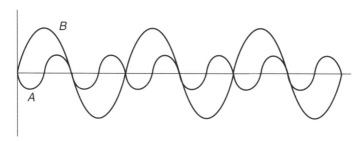

108. The period of wave A is
(1) 400 s   (2) 6 s   (3) $\dfrac{1}{4}$ s   (4) $\dfrac{1}{400}$ s

109. Under standard conditions of temperature and pressure, the wavelength in air of A is
(1) 2.5 m   (2) 12 m   (3) $\dfrac{331}{400}$ m   (4) 331×400 m

110. The frequency of wave B is
(1) 200 Hz   (2) 400 Hz   (3) 600 Hz   (4) 800 Hz

111. Sound waves produced by tuning forks are
(1) longitudinal   (2) hyperbolic   (3) torsional   (4) elliptical

112. Compared to the amplitude of wave B, the amplitude of wave A is
(1) less   (2) greater   (3) the same

113. Compared to the speed of wave A, the speed of wave B is
(1) less   (2) greater   (3) the same

Base your answers to questions 114 through 115 on the diagram and information below.

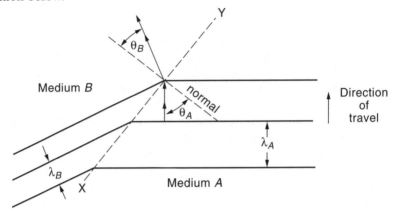

The diagram represents wave fronts traveling from medium A to medium B at boundary XY. The wave speed in medium A is 4.0 meters per second. The wave speed in medium B is 2.0 meters per second.

114. If $\sin \theta_A$ is 0.8, then $\sin \theta_B$ is
(1) 0.5   (2) 0.8   (3) 1.6   (4) 0.4

115. If wavelength λ_A is 2.0 m, then wavelength λB is
(1) 1.0 m   (2) 2.0 m   (3) 0.5 m   (4) 4.0 m

Base your answers to questions 116 through 119 on the diagram below, which represents a wave traveling to the right along an elastic medium.

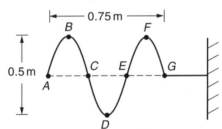

116. If the frequency of the wave is 0.25 hertz, the period of the wave is
(1) 1 s   (2) 2 s   (3) 0.25 s   (4) 4.0 s

117. What is the wavelength of the wave?
(1) 0.25 m   (2) 0.50 m   (3) 0.75 m   (4) 1.50 m

118. What is the amplitude of the wave?
(1) 1.0 m   (2) 0.75 m   (3) 0.50 m   (4) 0.25 m

119. Which point on the wave is in phase with point A?
(1) B (2) C (3) D (4) E

Chapter Thirteen	# LIGHT AND GEOMETRIC OPTICS

=== KEY IDEAS ===

Visible light is part of the electromagnetic spectrum of waves. Electromagnetic waves are transverse and have a constant speed in space. Because light is a periodic wave, it possesses the characteristics and properties of all periodic waves: reflection, refraction, interference, and diffraction, and it exhibits the Doppler effect.

The principal applications of reflected light involve the use of plane and curved mirrors. Lenses, prisms, and fiber-optic bundles are applications of refraction of light.

Diffraction and interference can be demonstrated by passing light through a single- or double-slit arrangement. These devices can be used to measure the wavelength of light. Interference also occurs with the reflected light from thin films and is responsible for the colors seen on soap bubbles and oil slicks.

If monochromatic light is generated so that all of the waves have a constant phase relationship, the light is said to be coherent. Lasers produce intense beams of coherent light.

KEY OBJECTIVES

At the conclusion of this chapter you will be able to:

- Define the term *polarization*, and explain why polarization distinguishes between transverse and longitudinal waves.
- Explain how reflection of light produces an image in a plane mirror, and describe the characteristics of such an image.
- Describe how images are produced by spherical mirrors, and use the mirror equations to solve problems relating to these images.
- Draw ray diagrams that illustrate image formation by plane and curved mirrors.
- Explain how light refracts as it passes from one medium to another.
- Define the term *absolute index of refraction*, and solve problems using this concept.
- State Snell's law in terms of absolute indices of refraction, and solve problems using this equation.
- Define the terms *critical angle* and *total internal reflection*, and relate them to Snell's law.

- Define the term *dispersion.*
- Explain how curved surfaces refract light.
- Describe how images are produced by spherical lenses, and use the lens equations to solve problems relating to these images.
- Define the terms *chromatic aberration* and *spherical aberration,* and describe these defects.
- Describe the patterns produced when monochromatic light passes through a double-slit arrangement, and explain how these patterns are formed.
- Apply the double-slit equation to the solution of problems.
- Explain the difference between the pattern produced by a double-slit arrangement and that produced by a single-slit arrangement.
- Define the term *thin-film interference,* and explain why soap bubbles and oil slicks produce colored patterns when illuminated by white light.
- Define the term *laser,* and explain how laser light differs from ordinary light.

13.1   INTRODUCTION

Light is an electromagnetic wave. The different forms of light constitute the *electromagnetic spectrum,* diagramed below, of which visible light is only a very small part.

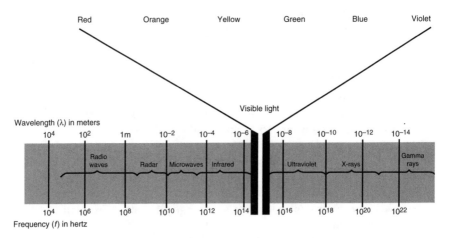

Electromagnetic waves differ in their frequency and in the sources used to produce them.

13.2 POLARIZATION OF LIGHT

Polarization is the separation of a beam of light so that the vibrations are in one plane. It is an exclusive property of transverse waves. We know that light is a transverse wave because it can be *polarized*. When a light wave is produced, it vibrates in many directions, as shown below.

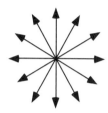

Unpolarized Light Beam

If, however, a beam of light passes through a polarizing filter, the beam that emerges will vibrate in one plane onlyand is said to be *plane polarized*.

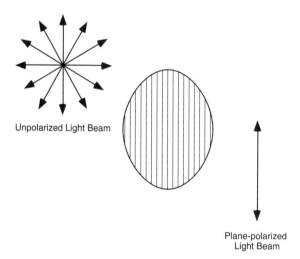

Unpolarized Light Beam

Plane-polarized
Light Beam

Since longitudinal waves, such as sound, vibrate parallel to the direction of motion, they cannot be polarized; therefore, polarization distinguishes between transverse and longitudinal waves. If a second polarizing filter is placed at a right angle to the plane of polarized light it will block nearly all of the light, as shown in the diagram.

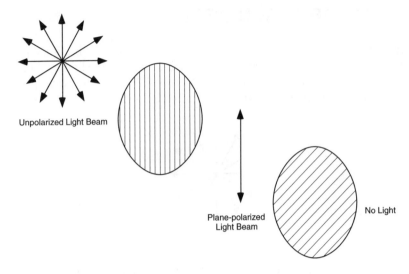

Unpolarized Light Beam

Plane-polarized
Light Beam

No Light

When light is reflected from a nonmetallic surface, it is polarized. Polarized sunglasses can be used to eliminate the glare associated with this type of reflection.

13.3 SPEED OF LIGHT

Anyone who has observed lightning or fireworks from a distance knows that the flash of light appears nearly instantaneously, while the explosion is heard somewhat later. We know that sound travels in air at approximately 330 meters per second (about 750 mi/h), but how fast does light travel?

This question was first answered in 1675 when the Dutch astronomer Olaus Roemer used his observations of Jupiter and the eclipse of one of its moons to measure the speed of light. In the nineteenth century the American physicist Albert Michaelson used sunlight and rotating mirrors to obtain more precise measurements.

As a result of Einstein's special theory of relativity, discussed in Chapter 15, we know that the speed of light in a vacuum is constant under all circumstances. It has been set at the value 2.99792458×10^8 meters per second (approximately 186,000 mi/s). The letter c is used to represent the speed of light in a vacuum.

The speed of light is less in a material medium than in a vacuum and depends on the nature of the medium and the frequency of the light.

13.4 VISIBLE LIGHT

The visible portion of the electromagnetic spectrum ranges from red to violet. The following table indicates the approximate ranges for visible light in a vacuum.

Wavelengths of Light in a Vacuum	
Violet	$4.0 - 4.2 \times 10^{-7}$ m
Blue	$4.2 - 4.9 \times 10^{-7}$ m
Green	$4.9 - 5.7 \times 10^{-7}$ m
Yellow	$5.7 - 5.9 \times 10^{-7}$ m
Orange	$5.9 - 6.5 \times 10^{-7}$ m
Red	$6.5 - 7.0 \times 10^{-7}$ m

Monochromatic light consists of light of a single color, that is, light of a single wavelength (or frequency). If all the colors of visible light are mixed together, the result is white light. Black is the complete absence of visible light.

13.5   REFLECTION

Law of Reflection

When light is reflected from a surface, the angle that the incident ray makes with the normal to the surface is equal to the angle that the reflected ray makes with the normal to the surface. This statement is known as the *law of reflection*.

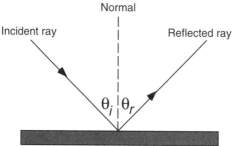

Polished surfaces such as plane mirrors produce *regular* reflection.

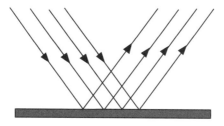

An irregular surface such as a windblown water surface or the paper on which this book is printed produces *diffuse* reflection.

343

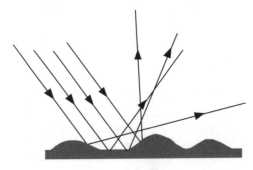

We note that, while the reflected rays emerge at different angles in the diagram above, the law of reflection holds for each individual pair of incident and reflected rays.

Mirrors

One primary application of reflected light is the image formed by a mirror. Mirrors may be plane or curved.

PLANE MIRRORS

When an object is viewed in a plane mirror, the image that is formed is erect (upright), left-right reversed, and the same size as the object. The object and image distances from the mirror are equal. These relationships can be proved by using the law of reflection and simple geometry.

The image produced by a plane mirror is *virtual* because light does not actually pass through the mirror to form the image; it only appears to be located inside the mirror.

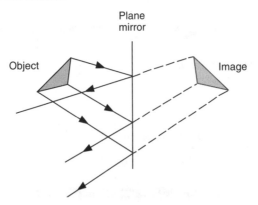

This phenomenon occurs because light appears to travel in straight lines, and a device such as a human eye searches for the apparent origin of the light.

PROBLEM

A beam of light enters and exits a hollow rectangular box. How could plane mirrors be placed in the box to produce the following patterns?

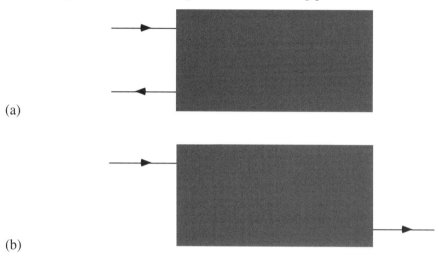

(a)

(b)

SOLUTION

The problem in each case is solved by orienting two plane mirrors at 45° angles in the upper and lower right corners of the box. Since the angles of incidence are all 45°, the incident and reflected rays will be perpendicular to one another, as illustrated in the diagrams below.

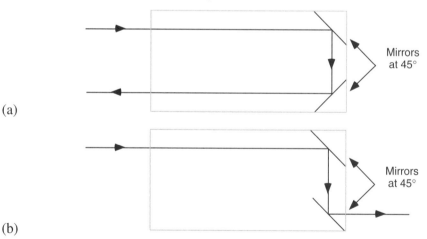

(a)

(b)

CURVED MIRRORS

When parallel rays of light strike the surface of a curved mirror, the light may either converge or diverge, as illustrated. Concave mirrors cause the light rays to *converge* to a point known as the *principal focus*. Convex mirrors cause light rays to *diverge*. If the diverging rays were projected backward, they would meet at a point called the *virtual focus*.

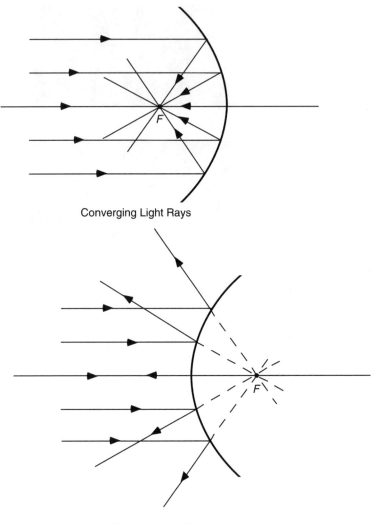

Converging Light Rays

Diverging Light Rays

Spherical Concave Mirrors

If an object is placed at various distances from a curved mirror, the mirror will produce different types of images.

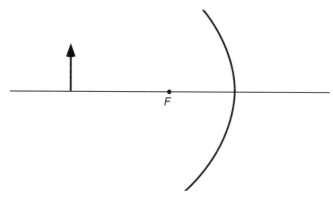

The diagram above represents an object placed in front of a spherical concave mirror. Four special rays emerging from the top of the object may be used to locate the image produced by the mirror.

If a ray parallel to the principal axis (the line passing perpendicularly through the center of the mirror) strikes the mirror, as illustrated below, the reflected ray will pass through the focus of the mirror. The distance between the surface of the mirror and the focus is known as the focal length of the mirror (f).

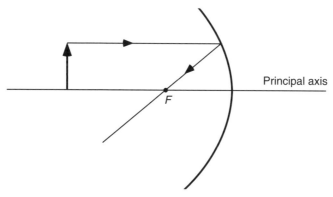

The center of curvature (C) is located at a point one radius from the surface of the mirror. This distance is equal to two focal lengths.

PHYSICS CONCEPTS

$$C = 2f$$

If a ray passes through the center of curvature of the mirror, it will strike the mirror perpendicularly to its surface and will be reflected back on itself.

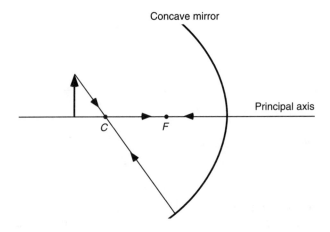

We note in the diagram above that, since the center of curvature lies on the principal axis, the principal axis is used as a ray to locate the base of the image.

If a ray passes through the principal focus, as shown below, the reflected ray will emerge parallel to the principal axis.

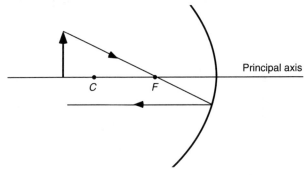

The following diagram shows that, if a ray from the object strikes the mirror at the principal axis, the principal axis acts as the normal and the ray will emerge at the same angle.

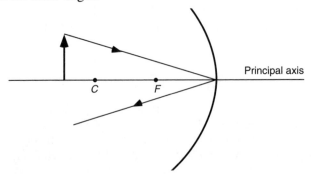

Any two rays along with the principal axis can be used to locate the image formed by a concave mirror. The following table and diagrams summarize the various types of images that can be produced by a spherical concave mirror.

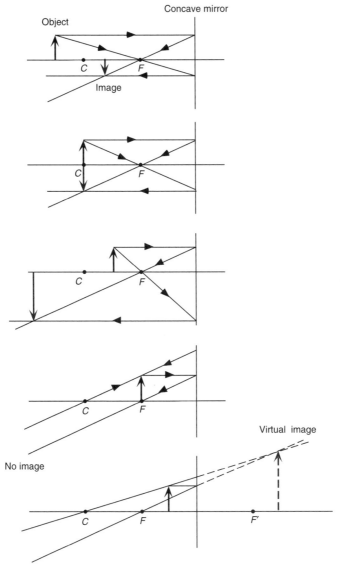

Object distance (d_o)	Image distance (d_i)	Size of Image (relative to size of object)	Description of image
infinity	F	point	real
beyond C	between F and C	smaller	real, inverted
2F	2F	equal	real, inverted
between F and C	beyond C	larger	real, inverted
F	infinity	no image	– – – –
less than F	behind mirror	larger	virtual, erect

Note that real, as well as virtual, images can be produced by this type of mirror.

Spherical Convex Mirrors

In every case, a spherical convex mirror produces an erect virtual, image that is located behind the mirror. The image is always smaller than the object. The diagram below shows how such an image is formed.

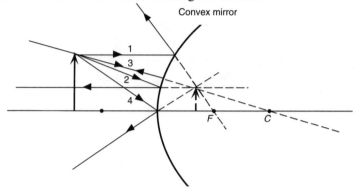

We note the following:
1. When a ray parallel to the principal axis strikes the surface of a convex mirror, it is reflected from the surface of the mirror in such a way that, if it is projected back behind the mirror, it passes through a virtual focus.
2. If a ray strikes the mirror at an angle such that, if it were projected as a straight line, it would pass through the virtual focus, it is reflected parallel to the principal axis.
3. If a ray strikes the mirror in such a way that, if it were projected as a straight line, it would pass through the center of curvature, the ray is reflected back on itself.
4. For a ray that strikes the surface of the mirror at the principal axis, the principal axis serves as a normal to the surface, and the ray is, therefore, reflected at the same angle.
   In all cases involving single curved mirrors, real images are inverted (upside down and left-right reversed), and virtual images are erect.

MIRROR EQUATIONS

We can calculate the sizes and distances of the images formed by curved mirrors by means of the following equations,

---
PHYSICS CONCEPTS

$$\frac{h_i}{h_o} = -\frac{d_i}{d_o}$$

$$\frac{1}{f} = \frac{1}{d_o} + \frac{1}{d_i}$$

---

where h_i and h_o represent the respective size (heights) of the image and object, and d_i and d_o represent the respective distances of the image and object from the mirror.

The following *sign conventions* are used for these equations: When the object, image, or focal point is "real" (i.e., is on the reflecting side of the mirror), the corresponding distance is considered positive; when the object, image, or focal point is "virtual" (i.e., behind the mirror), the distance is considered negative. (Virtual objects occur only in problems in which more than one mirror is used. Only then can an object be behind the mirror, that is, if it is the image produced by another mirror.)

We can use the following equation to calculate the magnification (m) of a mirror:

PHYSICS CONCEPTS

$$m = \frac{h_i}{h_o} = -\frac{d_i}{d_o}$$

The magnification is the ratio of the height of the image to the height of the object. The negative sign is inserted as a convention. Object and image heights are considered positive if they are above the axis and negative if below the axis. Thus the magnification is positive for an erect image and negative for an inverted image.

PROBLEM

An object whose height is 0.15 meter is placed 0.60 meter from a concave mirror whose focal length is 0.20 meter.
(a) Where is the center of curvature of the mirror?
(b) Where is the image located?
(c) What is the height of the image?

SOLUTION

(a) $C = 2f$
$= 2(0.20 \text{ m})$
$= 0.40 \text{ m}$

(c) $\dfrac{h_i}{h_o} = \dfrac{-d_i}{d_o}$

$h_i = -\dfrac{d_i h_o}{d_o}$

$= -\dfrac{(0.30 \text{ m})(0.15 \text{ m})}{0.60 \text{ m}}$

$= -0.075 \text{ m}$

(b) $\dfrac{1}{f} = \dfrac{1}{d_o} + \dfrac{1}{d_i}$

$\dfrac{1}{d_i} = \dfrac{1}{f} - \dfrac{1}{d_o}$

$= \dfrac{1}{0.20 \text{ m}} - \dfrac{1}{0.60 \text{ m}}$

$= \dfrac{3}{0.60 \text{ m}} - \dfrac{1}{0.60 \text{ m}}$

$d_i = 0.30 \text{ m}$

The image is real, is located 0.30 meter from the surface of the mirror, and is 0.075 meter in height, below the axis, inverted.

PROBLEM
Describe the image and its placement when a 0.10-meter object is placed 0.50 meter in front of a convex mirror whose focal length is 1.0 meter.

SOLUTION

$$\frac{1}{f} = \frac{1}{d_o} + \frac{1}{d_i}$$

$$\frac{1}{d_i} = \frac{1}{f} - \frac{1}{d_o}$$

$$= \frac{1}{-1.0 \text{ m}} - \frac{1}{0.50 \text{ m}}$$

$$= -\frac{3}{1.0 \text{ m}} = -3.0 \text{ m}$$

$$d_i = -0.33 \text{ m}$$

$$\frac{h_i}{h_o} = \frac{-d_i}{d_o}$$

$$h_i = -\frac{d_i h_o}{d_o}$$

$$= -\frac{(-0.33 \text{ m})(0.10 \text{ m})}{0.50 \text{ m}}$$

$$= 0.066 \text{ m}$$

The image is located 0.33 meter behind the mirror, is virtual, and is 0.066 meter in height (erect).

SPHERICAL ABERRATION

Spherical mirrors are subject to a deficiency known as *spherical aberration*. A spherical concave mirror will not focus the light exactly to one point. To correct this deficiency, a parabolic mirror must be used, as illustrated below.

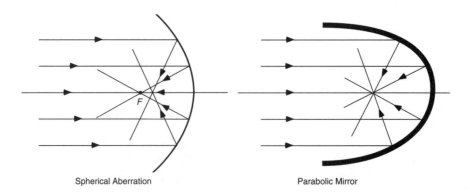

Spherical Aberration   Parabolic Mirror

13.6 REFRACTION

When monochromatic light travels between two media, there is a change in the speed of the light wave. If the light enters at an oblique angle, it will change direction as it passes into the second medium.

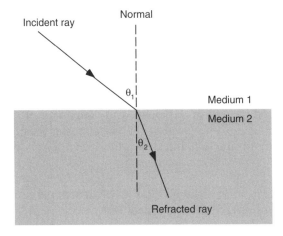

PHYSICS CONCEPTS

$$\frac{\sin \theta_1}{\sin \theta_2} = \frac{v_1}{v_2}$$

A result of the relationship expressed in this equation (Snell's law) is that the slower the speed, the smaller the angle.

PROBLEM

Monochromatic light passes between two media. If the angle in medium 1 is 45° and the angle in medium 2 is 30°, calculate the ratio of the light speeds between media 1 and 2.

SOLUTION

$$\frac{\sin \theta_1}{\sin \theta_2} = \frac{v_1}{v_2}$$

$$\frac{v_1}{v_2} = \frac{\sin 45°}{\sin 30°} = 1.4$$

$$v_1 = 1.4 v_2$$

353

Absolute Index of Refraction

To simplify refraction problems, a quantity known as the **absolute index of refraction** is defined as follows:

PHYSICS CONCEPTS

$$n = \frac{c}{v}$$

The equation states that the absolute index of refraction of a medium (n) is the ratio of the speed of light in a vacuum (c) to the speed of light in a medium (v). The absolute index of refraction is always greater than or equal to 1. The larger the index of refraction, the slower the speed of light in a medium.

PROBLEM
The speed of light in a medium is 2.4×10^8 meters per second. What is the absolute index of refraction of the medium?

SOLUTION

$$n = \frac{c}{v}$$

$$= \frac{3.0 \times 10^8 \text{ m/s}}{2.4 \times 10^8 \text{ m/s}}$$

$$= 1.25$$

The table below lists the indices of refraction for some common materials.

Absolute Indices of Refraction ($\lambda = 5.9 \times 10^{-7}$ m)	
Air	1.00
Alcohol	1.36
Benzene	1.50
Canada Balsam	1.53
Carbon Tetrachloride	1.46
Corn Oil	1.47
Diamond	2.42
Glass, Crown	1.52
Glass, Flint	1.61
Glycerol	1.47
Lucite	1.50
Quartz, Fused	1.46
Water	1.33

Snell's Law

In terms of the index of refraction, the relationship

$$\frac{\sin \theta_1}{\sin \theta_2} = \frac{v_1}{v_2}$$

becomes

$$\frac{\sin \theta_1}{\sin \theta_2} = \frac{\dfrac{c}{n_1}}{\dfrac{c}{n_2}} = \frac{n_2}{n_1}$$

We write this as follows:

PHYSICS CONCEPTS

$$n_1 \sin \theta_1 = n_2 \sin \theta_2$$

This relationship is an alternate form of Snell's law.

PROBLEM
A monochromatic light ray is incident on a surface boundary between air and corn oil at an angle of 60° to the normal. Calculate the refracted angle of the ray in the corn oil.

SOLUTION

$$n_1 \sin \theta_1 = n_2 \sin \theta_2$$

$$\sin \theta_2 = \frac{n_1 \sin \theta_1}{n_2}$$

$$= \frac{(1)(\sin 60°)}{1.47} = 0.59$$

$$\theta_2 = \sin^{-1}(0.59) = 36°$$

PROBLEM
A ray of monochromatic light in air is incident on a transparent block of material whose absolute index of refraction is 1.50, as shown in the diagram below.

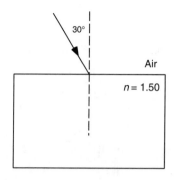

If the angle of incidence is 30°, trace the path of the light ray through the block and back into the air.

SOLUTION

We begin the solution by applying Snell's law in order to calculate the angle of refraction:

$$n_{air} \sin\theta_{air} = n_{block} \sin\theta_{block}$$

$$(1.00)(\sin 30°) = (1.50)(\sin\theta_{block})$$

$$\sin\theta_{block} = 0.333$$

$$\theta_{block} = 19.5°$$

We now extend the ray into the block at 19.5°. Using geometry (alternate interior angles), we find that the ray reaches the second surface at an angle of 19.5°.

If we applied Snell's law again, we would conclude that the ray would emerge in air at the original angle of 30°, as illustrated below.

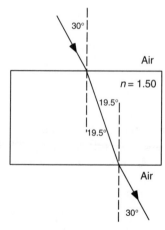

Critical Angle

If monochromatic light passes from medium 1 in which its speed is slower to medium 2 in which its speed is faster, we can draw a ray diagram as follows:

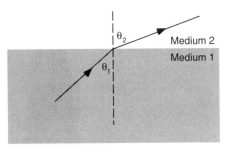

As angle θ_1 is made larger, angle θ_2 also becomes larger ($n_1\sin\theta_1 = n_2\sin\theta_2$).

There is one unique angle in medium 1 that will produce an angle of 90° in medium 2, as illustrated.

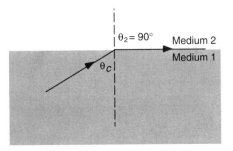

This unique angle is called the **critical angle** (θ_c). We can derive an equation for the critical angle using Snell's law:

$$n_1\sin\theta_c = n_2\sin90°$$

Then, since $\sin 90° = 1$,

$$n_1\sin\theta_c = n_2$$

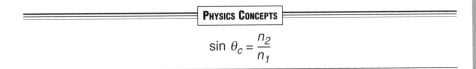

PHYSICS CONCEPTS

$$\sin \theta_c = \frac{n_2}{n_1}$$

PROBLEM

Calculate the critical angle between diamond and Lucite.

SOLUTION

$$\sin \theta_c = \frac{n_2}{n_1}$$

$$\theta_c = sin^{-1}\left(\frac{n_2}{n_1}\right) = sin^{-1}\left(\frac{1.50}{2.42}\right)$$

$$= 38°$$

In the event that medium 2 is a vacuum or air ($n_2 = 1$), the above equation simplifies to this expression:

PHYSICS CONCEPTS

$$\sin \theta_c = \frac{1}{n_1}$$

Total Internal Reflection

What is so special about the critical angle? At 90° the maximum refracted angle possible, light would skim the surface boundary between the two media. If the critical angle were exceeded, the light could no longer escape but would be reflected back into the medium. In this case, the law of reflection would hold as with any other pair of incident and reflected rays.

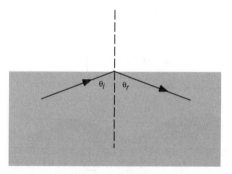

We note that the equation $\sin\theta_c = 1/n_1$, given above, implies that larger values of n yield smaller critical angles. A material such as diamond ($n = 2.42$) has a critical angle of only 24°. Because of this small value, much of the light inside a diamond is totally internally reflected. This **total internal reflection** is responsible for much of the diamond's sparkle.

PROBLEM

A monochromatic ray of light in air enters a triangular prism whose absolute index of refraction is 1.50. The prism is in the shape of an isosceles right triangle, and the incident ray is perpendicular to the surface as shown in the diagram.

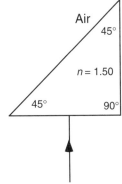

Complete the path of the light ray as it enters the prism.

SOLUTION

We begin by noting that the angle of incidence is 0°; consequently the angle of refraction will also be 0°. The ray will enter the prism without bending and will make an angle of 45° as it strikes the second surface (hypotenuse) of the prism.

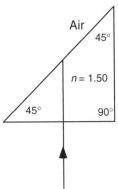

We now must ask: Will the critical angle of the block-air combination be exceeded?

$$\sin \theta_c = \frac{n_{air}}{n_{block}}$$

$$= \frac{1.00}{1.50} = 0.667$$

$$\theta_c = 41.8°$$

Since the angle of the ray is 45°, the critical angle is exceeded and total internal reflection results. The ray continues through the prism and exits into the air as shown below.

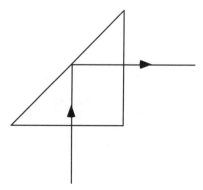

Dispersion

If white (polychromatic) light is passed through a prism, it is separated into its component colors as shown in the diagram.

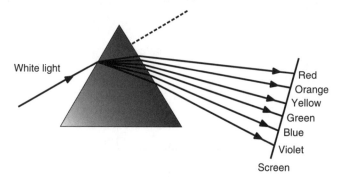

This phenomenon is known as **dispersion**. When water droplets in the air disperse light, rainbows may be produced.

Dispersion occurs because the speed of light inside the prism depends on the color of the light. From the diagram we can see that red light travels fastest because it has the largest refracted angle (that is, it is bent the least).

Lenses

One important practical application of refraction is the construction of lenses. As light passes through a lens, it can either converge or diverge, depending on the shape of the lens.

Lenses that are thicker in the middle and thinner at the ends are known as *converging lenses*. Samples of converging lenses are shown below.

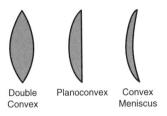

Double
Convex

Planoconvex

Convex
Meniscus

Lenses that are thinner in the middle and thicker at the ends are called *diverging lenses* and are illustrated below.

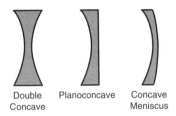

Double
Concave

Planoconcave

Concave
Meniscus

In this book we will assume that all of the lenses we discuss are very thin. This assumption allows us to use much simpler mathematical relationships.

CONVERGING (CONVEX) LENSES

If parallel monochromatic light rays strike a converging lens, the light passes through the lens and converges to a point known as the *focus*. The diagram below illustrates this situation.

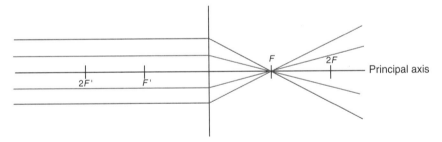

Note that the principal axis is also one of the rays of light. It does not bend because it strikes the lens along a normal to the surface. The distance between the center of the lens and the principal focus is known as the *focal length* of the lens.

The degree to which a lens refracts light depends on its index of refraction and its curvature. Large indices of refraction and highly curved surfaces produce more powerful lenses. (Eye-care professionals use a unit called the *diopter* to measure the refractive ability of a lens. The diopter is the reciprocal of the focal length and has the unit m^{-1}.)

If an object is placed at various distances from a converging lens, the lens will produce different types of images. Three special rays emerging from the top of the object may be used to locate the image produced by the lens. The diagram below represents an object placed in front of a converging lens.

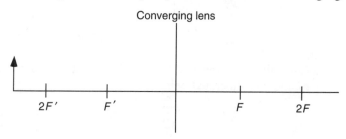

We substitute a straight line for our converging lens because, as stated above, we consider the lens to be extremely thin.

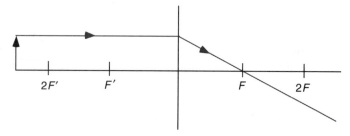

A ray that enters the lens parallel to the principal axis will be refracted so that it passes through the principal focus, as illustrated above.

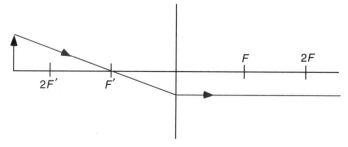

A ray that strikes the principal axis at one focal length from the lens will emerge parallel to the principal axis as illustrated above.

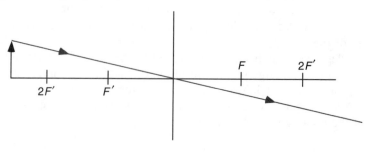

Any ray that passes through the optical center of the lens will pass through the lens without being refracted. We note in the preceding diagram that the principal axis is one such ray and can be used in locating the base of the object.

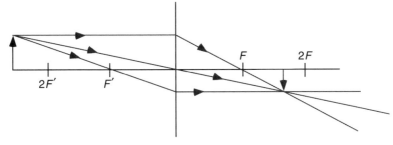

Any two rays along with the principal axis can be used to locate the image. The table and diagram below summarize the various types of images that can be produced by a converging lens. We note that real as well as virtual images can be formed by this type of lens.

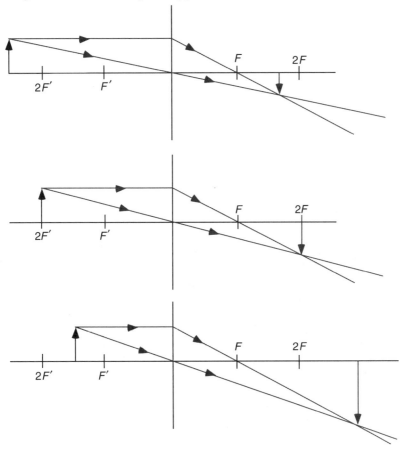

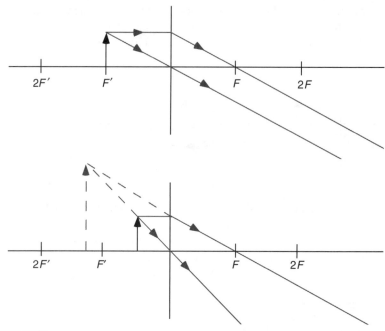

Object distance (d_o)	Image distance (d_i)	Size of Image (relative to size of object)	Description of image
infinity	F	point	real
beyond 2F	between F and 2F	smaller	real, inverted
2F	2F	equal	real, inverted
between F and 2F	beyond 2F	larger	real, inverted
F	infinity	no image	– – – –
less than F	behind object	larger	virtual, erect

DIVERGING (CONCAVE) LENSES

In every case, a diverging concave lens produces an erect, virtual image that is located in front of the object and on the same side of the lens as the object. The image is always smaller than the object. The diagram below illustrates how such an image is formed.

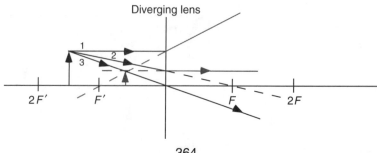

We note the following:

1. A ray entering the lens parallel to the principal axis is refracted at an angle such that, if we project it back through the lens, it passes through a *virtual focus.*

2. A ray entering the lens at an angle such that, if it were projected as a straight line, it would pass through point F on the diagram, is refracted parallel to the principal axis.

3. A ray that passes through the optical center of the lens passes straight through and is not refracted.

In all cases involving single lenses, real images are inverted and virtual images are erect.

LENS EQUATIONS

We can calculate the sizes and distances of images formed by lenses by means of the following equations, which are identical to the equations used for curved mirrors:

PHYSICS CONCEPTS

$$\frac{h_i}{h_o} = -\frac{d_i}{d_o}$$

$$\frac{1}{f} = \frac{1}{d_o} + \frac{1}{d_i}$$

The following *sign conventions* are used for these equations: The focal length is considered positive for converging lenses and negative for diverging lenses. The object distance is considered positive if it is on the side of the lens from which light is coming, that is, if the object is real, as usually the case. (Exceptions are some situations involving more than one lens where the object of one lens is the image of another and is located on the opposite side of the lens from which light is coming, that is, the object is virtual. In that case the object distance is considered negative.) The image distance is considered positive if it is on the opposite side of the lens from which light is coming, that is, if the image is real; if the image distance is on the same side of the lens, that is, if the image is virtual, the distance is negative.

The equation on page 352 for the magnification of a (m) mirror can be used also to calculate the magnification of a lens:

PHYSICS CONCEPTS

$$m = \frac{h_i}{h_o} = -\frac{d_i}{d_o}$$

The magnification is the ratio of the height of the image to the height of the object. The negative sign is inserted as a convention. Object and image heights are considered positive if they are above the axis and negative if below the axis. Thus the magnification is positive for an erect image and negative for an inverted image.

PROBLEM

A 0.70-meter object is placed 0.50 meter from a converging lens whose focal length is 0.25 meter. Describe the image and its placement.

SOLUTION

$$\frac{1}{f} = \frac{1}{d_o} + \frac{1}{d_i}$$

$$\frac{1}{d_i} = \frac{1}{f} - \frac{1}{d_o}$$

$$= \frac{1}{0.25 \text{ m}} - \frac{1}{0.50 \text{ m}}$$

$$= \frac{1}{0.50 \text{ m}}$$

$$d_i = 0.50 \text{ m}$$

$$\frac{h_i}{h_o} = \frac{-d_i}{d_o}$$

$$h_i = -\frac{d_i h_o}{d_o}$$

$$= -\frac{(0.50 \text{ m})(0.70 \text{ m})}{0.50 \text{ m}}$$

$$= -0.70 \text{ m}$$

The image is located 0.50 meter from the lens and is real. It is the same height as the object and is below the principal axis, that is, inverted.

PROBLEM

A 0.70-meter object is placed 0.50 meter from a diverging lens whose focal length is 0.25 meter. Describe the image and its placement.

SOLUTION

$$\frac{1}{f} = \frac{1}{d_o} + \frac{1}{d_i}$$

$$\frac{1}{d_i} = \frac{1}{f} - \frac{1}{d_o}$$

$$= \frac{1}{-0.25 \text{ m}} - \frac{1}{0.50 \text{ m}}$$

$$= -\frac{3}{0.50 \text{ m}}$$

$$d_i = -0.17 \text{ m}$$

$$\frac{h_i}{h_o} = \frac{-d_i}{d_o}$$

$$h_i = -\frac{d_i h_o}{d_o}$$

$$= -\frac{(-0.17 \text{ m})(0.70 \text{ m})}{0.50 \text{ m}}$$

$$= 0.24 \text{ m}$$

The image is located 0.17 meter away from the lens and on the same side of the lens as the object, and is therefore virtual. It is 0.24 meter in height above the principal axis; therefore, it is erect.

LENS DEFECTS

Two types of defects are common to all types of simple lenses. **Chromatic aberration** occurs because different colors of light do not focus at the same point. In cameras, this type of lens defect can be reduced by using combinations of lenses made of different types of glass.

Spherical aberration occurs because the spherical shape of the lens is not ideal for converging the light to a single point. Spherical aberration may be reduced (in cameras, for example) by restricting the light beam close to the center of the lens. This restriction is accomplished by reducing the size of the lens opening (the *aperture*).

13.7 DIFFRACTION AND INTERFERENCE OF LIGHT

The Double-Slit Experiment

If light is passed through a pair of closely spaced slits, an alternating pattern of bright and dark bands will appear on a distant screen, as shown below.

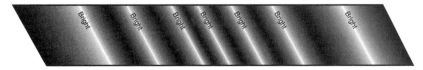

To explain this phenomenon we must first consider how a wave can travel from one point to another. Danish scientist Christian Huygens developed a principle that states that every point on a wave front can be considered a point source of secondary spherical waves (called *wavelets*). After a time, the new location of the wave is found by drawing a common tangent to these wavelets.

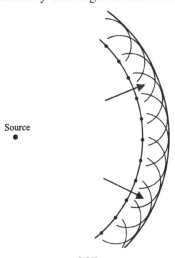

Source

367

In the double-slit arrangement illustrated below, each slit serves as a point source of waves. We see that, as the waves grow, the wave fronts interfere with one another. Points that interfere constructively will ultimately produce the bright bands, while points that interfere destructively will produce the dark bands, as shown in the diagram below.

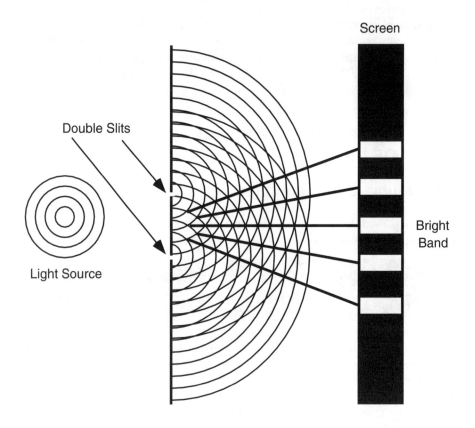

We note that the bright bands are nearly evenly spaced. For very closely spaced slits the distance between any two bright bands (X) depends on the wavelength of light used (λ), the distance between the slits (d), and the distance between the slits and the screen (L). In the diagram above, the central bright band is known as the zero-order band, and consequent bright bands on either side of the central band are referred to as the first-order band, the second-order band, and so on. The numerical relationship among the various factors is expressed by the following equation:

$$X = \frac{\lambda L}{d}$$

As the wavelength of light increases, the distance between the bright bands increases. Also, as the distance between the slits and the screen increases, the distance between the bright bands increases. However, as the distance between the slits increases, the distance between the bright bands decreases. This relationship can be used to measure the wavelength of monochromatic light. In this case we rewrite the equation as follows:

PHYSICS CONCEPTS

$$\lambda = \frac{dX}{L}$$

PROBLEM
Monochromatic light is incident on a pair of slits 1.95×10^{-5} meter apart. The distance between the first two bright bands is 2.11×10^{-2} meter. If the distance between the slits and the screen is 0.600 meter, (a) calculate the wavelength and (b) state the color of the light.

SOLUTION

(a) $\lambda = \frac{dX}{L}$

$= \frac{(1.90 \times 10^{-5} \text{ m})(2.11 \times 10^{-2} \text{ m})}{0.600 \text{ m}}$

$= 6.68 \times 10^{-7} \text{ m}$

(b) This wavelength corresponds to red light.

Single-Slit Diffraction

A single slit can also be used to produce a diffraction pattern. The width of the slit must have the same order of magnitude as the wavelength of the light used. The pattern produced by single-slit diffraction is different from the pattern obtained with double-slit diffraction in two respects: (1) the central bright band is much wider than any of the other bright bands, and (2) the intensity of the central band is greater than the intensity of any of the other bright bands.

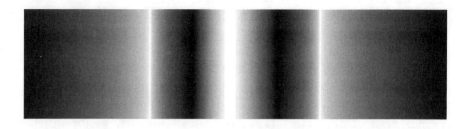

Thin-Film Interference

If a person observes an oil slick or a soap bubble that is illuminated with white light, he or she will see a series of colors on the surface of the slick or bubble. This phenomenon, known as **thin-film interference** and diagramed below, results when light waves reflected from the top of the film interfere with light waves reflected from the bottom of the film.

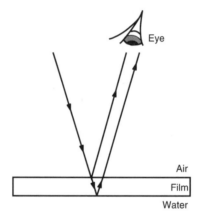

The colors observed depend on the nature of the film (its refractive index) and the thickness of the film.

Lasers

When light (even monochromatic light) is emitted by a source, the light waves have no special (phase) relationship with one another. As a result, the light beam spreads and loses its intensity (brightness) quickly. This light is said to be *incoherent*.

A **laser** is a device that produces monochromatic light in which nearly all of the waves are in phase. The word *laser* is an acronym for light amplification by stimulated emission of radiation. The beam produced by a laser spreads very little and is extraordinarily intense. This light is called *coherent*.

Because of its properties, laser light can be used for a variety of applications, from bar-code scanning to eye surgery.

The color of laser light depends on the sources that produce the light. Helium-neon lasers, for example, produce red light.

QUESTIONS

1. Which phenomenon is associated only with transverse waves?
   (1) interference   (2) dispersion   (3) refraction   (4) polarization

2. Which characterizes a polarized wave?
   (1) transverse and vibrating in one plane
   (2) transverse and vibrating in all directions
   (3) circular and vibrating at random
   (4) longitudinal and vibrating at random

3. A longitudinal wave cannot be
   (1) polarized   (2) diffracted   (3) refracted   (4) reflected

4. If the electromagnetic waves received on Earth from a source in outer space appear to be increasing in frequency, the distance between the source and the Earth is probably
   (1) decreasing   (2) increasing   (3) remaining the same

5. A star recedes rapidly from the Earth. The frequencies observed from the Earth compared to the frequencies of light emitted by the star are
   (1) lower   (2) higher   (3) the same

6. The wavelength of a spectral line emitted from a distant star is shifted toward a longer wavelength. The star is assumed to be
   (1) stationary relative to the Earth
   (2) moving in a circle around the Earth
   (3) moving away from the Earth
   (4) approaching the Earth

7. Which electromagnetic wave has the highest frequency?
   (1) radio   (2) infrared   (3) X-ray   (4) visible

8. Which electromagnetic radiation has a wavelength shorter than that of visible light?
   (1) ultraviolet waves   (3) radio waves
   (2) infrared waves   (4) microwaves

9. Which is not in the electromagnetic spectrum?
   (1) light waves   (2) radio waves   (3) sound waves   (4) X-rays

10. Which of the following electromagnetic waves has the lowest frequency?
    (1) violet light   (2) green light   (3) yellow light   (4) red light

11. In a vacuum, the wavelength of ultraviolet light is greater than that of
    (1) X-rays   (3) blue light
    (2) radio waves   (4) infrared light

12. In a vacuum, light waves and radio waves have the same
    (1) frequency   (2) period   (3) speed   (4) wavelength

13. If the frequency of a light wave in a vacuum is increased, its wavelength
    (1) decreases   (2) increases   (3) remains the same

14. Light of frequency 5.0×10^{14} hertz has a wavelength of 4.0×10^{-7} meter while traveling in a certain material. The speed of light in the material is
    (1) 1.3×10^7 m/s   (3) 3.0×10^8 m/s
    (2) 2.0×10^8 m/s   (4) 1.3×10^{21} m/s

15. What is the wavelength of X-rays with a frequency of 1.5×10^{18} hertz traveling in a vacuum?
    (1) 4.5×10^{26} m   (3) 5.0×10^{-10} m
    (2) 2.0×10^{-10} m   (4) 5.0×10^9 m

16. What is the color of a light wave with a frequency of 5.65×10^{14} hertz?
    (1) red   (2) yellow   (3) green   (4) blue

17. Which formula represents a constant for light waves of different frequencies in a vacuum?
    (1) $f\lambda$ (2) $\dfrac{f}{\lambda}$ (3) $\dfrac{\lambda}{f}$ (4) $f + \lambda$

18. A light beam from Earth is reflected by an object in space. If the round trip takes 2.0 seconds, then the distance of the object from Earth is
    (1) 6.7×10^7 m   (2) 1.5×10^8 m   (3) 3.0×10^8 m   (4) 6.0×10^8 m

19. Which set of electromagnetic radiations is arranged in order of increasing frequency?
(1) radio, ultraviolet, visible, gamma
(2) gamma, radio, visible, ultraviolet
(3) radio, visible, ultraviolet, gamma
(4) visible, ultraviolet, gamma. radio

20. The time required for light to travel a distance of 1.5×10^{11} meters is closest to
(1) 5.0×10^2 s   (2) 2.0×10^{-3} s   (3) 5.0×10^{-1} s   (4) 4.5×10^{19} s

21. The observed color of light depends on the light's
(1) speed   (2) amplitude   (3) intensity   (4) frequency

22. An observer at point O sees the reflected light ray as shown in the diagram below.

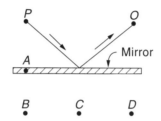

The image of point P that the observer sees is located at
(1) A (2) B (3) C (4) D

23. The diagram below shows two rays of light striking a plane mirror.

Which diagram below best represents the reflected rays?

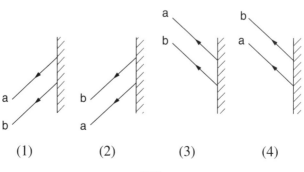

(1)   (2)   (3)   (4)

24. Which property of light is illustrated by the diagram below?

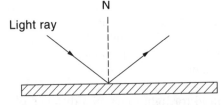

(1) diffraction   (2) dispersion   (3) refraction   (4) reflection

25. Which phenomenon of light is illustrated by the diagram below?

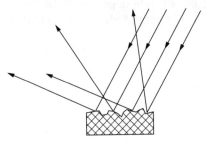

(1) refraction   (3) regular reflection
(2) dispersion   (4) diffuse reflection

26. Which diagram below best represents the image of the object formed by the plane mirror in the diagram below?

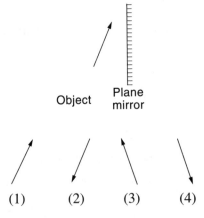

27. Which graph best represents the relationship between the image size and the object size for an object reflected in a plane mirror?

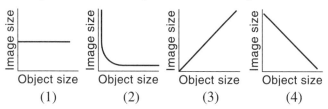

28. The diagram below shows a light ray being reflected from a plane mirror. What is the angle of reflection?

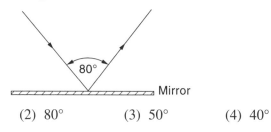

(1) 90°   (2) 80°   (3) 50°   (4) 40°

29. A man stands in front of a vertical plane mirror 2.0 meters high as shown in the diagram below.

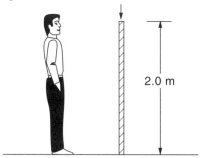

A ray of light reflects off the mirror, allowing him to see his foot. Approximately how far up the mirror from the floor does this ray strike the mirror?

(1) 1.0 m   (2) 2.0 m   (3) 0.25 m   (4) 0 m

30. Which diagram best represents the image of the letter "L" formed by the plane mirror *MM'*?

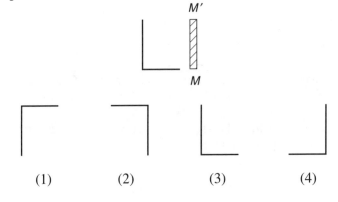

(1)   (2)   (3)   (4)

31. When a ray of light strikes a mirror perpendicularly to its surface, the angle of reflection is
(1) 0°   (2) 45°   (3) 60°   (4) 90°

32. If an observer at point *O* looks at the mirror, *M*, fixed in position as shown in the diagram below, which point will be seen?

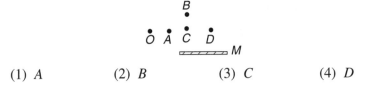

(1) *A*   (2) *B*   (3) *C*   (4) *D*

33. When a light ray is reflected from a surface, the ratio of the angle of incidence to the angle of reflection is
(1) less than 1
(2) greater than 1
(3) equal to 1

34. A light ray is incident upon a plane mirror. As the angle of incidence increases, the angle between the incident ray and the mirror's surface
(1) decreases   (2) increases   (3) remains the same

35. As a person approaches a plane mirror, the size of his or her image
(1) decreases   (2) increases   (3) remains the same

36. The diagram below represents an object in front of a concave mirror.

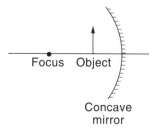

Focus   Object

Concave
mirror

The image of the object formed by the mirror is
(1) real and larger than the object
(2) real and smaller than the object
(3) virtual and larger than the object
(4) virtual and smaller than the object

37. The diagram below represents a spherical mirror with its center of curvature at *C* and focal point at *F*.

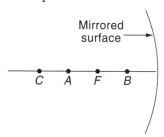

Mirrored
surface

C   *A*   *F*   *B*

At which position must a point source of light be placed to produce a parallel beam of reflected light?
(1) *A*   (2) *B*   (3) *C*   (4) *F*

38. If the distance between an object and a concave (converging) mirror is more than twice the focal length of the mirror, the image formed will be
(1) real and behind the mirror   (3) virtual and behind the mirror
(2) real and in front of the mirror   (4) virtual and in front of the mirror

39. An object is placed in front of a convex (diverging) mirror. The image of that object will be
(1) nonexistent   (3) virtual and smaller
(2) real and smaller   (4) virtual and larger

40. In the diagram below of a concave mirror, which point represents the center of curvature?

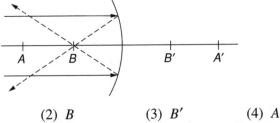

(1) *A* (2) *B* (3) *B'* (4) *A'*

41. Which diagram is a correct representation of a light ray reflected from a spherical surface? [Point *C* is the center of curvature, and point *F* is the focal point.]

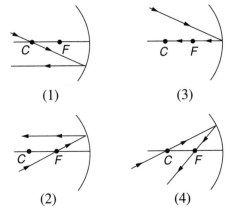

42. An object is located 0.12 meter in front of a concave (converging) mirror of 0.16-meter radius. What is the distance between the image and the mirror?
(1) 0.07 m   (2) 0.20 m   (3) 0.24 m   (4) 0.48 m

43. In the diagram below, an object is located in front of a convex (diverging) mirror. *F* is the virtual focal point of the mirror, and *C* is its center of curvature. Ray *R* is parallel to the principal axis.

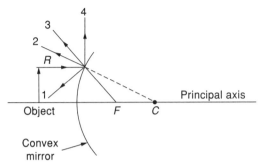

Ray *R* will most likely be reflected along path
(1) 1   (2) 2   (3) 3   (4) 4

44. A candle is located beyond the center of curvature, *C*, of a concave spherical mirror having principal focus *F*, as shown in the diagram below.

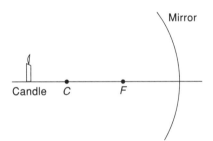

Where is the candle's image located?
(1) beyond *C*   (3) between *F* and the mirror
(2) between *C* and *F*   (4) behind the mirror

45. The diagram below represents a spherical mirror with three parallel light rays approaching. Which light ray will be reflected normal to the surface of the mirror?

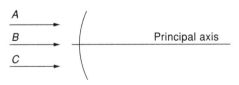

(1) *A*, only   (3) *C*, only
(2) *B*, only   (4) all of the rays

46. The convex spherical mirror found on the passenger side of many cars contains the warning: "Objects are closer than they appear." Which of the following best describes the image of an object viewed in such a mirror?
(1) real and smaller than the object
(2) real and larger than the object
(3) virtual and smaller than the object
(4) virtual and larger than the object

47. A pencil 0.10 meter long is placed 1.0 meter in front of a concave (converging) mirror whose focal length is 0.50 meter. The image of the pencil is
(1) erect and 0.030 meter long   (3) inverted and 0.030 meter long
(2) erect and 0.10 meter long   (4) inverted and 0.10 meter long

Base your answers to questions 48 through 51 on the diagram below, which shows the principal axis of a concave spherical mirror. The focal point is *F*, and *C* is the center of curvature of the mirror. The focal length of the mirror is 0.10 meter.

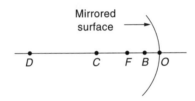

48. If an object is placed at point *B*, its image is
(1) virtual and inverted   (3) real and inverted
(2) virtual and erect   (4) real and erect

49. If an object is placed at point *D*, its image will be located at a position between points
(1) *O* and *F*   (2) *O* and *B*   (3) *C* and *D*   (4) *F* and *C*

50. The radius of curvature of the mirror *OC* is
(1) 0.30 m   (2) 0.20 m   (3) 0.10 m   (4) 0.050 m

51. An object 0.20 meter tall is placed 0.15 meter to the left of the mirror. If the image is located 0.30 meter to the left of the mirror, the size of the image will be
(1) 0.10 m   (2) 0.20 m   (3) 0.30 m   (4) 0.40 m

Base your answers to questions 52 through 55 on the diagram below, which shows four rays of light from object *AB* incident upon a spherical mirror whose focal length is 0.04 meter. Point *F* is the principal focus of the mirror, point *C* is the center of curvature, and point *O* is located on the principal axis.

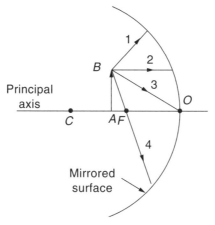

52. Which ray of light will pass through *F* after it is reflected from the mirror?
(1) 1   (2) 2   (3) 3   (4) 4

53. If object *AB* is located 0.05 meter from point *O*, its image will be located
(1) farther from the mirror than *C*   (3) between *F* and the mirror
(2) between *C* and *F*   (4) behind the mirror

54. As object *AB* is moved from its present position toward the left, the size of the image produced
(1) decreases   (2) increases   (3) remains the same

55. If the mirror's radius of curvature could be increased, the focal length of the mirror would
(1) decrease   (2) increase   (3) remain the same

56. Which ray diagram illustrates refraction?

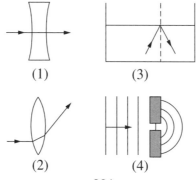

57. The diagram below shows light ray *R* entering air from water.

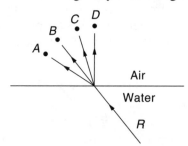

Through which point is the ray most likely to pass?
(1) *A*   (2) *B*   (3) *C*   (4) *D*

58. The ray *R* of monochromatic yellow light shown in the diagram is incident upon a glass surface at an angle of θ.

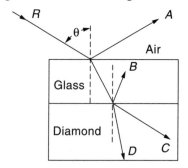

Which resulting ray is *not* possible?
(1) *A*   (2) *B*   (3) *C*   (4) *D*

59. Light travels from medium *A*, where its speed is 2.5×10^8 meters per second, into medium *B*, where its speed is 2.0×10^8 meters per second. Compared to the absolute index of refraction of medium *A*, the absolute index of refraction of medium *B* is
(1) less   (2) greater   (3) the same

60. If the speed of light in a medium is 1.5×10^8 meters per second, the index of refraction of the medium is
(1) 0.5   (2) 2.0   (3) 0.67   (4) 1.5

61. The diagram shows a transparent cube made of a substance that has an index of refraction (n) of 1.73 in air.

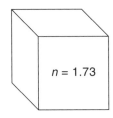

If a ray of light strikes a side of the cube with an angle of incidence of 60°, the angle of refraction will be
(1) smaller than 30°   (3) 45°
(2) 30°   (4) greater than 45°

62. Which graph best shows the relationship between the sine of the angle of incidence (sin i) and the sine of the angle of refraction (sin r) for light moving from air to water?

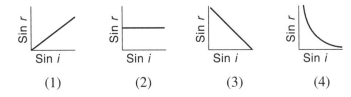

    (1)   (2)   (3)   (4)

63. As a light wave passes from Lucite into flint glass, the frequency of the wave will
(1) decrease   (2) increase   (3) remain the same

64. For a given medium, the index of refraction (n) for a given light wave is equal to
(1) $\dfrac{c}{v}$ (2) cf (3) $\dfrac{\lambda}{c}$ (4) cv

65. In which of the following materials is the speed of light the greatest?
(1) quartz   (2) alcohol   (3) glycerol   (4) benzene

Base your answers to questions 66 through 71 on the diagram below, which represents two media with parallel surfaces in air and a ray of light passing through them.

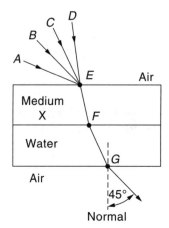

66. What is the approximate speed of light in water?
    (1) 4.4×10^7 m/s   (3) 3.0×10^8 m/s
    (2) 2.3×10^8 m/s   (4) 4.0×10^8 m/s

67. If the angle of refraction in air is 45°, what is the sine of the angle of incidence in water?
    (1) 0.53   (2) 0.71   (3) 0.94   (4) 1.33

68. Which line best represents the incident ray in the air?
    (1) *AE*   (2) *BE*   (3) *CE*   (4) *DE*

69. What is the sine of the critical angle for light passing from water into air?
    (1) 0.50   (2) 0.75   (3) 0.87   (4) 1.33

70. Compared to the speed of light in water, the speed of light in medium *X* is
    (1) lower   (2) higher   (3) the same

71. Ray *EFG* would be a straight line if the index of refraction for medium *X* were
    (1) less than 1.33
    (2) greater than 1.33
    (3) equal to 1.33

Base your answers to questions 72 through 76 on the information and diagram below.

This diagram represents three transparent media arranged one on top of the other. A light ray in air is incident on the upper surface of layer A.

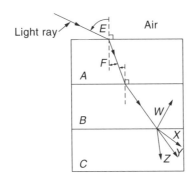

72. If layers B and C both have the same index of refraction, in which direction will the light ray travel after reaching the boundary between layers B and C?
    (1) W (2) X (3) Y (4) Z

73. If layer A were Lucite, layer B could be
    (1) water   (2) diamond   (3) benzene   (4) flint glass

74. If angle E is 60° and layer A has an index of refraction of 1.61, the sine of angle F will be closest to
    (1) 1.00   (2) 0.866   (3) 0.538   (4) 0.400

75. If angle E were increased, angle F would
    (1) decrease   (2) increase   (3) remain the same

76. Compared to the apparent speed of light in layer A, the apparent speed of light in layer B is
    (1) greater   (2) the same   (3) less

77. As the index of refraction of an alcohol-water mixture increases, the critical angle for the mixture
    (1) decreases   (2) increases   (3) remains the same

78. If the critical angle of glass is 45°, the light ray shown in the diagram will be

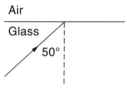

(1) reflected   (2) refracted   (3) dispersed   (4) diffracted

79. The critical angle for light traveling from a given medium to air is 38°. If light is traveling from another medium with a greater index of refraction to air, the critical angle is
(1) less than 38°
(2) greater than 38°
(3) equal to 38°

80. When a light ray passes from medium 1 to medium 2, its speed decreases. Which arrow best represents the path of the ray in medium 2?

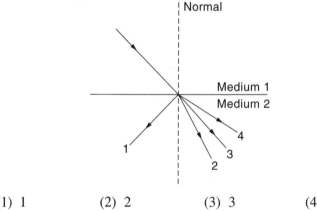

(1) 1   (2) 2   (3) 3   (4) 4

81. Which diagram shows the path that a monochromatic ray of light will travel as it passes through air, benzene, and Lucite and then back into air?

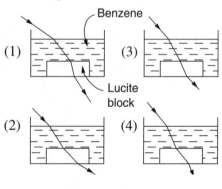

Base your answers to questions 82 through 85 on the diagram below, which represents a ray of monochromatic light ($f = 5.08 \times 10^{14}$ Hz) passing from air to benzene, through material X, and back into air.

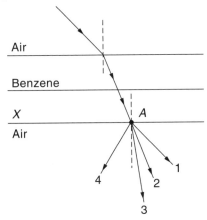

82. What is the speed of the light in the benzene?
(1) 0.50×10^8 m/s   (3) 3.0×10^8 m/s
(2) 2.0×10^8 m/s   (4) 4.5×10^8 m/s

83. The wavelength of the light in air is approximately
(1) 2.9×10^{-7} m   (3) 5.9×10^{-7} m
(2) 3.9×10^{-7} m   (4) 7.9×10^{-7} m

84. Material X could be
(1) diamond   (2) water   (3) flint glass   (4) Lucite

85. Which line represents the path of the light ray after it reenters the air at point A?
(1) 1   (2) 2   (3) 3   (4) 4

86. The diagram shows a ray of light (R) incident upon a surface at an angle greater than the critical angle. Through which point is the ray most likely to pass?

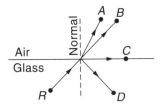

(1) A (2) B (3) C (4) D

87. As a beam of monochromatic light passes from water into air, the frequency of the light
    (1) decreases   (2) increases   (3) remains the same

88. The critical angle is the angle of incidence that produces an angle of refraction of
    (1) 0°   (2) 45°   (3) 60°   (4) 90°

89. The index of refraction of a transparent material is 2.0. Compared to the speed of light in air, the speed of light in this material is
    (1) less   (2) greater   (3) the same

90. As the index of refraction of a medium increases, the critical angle between the medium and air θ_C
    (1) decreases   (2) increases   (3) remains the same

91. The critical angle for a monochromatic light ray traveling from a dispersive material into air is 45°. What is the index of refraction for the material?
    (1) 2.41   (2) 1.71   (3) 1.41   (4) 0.707

92. If a ray of light in glass is incident upon an air surface at an angle greater than the critical angle, the ray will
    (1) reflect, only   (3) partly refract and partly reflect
    (2) refract, only   (4) partly refract and partly diffract

93. A light ray travels from an unknown material into air. If the sine of the critical angle is 0.685, the substance could be
    (1) alcohol   (3) flint glass
    (2) diamond   (4) carbon tetrachloride

94. The critical angle of glass is 43°. In which diagram below will the ray of light be refracted?

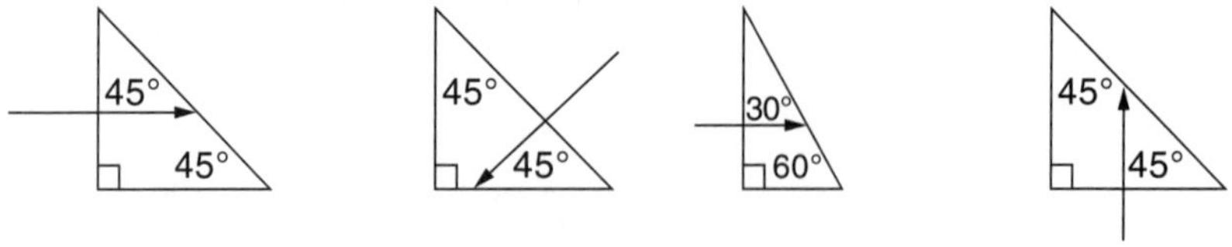

     (1)   (2)   (3)   (4)

95. In which diagram will the ray of light most likely undergo total internal reflection?

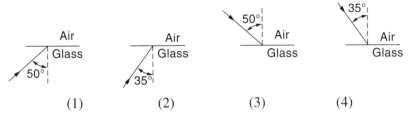

(1)   (2)   (3)   (4)

Base your answers to questions 96 through 100 on the accompanying diagram, which represents a ray of monochromatic light incident upon the surface of plate *X*. The values of *n* in the diagram represent absolute indices of refraction.

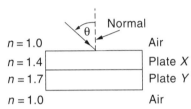

96. What is the relative index of refraction of the light going from plate *X* to plate *Y*?
(1) $\dfrac{1.0}{1.7}$ (2) $\dfrac{1.0}{1.4}$ (3) $\dfrac{1.7}{1.4}$ (4) $\dfrac{1.4}{1.7}$

97. The speed of the light ray in plate *X* is approximately
(1) 1.8×10^8 m/s   (3) 2.5×10^8 m/s
(2) 2.1×10^8 m/s   (4) 2.9×10^8 m/s

98. What is the sine of the critical angle for the light ray at the boundary between plate *Y* and the air?
(1) 1.0   (2) 0.83   (3) 0.71   (4) 0.59

99. Compared to angle θ, the angle of refraction of the light ray in plate *X* is
(1) smaller   (2) greater   (3) the same

100. Compared to angle θ, the angle of refraction of the ray emerging from plate *Y* into air will be
(1) smaller   (2) greater   (3) the same

101. In the diagram below, a monochromatic light ray is passing from medium *A* into medium *B*. The angle of incidence, θ, is varied by moving the light source, *S*.

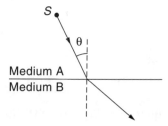

When angle θ becomes the critical angle, the angle of refraction will be
(1) 0°   (3) greater than θ, but less than 90°
(2) θ   (4) 90°

102. When a ray of white light is refracted, which component color has the greatest change in direction?
(1) orange   (2) red   (3) green   (4) violet

103. Compared to the wavelength of a wave of green light in air, the wavelength of the same wave of green light in Lucite is
(1) less   (2) greater   (3) the same

104. Monochromatic light cannot be
(1) dispersed   (2) absorbed   (3) reflected   (4) refracted

105. Compared to the speed of light in a vacuum, the speed of light in a dispersive medium is
(1) less   (2) greater   (3) the same

106. The separation of white light into its component colors as it passes through a prism is called
(1) convergence   (3) diffraction
(2) dispersion   (4) total internal reflection

107. Which terms best describe the phenomenon illustrated by the diagram below?

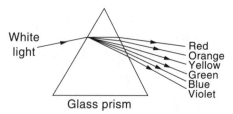

(1) scattering and diffraction   (3) transmission and Doppler effect
(2) reflection and interference   (4) refraction and dispersion

Base your answers to questions 108 through 112 on the diagram below, which represents a ray of monochromatic green light incident upon the surface of a glass prism.

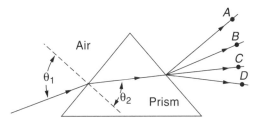

108. The index of refraction of the glass prism for green light equals

(1) $\dfrac{\theta_1}{\theta_2}$ (2) $\dfrac{\theta_2}{\theta_1}$ (3) $\dfrac{\sin \theta_2}{\sin \theta_1}$ (4) $\dfrac{\sin \theta_1}{\sin \theta_2}$

109. After the ray leaves the prism, it will most likely pass through point

(1) *A* (2) *B* (3) *C* (4) *D*

110. If the monochromatic green ray is replaced by a monochromatic red ray, θ_2 will

(1) decrease (2) increase (3) remain the same

111. Compared to the speed of the monochromatic green light in the prism, the speed of the monochromatic red light in the prism is

(1) less (2) greater (3) the same

112. Compared to the frequency of the green light in the prism, the frequency of the red light in the prism is

(1) less (2) greater (3) the same

Base your answers to questions 113 through 117 on the diagram below, which shows a ray of white light incident upon a surface of a triangular glass prism in air.

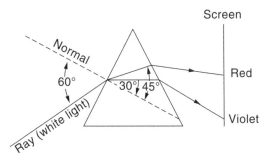

113. The index of refraction for the violet light ray is approximately
    (1) 0.577   (2) 2.00   (3) 1.24   (4) 1.73

114. If the index of refraction for the red light is approximately 1.22, the speed of the red light in glass is approximately
    (1) 1.22×10^8 m/s   (3) 3.00×10^8 m/s
    (2) 2.46×10^8 m/s   (4) 4.07×10^8 m/s

115. Which color of light has the greatest speed in the prism?
    (1) red   (2) green   (3) blue   (4) violet

116. The continuous spectrum produced by the prism is caused by
    (1) diffraction   (2) interference   (3) refraction   (4) reflection

117. Assume that the prism is made of Lucite and is immersed in benzene. After the ray of white light strikes the prism, the ray will be
    (1) totally reflected
    (2) refracted, only
    (3) refracted and then dispersed
    (4) transmitted without change in direction or speed

118. Which device may form both real and virtual images?
    (1) a diverging lens   (3) a double-slit arrangement
    (2) a converging lens   (4) a plane mirror

119. Which diagram below correctly represents rays of light passing through a glass lens?

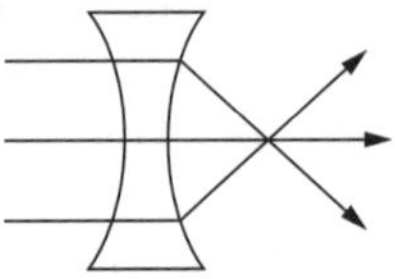

        (1) 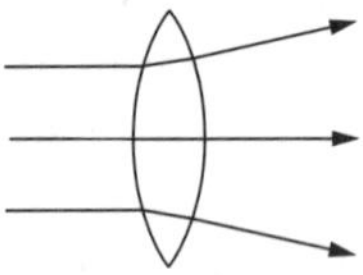   (3)

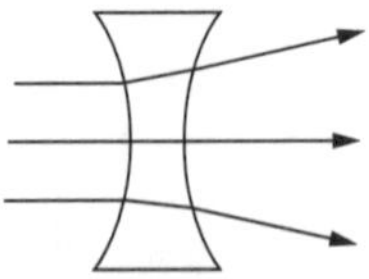

        (2) 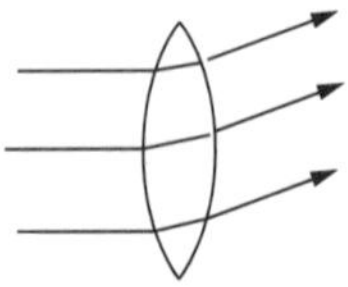   (4)

120. Which diagram most accurately represents the behavior of light rays passing through a glass lens in air?

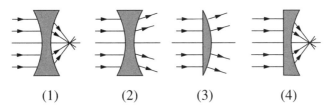

     (1)   (2)   (3)   (4)

121. Which optical device shown below should be placed in the box indicated by the dotted lines in the diagram to cause the parallel light rays to diverge?

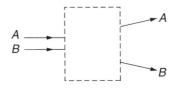

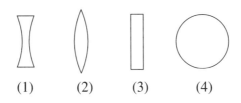

     (1)   (2)   (3)   (4)

122. Virtual images can be formed by
    (1) converging lenses, only
    (2) diverging lenses, only
    (3) neither converging nor diverging lenses
    (4) either converging or diverging lenses

123. The image formed by a diverging lens is
    (1) enlarged   (2) inverted   (3) real   (4) virtual

Base your answers to questions 124 and 125 on the diagram below.

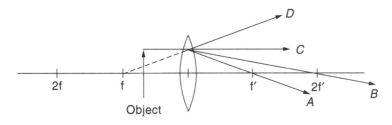

124. A ray of light from the object parallel to the principal axis will be refracted by the lens along path

(1) *A*   (2) *B*   (3) *C*   (4) *D*

125. The image formed will be

(1) real, inverted, and smaller   (3) virtual, erect, and smaller
(2) real, erect, and enlarged   (4) virtual, erect, and enlarged

126. As an object is moved closer to a convex lens, the distance of the real image from the lens

(1) decreases   (2) increases   (3) remains the same

127. Which diagram best represents a lens being used to produce a real, enlarged image of object *O*?

(1)

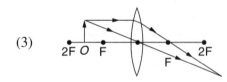

(2)

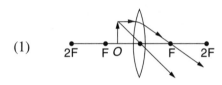

(3)

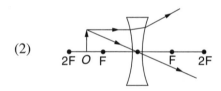

(4)

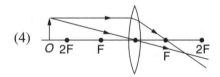

128. When ray *I* emerges from the lens shown in the diagram, it will travel along a path toward point

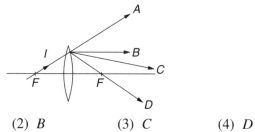

    (1) *A*   (2) *B*   (3) *C*   (4) *D*

129. A hollow convex lens contains water. If the water is replaced by benzene, the focal length of the lens will
    (1) decrease   (2) increase   (3) remain the same

130. If a converging Lucite lens is made more curved, the index of refraction of the Lucite will
    (1) decrease   (2) increase   (3) remain the same

131. Which diagram correctly shows how wave fronts of light will move as they pass through a glass lens in air?

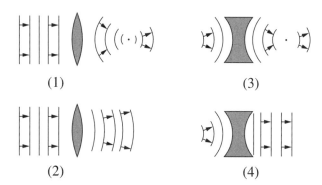

132. In the diagram below, ray *XO* is incident upon the concave (diverging) lens.

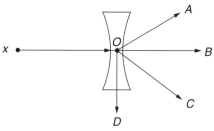

Along which path will the ray continue?
    (1) *OA*   (2) *OB*   (3) *OC*   (4) *OD*

133. The lens shown in the diagram will form an image of the arrow that is

(1) enlarged and virtual   (3) enlarged and real
(2) reduced and virtual   (4) reduced and real

134. An object placed 0.07 meter from a converging lens produces a real image 0.42 meter from the lens. The focal length of the lens is
(1) 0.03 m   (2) 0.06 m   (3) 0.07 m   (4) 2.4 m

135. Compared to the object, real images formed by a lens are always
(1) larger   (2) smaller   (3) inverted   (4) erect

136. An object is located 0.12 meter to the left of a converging lens of focal length 0.080 meter. The image will be located
(1) 0.24 m to the right of the lens   (3) 0.12 m to the right of the lens
(2) 0.24 m to the left of the lens   (4) 0.20 m to the left of the lens

137. A 0.20-meter-high image is formed 0.60 meter to the right of a converging lens when the object distance is 0.30 meter to the left of the lens. What is the size of the object?
(1) 0.10 m   (2) 0.20 m   (3) 0.30 m   (4) 0.40 m

Base your answers to questions 138 through 141 on the information below.

A converging lens produces an image of a large, distant object. The object distance is 5000 meters, and the image distance is 0.50 meter.

138. What is the approximate focal length of the lens?
(1) 10 m   (2) 0.75 m   (3) 0.50 m   (4) 0.25 m

139. If the object size is 1000 meters, what is the image size?
(1) 0.050 m   (2) 0.10 m   (3) 0.15 m   (4) 0.20 m

140. If the size of the object were increased, the size of the image would
(1) decrease   (2) increase   (3) remain the same

141. If the index of refraction of the lens were increased, the focal length of the lens would be
(1) smaller   (2) the same   (3) larger

142. Compared to the focal length of a crown glass lens, the focal length of a flint glass lens of the same dimensions would be
(1) less   (2) greater   (3) the same

Base your answers to questions 143 through 146 on the diagram below, which represents a converging crown glass lens with a focal length of 8.0 centimeters. Ray A represents one of the light rays incident upon the lens.

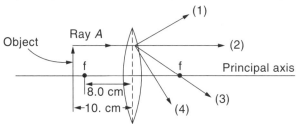

143. Which ray best represents the path of ray A after it emerges from the lens?
(1) 1   (2) 2   (3) 3   (4) 4

144. The image of the object formed by the lens is
(1) virtual and erect   (3) real and erect
(2) virtual and inverted   (4) real and inverted

145. How far from the lens is the image formed?
(1) 18 cm   (2) 2 cm   (3) 40 cm   (4) 80 cm

146. Which phenomenon best explains why the lens produces an image?
(1) diffraction   (2) dispersion   (3) reflection   (4) refraction

Base your answers to questions 147 through 149 on the diagram below, which shows a crown glass lens mounted on an axis. The focal length (f) of the lens is 0.30 meter.

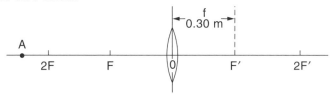

147. If the object is placed between the focal point and the lens, the image formed will be
(1) virtual, enlarged, erect   (3) real, smaller, inverted
(2) virtual, smaller, erect   (4) real, enlarged, erect

148. An object placed 0.40 meter from the lens produces an image located 1.20 meters from the lens. If the size of the object is 0.05 meter, the size of the image will be
(1) 0.03 m   (2) 0.05 m   (3) 0.15 m   (4) 0.20 m

149. If an object is placed at point A, an image will be formed
    (1) between O and F′
                    (3) between F′ and 2F′
    (2) at F′
                          (4) beyond 2F′

Base your answers to questions 150 through 154 on the diagram below, which shows a lens that has a focal length of 2 meters. The principal focus of the lens is at F. Points 2F, A, and B are other points on the principal axis.

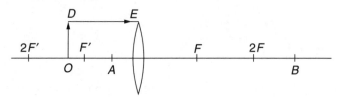

150. Ray DE would most likely cross the principal axis at
    (1) A   (2) B   (3) F   (4) 2F

151. The distance between the lens and the image is
    (1) between 0 and 2 m
                    (3) greater than 4 m
    (2) between 2 and 4 m
                    (4) 4 m

152. The object is placed at point A. Compared to the object, the image formed will be
    (1) smaller and inverted
                    (3) larger and inverted
    (2) smaller and upright
                    (4) larger and upright

153. As the object is moved from 2F′ to F′, the image size will
    (1) decrease   (2) increase   (3) remain the same

154. As the object is moved from F′ toward the lens, the distance of the image from the lens will
    (1) decrease   (2) increase   (3) remain the same

Base your answers to questions 155 through 157 on the diagram below, which shows a crown glass lens of focal length f in air. Monochromatic red light from an object placed on the left side of the lens passes through the lens and forms a real, inverted image on the right side of the lens. The image size is 0.04 meter, and the image is smaller than the object. The image forms at a distance of 0.1 meter from the center of the lens.

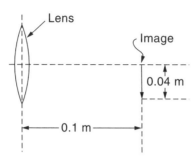

155. If the size of the object is 0.08 meter, the distance from the object to the center of the lens is
(1) 0.1 m   (2) 0.2 m   (3) 0.3 m   (4) 0.5 m

156. The distance from the object to the center of the lens must be
(1) greater than $2f$ (3) between f and $2f$
(2) equal to $2f$ (4) less than f

157. A flint glass lens of identical curvature is substituted for the crown glass lens. Compared to the focal length of the crown glass lens, the focal length of the flint glass lens is
(1) shorter   (2) longer   (3) the same

Base your answers to questions 158 through 161 on the diagram below which represents an object placed at point O, located 1.0 meter from a lens of focal length 0.50 meter.

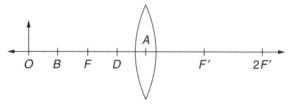

158. The distance between the lens and the image is
(1) 1.0 m   (2) 0.50 m   (3) 0.25 m   (4) 1.5 m

159. Compared to the size of the object, the size of the image is
(1) smaller   (2) larger   (3) the same

160. Compared to the image when the object is at point O, the image when the object is moved to point B will be
(1) smaller
(3) nearer the lens
(2) changed from inverted to erect
(4) farther from the lens

161. The original lens is replaced with one having a shorter focal length. If the object remains at point O, the new image distance, compared to the original image distance, will
(1) be smaller
(2) be larger
(3) remain the same

Base your answers to questions 162 through 165 on the diagram below which represents a converging lens with a focal length of 0.20 meter. DE represents a light ray parallel to the principal axis. OD is an object perpendicular to the principal axis.

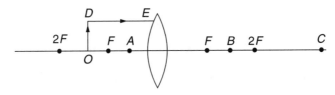

162. When ray DE emerges from the lens, it will most likely pass through point
(1) F
(2) B
(3) C
(4) $2F$

163. This lens cannot be used to form images that are
(1) real and enlarged
(3) virtual and enlarged
(2) real and reduced
(4) virtual and reduced

164. If the object is 0.25 meter from the lens, at what distance from the lens is the image located?
(1) 1.0 m
(2) 0.80 m
(3) 0.40 m
(4) 0.25 m

165. As the object is moved toward F, the size of the image will
(1) decrease
(2) increase
(3) remain the same

166. When white light is directed through a double slit, a stationary interference pattern is produced. Which color is diffracted the *least*?
(1) red
(2) green
(3) blue
(4) violet

167. Monochromatic light is passed through two slits 2×10^{-4} meter apart. The pattern produced illuminates a screen 4 meters from the slits. If the first-order maximum is 1×10^{-2} meter from the central maximum, the wavelength of the light is
(1) 2×10^{-7} m
(2) 5×10^{-7} m
(3) 2×10^{-4} m
(4) 5×10^{-2} m

Base your answers to questions 168 through 171 on the information below.

Light of wavelength 5.4×10^{-7} meter shines through two narrow slits 4.0×10^{-4} meter apart onto a screen 2.0 meters away from the slit.

168. What is the color of the light?
    (1) red   (2) orange   (3) green   (4) violet

169. The distance between the bright areas in the stationary interference pattern formed on the screen is
    (1) 1.1×10^{-10} m   (3) 5.4×10^{-3} m
    (2) 2.7×10^{-3} m   (4) 3.7×10^{2} m

170. Reducing the distance between the slits by one-half would cause the distance between the bright lines in the interference pattern to
    (1) remain the same   (3) halve
    (2) double   (4) quadruple

171. Changing the color of the light used to a color with a higher frequency would cause the distance between the bright lines in the interference pattern to
    (1) decrease   (2) increase   (3) remain the same

Base your answers to questions 172 through 177 on the diagram below, which shows the top view of a double-slit barrier. The distance between the slits (d) is 1.0×10^{-3} meter. The barrier is illuminated with light whose wavelength is 6.0×10^{-7} meter. An interference pattern is observed on the screen.

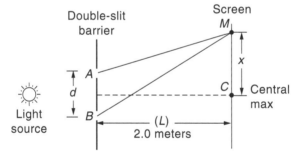

172. If point M on the screen is the location of a bright line, the path difference between AM and BM may be
    (1) one wavelength   (3) one-half wavelength
    (2) three-quarter wavelength   (4) one-quarter wavelength

173. What color is the light?
    (1) blue   (2) yellow   (3) orange   (4) green

174. If point M is the location of the first maximum, distance x is
    (1) 1.2×10^{-3} m (2) 2.4×10^{-3} m (3) 0.2 m   (4) 0.4 m

175. If light of a longer wavelength is used, distance x will
    (1) decrease   (2) increase   (3) remain the same

176. If the distance L between the barrier and the screen is decreased, the distance x between the central maximum and the first maximum will
    (1) decrease   (2) increase   (3) remain the same

177. If distance d between the slits is increased, distance x between the central maximum and the first maximum will
    (1) decrease   (2) increase   (3) remain the same

178. The diagram below represents light shining through a pair of slits 3.0×10^{-3} meter apart and onto a screen. What color of light produces a first-order bright band that is 2.0×10^{-4} meter from the center of the interference pattern?

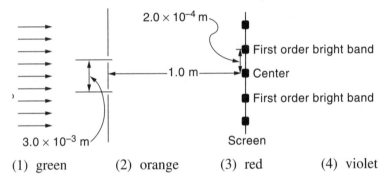

    (1) green   (2) orange   (3) red   (4) violet

179. Coherent light of wavelength 6.4×10^{-7} meter passes through two narrow slits, producing an interference pattern on a screen with the first-order bright bands 2.0×10^{-2} meter from the central maximum. The screen is 4.0 meters from the slits. What is the distance between the slits?
    (1) 2.6×10^{-6} m   (3) 1.3×10^{-4} m
    (2) 3.2×10^{-9} m   (4) 7.8×10^{3} m

180. The diagram below represents electromagnetic radiation of wavelength 8.0×10^{-7} meter passing through two narrow slits 1.0×10^{-4} meter apart and falling on a detector. How far from the central maximum is the first-order maximum detected?

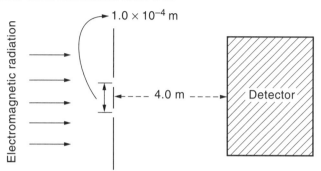

(1) 1.4×10^{-11} m (3) 1.4×10^{-2} m
(2) 3.2×10^{-6} m (4) 3.2×10^{-2} m

181. The results produced by coherent light passing through a double slit provide evidence to support the
(1) wave theory of light (3) quantum theory of light
(2) Newton model for light (4) particle theory of light

Base your answers to questions 182 through 186 on the diagram below, which represents monochromatic light incident upon a double slit in barrier A, producing an interference pattern on screen B.

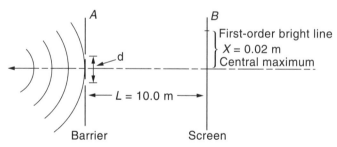

182. The observed phenomenon in this experiment is produced by
(1) refraction and reflection (3) diffraction and interference
(2) polarization and reflection (4) diffraction and polarization

183. If $x = 0.02$ meter, $L = 10.0$ meters, and the wavelength of the incident light is 5.0×10^{-7} meter, the distance d between the slits is
(1) 2.5×10^{-4} m (3) 2.5×10^{-2} m
(2) 2.0×10^{-2} m (4) 4.0×10^{-5} m

184. If the distance between the slits is increased, distance x will
    (1) decrease   (2) increase   (3) remain the same

185. If the distance from the slits to the screen is increased, distance x will
    (1) decrease   (2) increase   (3) remain the same

186. If the wavelength of the incident light is increased, distance x will
    (1) decrease   (2) increase   (3) remain the same

187. The diffraction pattern produced by a double slit will show greatest separation of maxima when the color of the light source is
    (1) red   (2) orange   (3) blue   (4) green

Base your answers to questions 188 through 193 on the information and diagram below.

The diagram represents a point light source incident upon a double slit that produces an interference pattern on a screen.

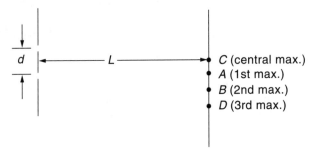

188. If distance d were halved, the distance between C and A would be
    (1) quartered   (2) halved   (3) unchanged   (4) doubled

189. If distance L were quadrupled, the distance between C and A would be
    (1) quartered   (2) unchanged   (3) doubled   (4) quadrupled

190. If the distance between the light source and the slits was halved, the distance between B and D would be
    (1) quartered   (2) halved   (3) unchanged   (4) doubled

191. Which graph best represents the relationship between the intensity of the maxima and their distance from the central maximum?

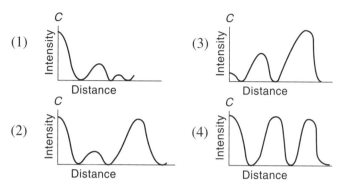

(1)

(2)

(3)

(4)

192. If the intensity of the light source is decreased, the distance between C and A will
(1) decrease (2) increase (3) remain the same

193. If the frequency of the light source is increased, the distance between C and A will
(1) decrease (2) increase (3) remain the same

194. Which pattern of bright and dark bands could have been produced by a beam of monochromatic light passing through a single narrow slit?

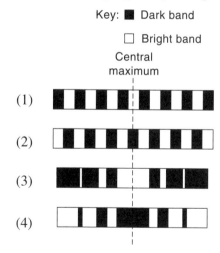

Key: ■ Dark band
□ Bright band
Central maximum

(1)

(2)

(3)

(4)

195. Interference experiments demonstrate the
(1) particle nature of light
(2) polarization of light
(3) intensity of light
(4) wave nature of light

196. Which diagram best represents the phenomenon of diffraction?

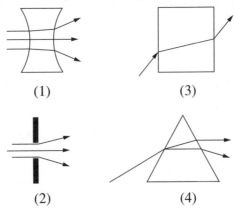

197. An interference pattern is produced on a screen by a green monochromatic light beam that has passed through a single narrow slit. Which diagram best represents this pattern?

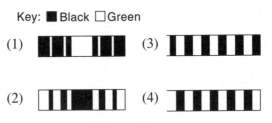

198. Which phenomenon is the best evidence for the wave nature of light?
(1) reflection
(2) photoelectric emission
(3) diffusion
(4) interference

199. Monochromatic light passes through a single narrow slit, forming a diffraction pattern on a screen. Which graph best represents the light intensity of the single-slit diffraction pattern for monochromatic light?

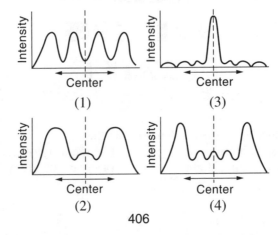

Base your answers to questions 200 through 205 on the diagram below, which represents parallel rays of light passing behind an opaque screen. For each diagram in questions 200 through 205, select the optical device that must be placed behind the screen in order to produce the emerging beam of light.

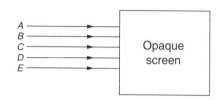

200.

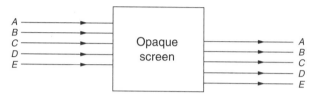

(1) converging lens
(2) diverging lens
(3) rectangular prism
(4) plane mirror

201.

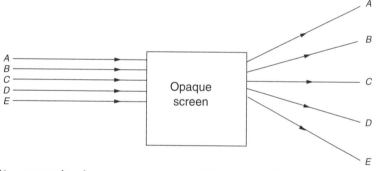

(1) converging lens
(2) diverging lens
(3) rectangular prism
(4) right-angle prism

202.

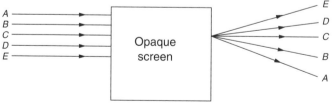

(1) converging lens
(2) diverging lens
(3) rectangular prism
(4) right-angle prism

407

203.

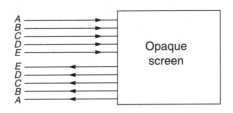

(1) one plane mirror   (3) two right-angle prisms
(2) one curved mirror   (4) one rectangular prism

204.

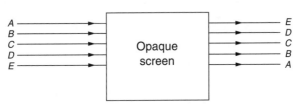

(1) two converging lenses   (3) converging lens
(2) two diverging lenses   (4) rectangular prism

205.

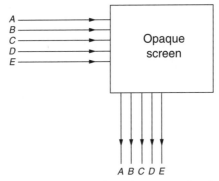

(1) curved mirror   (3) converging lens
(2) right-angle prism   (4) diverging lens

206. Which term best describes the light generated by a laser?
(1) diffused   (2) coherent   (3) dispersive   (4) longitudinal

408

207. The diagram below represents a group of light waves emitted simultaneously from a single source.

These light waves would be classified as
(1) coherent, but not monochromatic
(2) monochromatic, but not coherent
(3) both monochromatic and coherent
(4) neither monochromatic nor coherent

208. Two sources are coherent if the waves that they produce
(1) have the same amplitude   (3) are both transverse in nature
(2) travel at the same speed   (4) have a constant phase relation

SOLID-STATE PHYSICS

Chapter
Fourteen

KEY IDEAS

Semiconductors have unique electrical properties that make them useful for a wide range of applications. The band theory explains why semiconductors possess these properties.

There are two types of semiconductors (*N*-type and *P*-type), which are made by adding small amounts of impurities to semiconductor elements such as germanium and silicon. The impurities have the property of donating or accepting electrons.

The simplest semiconductive device is the junction diode, which is made by bringing *N*-type and *P*-type semiconductor material into contact. The diode allows charge to flow in one direction only. If a diode is part of an AC circuit, the current is changed from alternating to pulsed direct current. In this case the diode acts as a rectifier.

Bipolar transistors are semiconductor devices in which a thin slab of *N*-type or *P*-type semiconductor is "sandwiched" between two slabs of the opposite type of semiconductor. Transistors act as amplifiers in electric circuits.

Integrated circuits are miniaturized combinations of many semiconductor devices and are used in applications such as computers.

KEY OBJECTIVES

At the conclusion of this chapter you will be able to:

- Define the term *conductivity*, and classify solids by their abilities to conduct electricity.
- Describe how the electron band theory of conduction distinguishes among conductors, insulators, and semiconductors.
- Define the term *intrinsic semiconductor*, and explain how temperature affects the conductivities of semiconductors.
- Define the term *extrinsic semiconductor*, and explain how the process of doping enhances the conductivity of materials such as silicon and germanium.
- Define the terms *donor element, N-type semiconductor, acceptor element, positive hole,* and *P-type semiconductor.*
- Describe the operation of a junction diode when it is forward biased and reversed biased.

410

- Define the terms *P-N junction, electric field barrier, biasing,* and *avalanche.*
- Describe how a diode can be used to rectify alternating current.
- Describe the construction of *N-P-N* and *P-N-P* bipolar transistors and their uses in semiconductor circuits.
- Define the term *integrated circuit* or *chip.*

14.1   INTRODUCTION

Solid-state physics is the study of semiconductors (such as silicon and germanium) and the electronic devices in which they are used. We study this topic because virtually every electronic device, from computers to talking toys, is made with circuits containing semiconductor devices. To understand semiconductors, we need first to understand how solid substances conduct electricity.

14.2   CONDUCTION IN SOLIDS

A solid can be classified as a conductor, an insulator, or a semiconductor, depending on how well it conducts electricity, that is, how well it carries an electric current. *Conductors* have low resistances and, consequently, carry currents easily. Generally, metals such as silver and copper are the best conductors. *Insulators* have high resistances and do not carry currents well. Nonmetals such as wood and glass are insulators. *Semiconductors* vary widely in their abilities to conduct electricity, but these abilities are usually intermediate between those of conductors and those of insulators. Silicon and germanium are typical semiconductors.

In Chapter 10, we learned that a quantity known as *resistivity* (ρ) is used to compare the resistive properties of substances. The reciprocal of resistivity is known as **conductivity** (σ) and we use this quantity to compare how well substances conduct electricity. The table below lists the conductivities of a conductor, an insulator, and a semiconductor.

Substance	Conductivity $(\Omega^{-1} \cdot m^{-1})$
Aluminum (conductor)	3.7×10^7
Germanium (semiconductor)	2.2
Phosphorus (insulator)	1.0×10^{-9}

14.3 THEORIES OF CONDUCTION IN SOLIDS

Electron Sea Model

The simplest theory of conduction is known as the *electron sea* model. In this model, the ability of a solid to conduct electricity depends on the number of valence electrons that can move freely through the solid. (*Valence* electrons are the electrons that occupy the outermost energy level of an atom.) Although this model is useful in explaining conduction in metals, it does not adequately explain the properties of semiconductors, particularly why the conductivity of a semiconductor *increases* with a rise in temperature.

Electron Band Model

The second theory is known as the *electron band* model. In this model, the atoms of the solid form a series of energy bands that may be populated by electrons. (These bands are formed by the merging of individual *atomic orbitals* that have similar energies.) For our purposes, the two most important bands are the *valence band* and the *conduction band*.

A valence band contains the electrons in the outermost energy levels of the atoms comprising the solid. However, these electrons must be "promoted" to the conduction band before they can serve as current carriers. Between the valence band and the conduction band is an *energy gap*. The size of this gap determines how many electrons will be promoted to the conduction band and, as a result, how well the solid will conduct electricity. The diagram below illustrates the valence and conduction bands in conductors, semiconductors, and insulators:

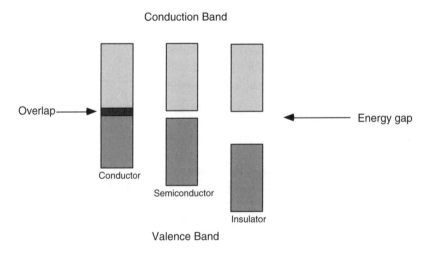

In the conductor, the valence and conduction bands overlap, as shown in the diagram. As a result, electrons flow easily through the conductor. Metals with small numbers of valence electrons (sodium, copper, silver) conduct especially well. When the temperature is raised, the electrons collide with the atoms in the conductor and kinetic energy is lost as heat. These collisions reduce the current in the conductor (i.e., they raise its resistance).

In the insulator, the energy gap is large enough to prevent a significant number of electrons from being promoted to the conduction band. As a result, the resistance of an insulator is very high, and this device carries only a minute current under normal circumstances.

The semiconductor has a small energy gap. At low temperatures, this device behaves as an insulator. At higher temperatures, however, the increase in the kinetic energy of the electrons allows some of them to be promoted to the conduction band. As a result, the conductivity is increased dramatically. A material that behaves this way is known as an **intrinsic semiconductor**, and the diagram below illustrates this process.

Conduction Band

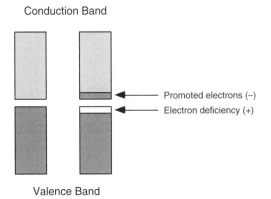

Valence Band

In the conduction band, the promoted electrons act as *negative* charge carriers. The promotion creates an electron deficiency in the valence band. This deficiency can serve as the equivalent of *positive* charge carriers, and conduction can take place in *both* bands.

14.4   EXTRINSIC SEMICONDUCTORS: DOPING

Another way of increasing the conductivity of materials such as silicon and germanium is to add very small amounts of other elements. This process is known as *doping,* and semiconductors made in this way are called **extrinsic semiconductors.** The ratio of doping atoms to semiconductor atoms is on the order of 1 to 100,000,000. If more of the doping material were added, the conductivity of the silicon (or germanium) would become too high for the element to function as a semiconductor. Two types of semiconductors can be made by the doping process: *N-type* and *P-type*.

N-Type Semiconductors

A silicon or germanium atom contains four valence electrons. If a doping atom contains *five* valence electrons, as does *arsenic* or *antimony,* it will contribute an electron to the conduction band. For this reason, the doping material is known as a **donor element**.

The doping atom remains in the crystal as a *positive* ion; therefore, most of the charge carriers in the semiconductor are electrons. This type of material is known as an **N-type semiconductor**. The diagram below illustrates how the current is carried by *N*-type material in the presence of a potential difference:

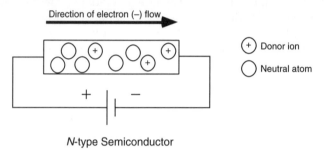

N-type Semiconductor

P-Type Semiconductors

If a doping atom contains *three* valence electrons, as does *gallium* or *aluminum,* it will accept an electron from the conduction band. For this reason, the doping material is known as a **acceptor element**.

The doping atom remains in the crystal as a *negative* ion; therefore, an electron deficiency remains in the conduction band. Since these *holes* behave as *positive* charges, most of the charge carriers in the semiconductor can be considered to be **positive holes**. This type of material is known as a **P-type semiconductor.** The diagram below illustrates how the current is carried by *P*-type material in the presence of a potential difference:

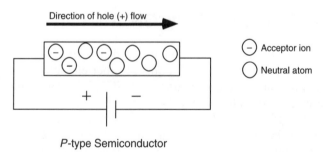

P-type Semiconductor

14.5 JUNCTION DIODES

The junction diode is the simplest semiconductor device. A diode is a "one-way valve," which permits the passage of current in only one direction. A junction diode is made by joining *P*-type and *N*-type semiconductors; the boundary between the two materials is known as the **P-N junction**. The diode's unique characteristics are due to the presence of this boundary. The diagram below represents a junction diode.

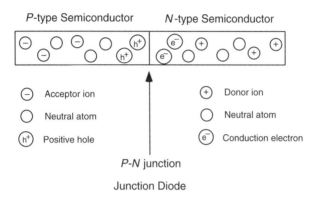

Junction Diode

The electrical symbol for a diode is as follows:

In a junction diode, the anode comprises the *P* side of the diode and the cathode comprises the *N* side. The arrow points in the direction in which the diode will allow the flow of positive charges.

In the narrow region at the *P-N* junction, the (negative) conduction electrons drift into the (positive) holes, producing neutral atoms. The positive and negative ions that remain establish an electric field that acts as a barrier to further migration. This **electric field barrier** is illustrated below.

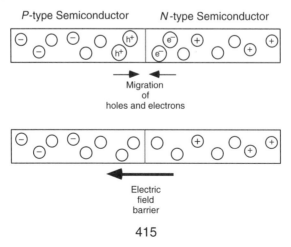

415

Biasing

When a potential difference is applied across a diode, the direction of the potential difference will either weaken or strengthen the electric field barrier. This effect is known as **biasing**.

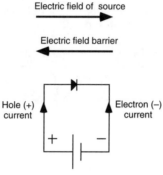

Forward-Biased Diode

In the simple circuit shown above, the anode is connected to the positive terminal and the cathode to the negative terminal. The electric field produced by the source overcomes the electric field barrier of the diode and allows current in the circuit. Electrons will flow from the negative terminal, and holes will flow from the positive terminal, into the diode. In this case, the diode is said to be *forward biased.*

In contrast, in the circuit shown below, the anode and cathode are connected to the negative and positive terminals, respectively, of the source.

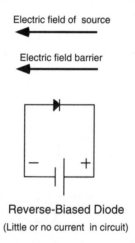

Reverse-Biased Diode

(Little or no current in circuit)

The electric field barrier is *strengthened* by the electric field of the source. As a result, little or no current is present in the circuit because the electric field barrier severely inhibits charge flow *within the diode*. In this case, the diode is said to be *reverse biased.*

A practical application of the diode is the conversion of alternating current to direct current. When used in this way, the diode acts as a *half-wave rectifier* because the direction of the alternating current is reversed every half-cycle. When an AC source is connected to a junction diode, the diode will alternate between being forward biased and being reverse biased. Current will be present only during the forward-biased phase and therefore will always be in the same direction. The diagram below illustrates half-wave rectified alternating current.

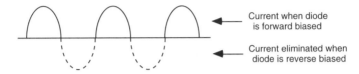

Current when diode
is forward biased

Current eliminated when
diode is reverse biased

The graph below illustrates the characteristics of currents in a junction diode.

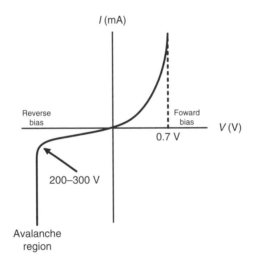

If we examine the portion of the graph in which the diode is forward biased, we can see that a semiconductor diode does *not* obey Ohm's law. When the diode is reverse biased, little current is present until the potential difference is large enough (200–300 V) to cause the diode to break down or **avalanche**. In the avalanche region the potential difference hardly changes, as shown in the graph.

Zener diodes operate close to their avalanche regions and are used to maintain constant potential differences. In other words, they are *voltage regulators*.

14.6 TRANSISTORS

The term *transistor* is a combination of the word parts *transf*er and res*istor*. A transistor serves as an amplification device; a small amount of power input to a transistorized circuit is accompanied by a large power output. Microphone-speaker circuits, for example, make use of this amplification property.

The simplest transistor, the *bipolar* transistor, is just two junction diodes that share a thin, common semiconductor material between them. The diagrams below illustrates how bipolar transistors may be constructed.

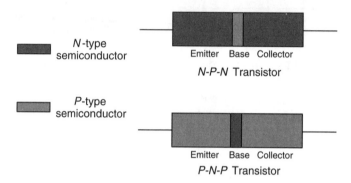

N-type semiconductor

Emitter Base Collector

N-P-N Transistor

P-type semiconductor

Emitter Base Collector

P-N-P Transistor

The *emitter* serves as the primary source of the charge carriers in the circuit. In an *N-P-N* transistor, these charge carriers are negative electrons; in a *P-N-P* transistor, they are positive holes. The *base* is the thin (about 100 nm), common layer of semiconductor material. The *collector* receives the charge carriers from the base.

The emitter is more heavily doped than either the collector or the base. As a result, the resistance at the collector-base junction is approximately 1000 times larger than the resistance at the emitter-base junction.

The electrical symbols for bipolar transistors are as follows:

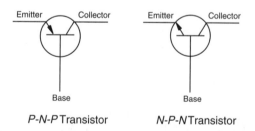

Emitter   Collector

Base

P-N-P Transistor

Emitter   Collector

Base

N-P-N Transistor

In each symbol, the arrow points in the direction of the conventional (positive) current.

The diagrams below illustrate an operating circuit containing an *N-P-N* transistor.

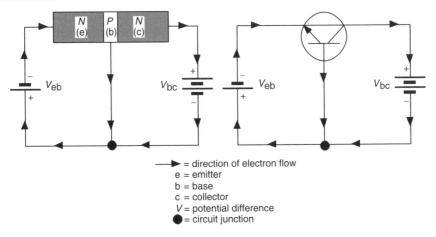

= direction of electron flow
e = emitter
b = base
c = collector
V = potential difference
● = circuit junction

The diagrams are equivalent, but the one on the right uses the transistor symbol as part of the circuit. When the transistor is placed in the circuit, the emitter-base pair is forward biased while the collector-base pair is reverse biased (see Section 14.5). Because of the extreme thinness of the base, approximately 98% of the electrons pass through the base to the collector. In addition, there is a small electron flow in the base due to the combination of electrons and holes within the base. The withdrawal of electrons creates new holes in the base.

If a *P-N-P* transistor were used, the polarities of the potential differences (V_{eb} and V_{bc}) would have to be reversed and the arrows in the diagram would represent positive-hole current rather than electron flow. This situation is illustrated in the diagram below.

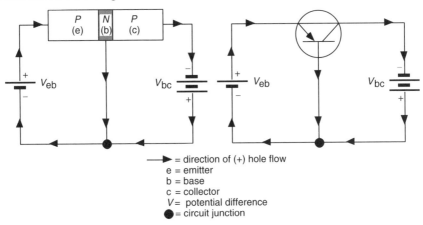

= direction of (+) hole flow
e = emitter
b = base
c = collector
V = potential difference
● = circuit junction

In either case, the result, as stated above, is that a low power-input signal (usually AC) becomes an amplified output signal.

419

14.7 INTEGRATED CIRCUITS

Because it is possible to miniaturize semiconductor devices, many of these components can be packaged together as an **integrated circuit** (also known as a semiconductor **chip**). Integrated circuits are essential parts of computers (microprocessor and logic chips) and of virtually every other modern electronic device.

QUESTIONS

1. As the temperature of a semiconductor increases, its conductivity
   (1) decreases   (2) increases   (3) remains the same

2. Which quantity is equal to the reciprocal of resistivity?
   (1) conductivity (2) current   (3) resistance   (4) voltage

3. A large number of electrons move freely throughout a solid material. This material is best classified as
   (1) a conductor   (3) an N-type semiconductor
   (2) an insulator   (4) a P-type semiconductor

4. Compared to an insulator, a conductor of electric current has
   (1) more free electrons per unit volume
   (2) fewer free electrons per unit volume
   (3) more free atoms per unit volume
   (4) fewer free atoms per unit volume

5. In each diagram below of a band model element, C is the conduction band and V the valence band. Which diagram best represents a conductor?

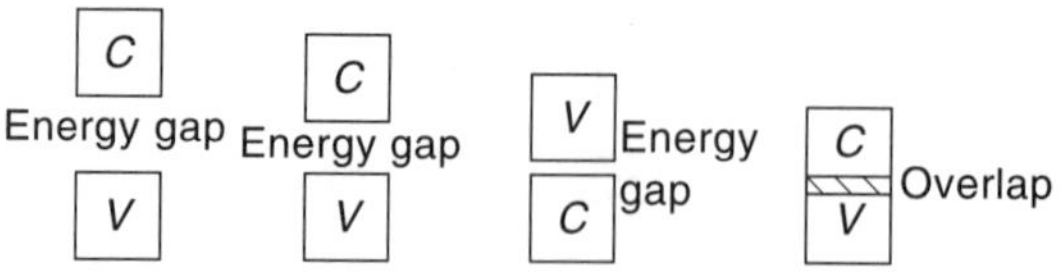

   (1)   (2)   (3)   (4)

6. What type of semiconductor is represented by the diagram below?

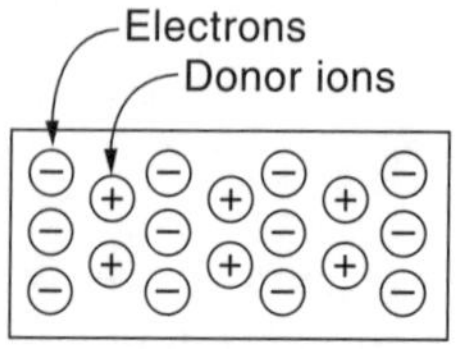

   (1) P-type   (2) N-type   (3) N-P-type   (4) P-N-type

7. Current carriers in a semiconductor are
   (1) electrons, only
   (2) protons, only
   (3) electrons and positive holes
   (4) protons and negative holes

8. Which graph best represents the relationship between the conductivity and the temperature of semiconductors?

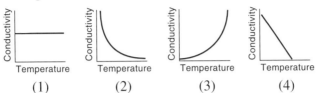

   (1) (2) (3) (4)

9. As the number of free charges per unit volume of a solid increases, its electrical conductivity
   (1) decreases (2) increases (3) remains the same

10. Which procedure would cause most of the free electrons in a conductor to move in the same direction?
    (1) heating the conductor
    (2) charging the conductor
    (3) maintaining a potential difference across the conductor
    (4) holding the conductor between the poles of a magnet

11. According to the energy band model, in which material is the energy gap between the conduction band and the valence band greatest?
    (1) a conductor
    (2) an insulator
    (3) an extrinsic semiconductor
    (4) an intrinsic semiconductor

12. A hole in the atoms of a semiconductor crystalline lattice structure can be defined as a
    (1) free electron
    (2) negative atom
    (3) region in which an electron is located
    (4) region from which an electron has vacated

13. A solid is determined to be a conductor, an insulator, or a semiconductor on the basis of
    (1) its melting point and boiling point
    (2) its atomic number and mass number
    (3) the number of electrons in the conduction band and the energy gap between bands
    (4) the number of electrons per square meter of cross-sectional area of the material

14. When a semiconductor is replaced in a circuit by an insulator, the resistance of that section of the circuit
  (1) decreases   (2) increases   (3) remains the same

15. Which is an important factor in determining whether two materials can be used to form a semiconductor?
  (1) One material must be able to donate electrons, and the other must be able to accept electrons.
  (2) Both materials must have the same number of electrons in their valence levels.
  (3) One material must have positive electrons and the other must have negative electrons.
  (4) Both materials must have more neutrons than protons in their nuclei.

16. A small amount of antimony is deposited on a crystal of silicon and then heated. The antimony diffuses inward to form a semiconducting material. This process is called
  (1) crystallization   (3) solidifying
  (2) conduction   (4) doping

17. Gallium accepts bound valence electrons in a semiconductor. The deficiency of conducting electrons provided by gallium causes an excess of
  (1) holes   (2) neutrons   (3) electrons   (4) protons

18. Germanium is doped with arsenic to form an N-type semiconductor. A majority of the charge carriers in this semiconductor are
  (1) electrons   (2) holes   (3) P-ions   (4) protons

19. A doping agent that adds electrons to a semiconducting material is called
  (1) an N-type semiconductor   (3) a donor
  (2) a P-type semiconductor   (4) an acceptor

20. The majority of charge carriers in a P-type semiconductor are holes. When a section of P-type semiconductor is connected across the terminals of a battery, where do the holes flow?
  (1) toward the negative terminal, only
  (2) toward the positive terminal, only
  (3) equally toward both the negative and positive terminals
  (4) toward neither terminal

21. The diode in the circuit diagram below is considered to be

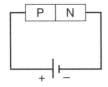

(1) open biased
(2) closed biased
(3) forward biased
(4) reverse biased

Questions 22–24 refer to the following diagram of the semiconductor device.

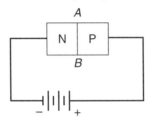

22. Which type of semiconductor device is shown?
(1) an emitter   (2) a resistor   (3) a diode   (4) a transistor

23. Line *AB* identifies the
(1) emitter   (2) base   (3) collector   (4) junction

24. As drawn, this device is
(1) forward biased
(2) reverse biased
(3) grounded
(4) open

25. The graph below represents the relationship between the current and the applied potential for a diode.

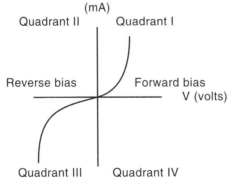

The avalanche region of the graph is in Quadrant
(1) I   (2) II   (3) III   (4) IV

26. The very narrow region where *P*-type and *N*-type semiconductors are joined is called a
(1) bias   (3) collector
(2) valence region   (4) junction

Base your answers to questions 27 through 29 on the diagram below, which represents a diode.

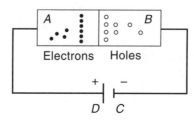

27. The *P-N* junction in the diagram is biased
(1) reverse   (2) forward   (3) *B* to *C*   (4) *A* to *D*

28. In the diagram, *B* represents the
(1) *N*-type silicon   (3) cathode
(2) *P*-type silicon   (4) diode

29. If the positive and negative wires of the circuit in the diagram were reversed, the current would
(1) decrease   (2) increase   (3) remain the same

30. What does the circuit diagram below represent?

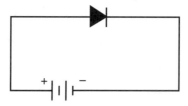

(1) a forward-biased diode   (3) a forward-biased transistor
(2) a reverse-biased diode   (4) a reverse-biased transistor

31. As the amount of doping material used in a diode increases, the potential difference needed to cause the avalanche
(1) decreases   (2) increases   (3) remains the same

32. Which would occur if the connections to the source of potential difference in the diagram below were reversed?

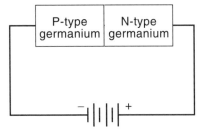

(1) The current in the circuit would decrease.
(2) The current in the circuit would increase.
(3) The current in the circuit would remain the same.

33. In which circuit diagram below is the light bulb most likely to light?

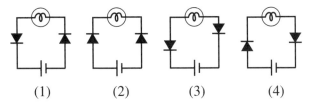

    (1)   (2)   (3)   (4)

34. The diagram below shows the AC input signal to a diode.

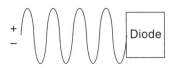

What will the output signal look like?

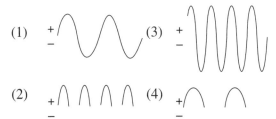

35. Why do current carriers have difflculty crossing a *P-N* junction?
(1) The junction area has a large positive charge.
(2) The junction area has a large negative charge.
(3) The junction acts as an electric field barrier.
(4) The resistance across the junction is extremely low.

36. The simplest circuit element that will allow electric current to pass through a circuit in only one direction is
(1) a transistor
(2) a diode
(3) an *N*-type semiconductor
(4) a *P*-type semiconductor

37. The graph below shows the relationship between current and applied potential difference for an electrical device.

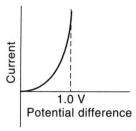

This device is most likely a
(1) forward-biased *P*-type semiconductor
(2) reverse-biased *P*-type semiconductor
(3) forward-biased *P-N* junction
(4) reverse-biased *P-N* junction

38. An incandescent light bulb is connected in series with a diode and a source of direct current. If the diode is forward biased, the light bulb will
(1) light and stay lighted
(2) flicker with a period dependent on the DC current
(3) not light at all
(4) light at first, then dim out

Base your answers to questions 39 and 40 on the graph below of current versus potential difference for a diode.

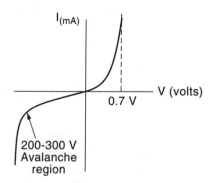

39. The avalanche occurs only if the diode circuit is
(1) open
(2) closed
(3) reverse biased
(4) forward biased

40. As the amount of doping material in the semiconductor increases, the resistance of the semiconductor
(1) decreases   (2) increases   (3) remains the same

41. How is current affected when a semiconductor diode is reverse biased?
(1) Current conduction increases because the junction electric field barrier increases.
(2) Current conduction increases because the junction electric field barrier decreases.
(3) Current conduction decreases because the junction electric field barrier increases.
(4) Current conduction decreases because the junction electric field barrier decreases.

42. A potential difference is applied to a *P*-type semiconductor as shown in the diagram below.

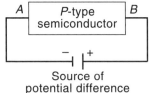

Current is conducted through this semiconductor by
(1) positive holes moving from end *A* to end *B*
(2) positive holes moving from end *B* to end *A*
(3) conducting protons moving from end *A* to end *B*
(4) conducting protons moving from end *B* to end *A*

43. In the circuit shown in the diagram below, ammeter A reads 4 milliamperes when connected to an *N*-type semiconductor.

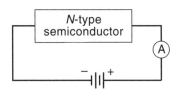

If the connections to the battery are reversed, the reading of ammeter A will be closest to
(1) 8 mA   (2) 2 mA   (3) 0 mA   (4) 4 mA

44. In an *N*-type semiconductor there is an excess of
(1) free electrons   (3) atoms
(2) protons   (4) ions

45. If silicon has a small amount of arsenic as an impurity in the crystal structure, the current in this *N*-type semiconductor is carried by
(1) electrons (2) neutrons (3) protons (4) ions

46. The diagram below shows an electron moving through a semiconductor.

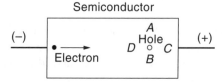

Toward which letter will the hole move?
(1) *A* (2) *B* (3) *C* (4) *D*

47. The circuit diagram below shows a *P*-type semiconductor in series with a lamp, a resistor, and a battery.

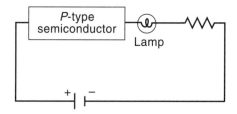

What procedure would increase the current in the circuit?
(1) increasing the resistance of the resistor
(2) reversing the battery polarity
(3) reversing the connections to the semiconductor
(4) increasing the temperature of the semiconductor

48. How is the semiconductor in the circuit shown below classified?

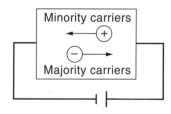

(1) *N*-type (2) *P*-type (3) a diode (4) a transistor

49. What permits current to flow through a semiconductor when it is connected to a battery, as shown in the diagram below?

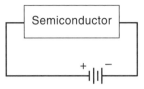

(1) holes moving toward the right, only
(2) electrons moving toward the left, only
(3) both electrons and holes moving toward the left
(4) electrons moving left and holes moving right

Base your answers to questions 50 through 54 on the diagram below, which represents a germanium device.

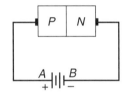

50. In the circuit diagram, the diode is forward biased when electron current flows from
(1) *N* to *B*   (2) *A* to *P*   (3) *P* to *N*   (4) *N* to *P*

51. In the circuit diagram, the electric field barrier is represented by the region between
(1) *A* and *P*   (2) *P* and *N*   (3) *N* and *B*   (4) *B* and *A*

52. In this type of circuit, holes migrate from
(1) *P* to *N*   (2) *A* to *B*   (3) *N* to *P*   (4) *B* to *A*

53. According to the circuit diagram, which symbol best represents the orientation of the device shown?

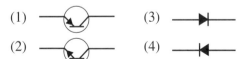

54. If an alternating potential difference is substituted for the battery, the result will be
(1) an alternating current   (3) a steady direct current
(2) a pulsating direct current   (4) no current flow

55. Which is the base of the *P-N-P* semiconductor device shown below?

(1) left *P*　　　(2) right *P*　　　(3) *N*　　　(4) *P-N* junction

56. The graph below represents the AC input signal to a transistor amplifier.

Which graph below best represents the amplified output signal from this transistor?

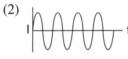

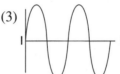

57. In an *N-P-N* transistor, the emitter-base combination is forward biased. In this transistor, there is a flow of
(1) electrons from the base to the emitter
(2) electrons from the emitter to the base
(3) holes from the base to the collector
(4) holes from the emitter to the base

58. Which device is represented by the diagram below?

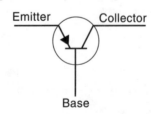

(1) an *N-P-N* transistor　　　(3) a zener diode
(2) a *P-N-P* transistor　　　(4) a solenoid

59. An *N-P-N* transistor is connected as shown in the diagram below.

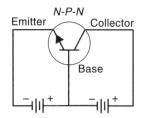

Within the transistor, a forward-biased current will flow from the
(1) collector to the base   (3) base to the emitter
(2) base to the collector   (4) emitter to the base

60. The diagram below shows an *N-P-N* semiconductor device.

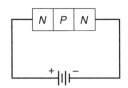

The base in the semiconductor shown is the
(1) right *N*   (2) left *N*   (3) *P*   (4) *P-N* junction

61. What is a basic difference between a transistor and a diode?
(1) the number of junctions between P-type and N-type material
(2) the size of the single junction between the P-type and N-type material
(3) the nature of the donor material used in the N-type semiconductor
(4) the amount of current applied to the semiconductor material

62. Which diagram shows both a forward and a reverse bias?

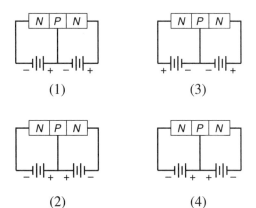

63. A small change in the emitter-base current in a transistor brings about a large change in the collector current. This current-increasing property of a transistor is called

(1) amplification   (3) Ohm's law

(2) biasing   (4) rectification

64. As the emitter-base current in a transistor increases, the base-collector current

(1) decreases   (2) increases   (3) remains the same

65. Which device could be used to amplify the input signal in a microphone-speaker circuit?

(1) a diode   (2) a resistor   (3) a transistor   (4) a solenoid

66. The semiconductor material located between the emitter and the collector of a transistor is called the

(1) base   (2) junction   (3) bias   (4) rectifier

67. In the diagram below, which part of the operating *N-P-N* transistor is the collector?

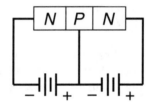

(1) the left *N*-type section   (3) the left *P-N* junction

(2) the right *N*-type section   (4) the right *P-N* junction

68. Which diagram below correctly represents the basic operating circuit of an *N-P-N* transistor?

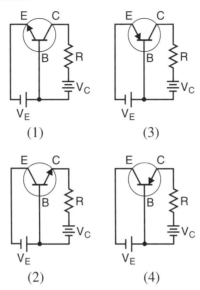

(1)   (3)

(2)   (4)

69. The diagram below represents an operating *N-P-N* transistor circuit. Ammeter A_c reads the collector current and ammeter A_b reads the base current.

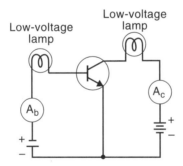

Compared to the reading of ammeter A_c, the reading of ammeter A_b is
(1) less   (2) greater   (3) the same

70. A student is designing a circuit to amplify a small voltage change into a larger voltage change. The electric circuit element best suited to this task is
(1) a transistor   (3) an *N*-type semiconductor
(2) a diode   (4) a *P*-type semiconductor

71. What is the name for a large group of electronic components that are interconnected on a single block of silicon inside a computer?
(1) a thermistor   (3) a transistor
(2) an integrated circuit   (4) a collector

MODERN PHYSICS

KEY IDEAS

Energy is not continuous; it occurs in a series of discrete bundles called quanta. Einstein's explanation of the photoelectric effect introduced the concept that light energy is quantized in units called photons. The energy of a photon is directly related to its frequency. Compton's experiments verified the photon nature of light. De Broglie proposed that matter such as electrons posses wavelike characteristics, and diffraction experiments verified this theory.

Rutherford established the nuclear model of the atom by analyzing the results when alpha particles were scattered by metallic foils. Bohr refined the model of hydrogen by proposing that the energy levels in the atom are quantized and that electrons can make only discrete transitions within these levels. The currently accepted model of the atom is based on the idea of probability, and each electron is viewed as a cloud rather than as a specific point in space.

Einstein's special theory of relativity is based on the ideas that the speed of light is constant in all frames of reference and that the laws of physics hold in all inertial frames. As a result of this theory, quantities such as mass, length, and time are not constant, but change as the velocity of an object changes. In addition, two events that occur simultaneously in one frame of reference may not appear to be simultaneous in another frame. Another result of special relativity is the famous $E = mc^2$ formula, which states that matter and energy are equivalent and that each can be transformed into the other.

KEY OBJECTIVES

At the conclusion of this chapter you will be able to:
- Define the term *quantum of energy*, and relate this term to Planck's constant.
- Describe the photoelectric effect and Einstein's explanation of it.
- Solve problems using Einstein's photoelectric equation.
- Explain how the Compton effect supports the photon theory of light.
- Calculate the momentum of a photon, given its frequency or wavelength.

- Define the term *matter wave*, and calculate the wavelength of a particle of matter when it is in motion.
- Describe Rutherford's experiments involving the scattering of alpha particles by metallic foils and the model of the atom he proposed as a result of those experiments.
- Explain why the Rutherford model did *not* provide a complete picture of the atom.
- State the hypotheses that Bohr used in developing his model of the hydrogen atom.
- Define the terms *ground state, excited state,* and *stationary state* as they apply to the Bohr model.
- Describe how Bohr was able to explain the existence of line spectra.
- Define the term *ionization*, and use the energy level diagrams for hydrogen and mercury to calculate the energies involved in various electron transitions.
- Define the term *electron cloud*, and state why the cloud model was needed to provide a more nearly complete picture of the atom.
- State the postulates of Einstein's special theory of relativity.
- Define the terms *simultaneity, time dilation, twin paradox,* and *length contraction*, and apply the corresponding equations to solve problems in special relativity.

15.1   INTRODUCTION

The title of this chapter is not completely accurate since the theories and discoveries we describe originated between 1895 and 1930. In this time period, two major scientific theories were advanced: quantum physics and Einstein's theory of relativity. Although both theories revolutionized the sciences and technology, they presented various aspects of nature in ways that defied the familiar, "commonsense" view of the world.

We begin the chapter with quantum physics and end with a section on Einstein's special theory of relativity.

15.2   BLACK-BODY RADIATION AND PLANCK'S HYPOTHESIS

When a solid is heated, it emits a variety of electromagnetic radiation. As the temperature of the solid is increased, the radiation shifts toward shorter wavelengths, a change that explains why heated solids begin to glow red, then orange, and finally white. To study this radiation effectively, physicists used a device called a *cavity radiator*, also known as an *ideal black body*. This device is a hollow solid with a small opening drilled in one of the walls, as shown in the following diagram.

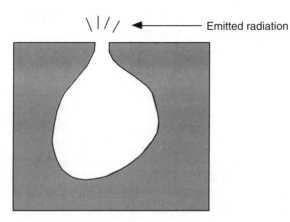

When the solid was heated, it was found that that the radiation emitted *through the opening* depended only on the temperature of the solid, not on the material of which the cavity radiator was made. The origin of the radiation was presumed to be the atoms of the solid. The heating caused the atoms to oscillate (just as springs do), and then the energy was released into the cavity as electromagnetic radiation. However, there was serious disagreement between the theory and the experimental results.

In 1900, German physicist Max Planck reported a startling discovery: the experimental results could be explained precisely if it was assumed that the atomic "oscillators" can have only certain energies given by this equation:

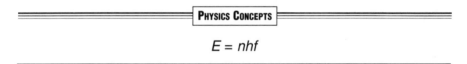

$$E = nhf$$

where E is the energy of the atomic oscillator, f is its frequency of oscillation, n is an integer (1, 2, 3, . . .) and h, known as *Planck's constant*, has the value of 6.6×10^{-34} joule · second. Each of these discrete values of energy is known as a *quantum* of energy. As a result, the radiation emitted from the cavity opening is also restricted to certain values.

This idea was revolutionary because physicists had always assumed that a particle could take on or emit any value of energy. (In fact, Planck himself was not comfortable with the concept of quantized energy!) As we will see in the next section, Planck's idea was reinforced in 1905 by German-American physicist Albert Einstein, who used it to explain the nature of light in an experiment involving the *photoelectric effect*.

15.3   THE PHOTOELECTRIC EFFECT

The photoelectric effect is based on a very simple premise: When light energy strikes the surface of certain materials, electrons are ejected, creating a small but measurable current. Today, photoelectric devices are in wide use in products such as cameras and automatic door openers. The diagram below illustrates how the photoelectric effect can be studied in a laboratory.

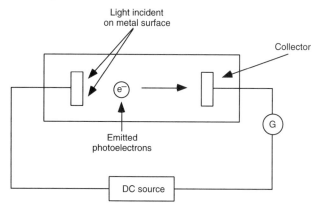

As monochromatic light strikes a metal surface, a potential difference causes the ejected photoelectrons (e⁻) to move onto a collector. The galvanometer (G) measures the current that is produced as a result. The arrangement illustrated above allows for changes in the (1) metal surface, (2) potential difference, (3) light intensity, and (4) color of the light used.

Such photoelectric experiments yielded results that did not agree with the idea that light behaves as a wave. For example, certain colors of light did not eject electrons, regardless of the intensity of the incident light. According to wave theory, very bright light (of any color) should possess a great deal of energy and be able to eject electrons from a metal surface. It was found that, when electrons are ejected, the *color* of the light determines how energetic the electrons are; the intensity determines the *rate* at which the electrons are ejected (i.e., the current).

The potential difference can also be made to *oppose* the movement of the electrons. There is a value, known as the *stopping potential*, that will stop even the most energetic electrons from reaching the collector. If we measure the value of the stopping potential (V_0), we can calculate the kinetic energy of the fastest electrons (KE_{max}). Since energy (work) is equal to potential difference multiplied by charge, it follows that:

PHYSICS CONCEPTS

$$KE_{max} = V_0 e$$

PROBLEM

In a photoelectric experiment, the stopping potential is 5.0 volts. Calculate the maximum kinetic energy of the photoelectrons.

SOLUTION

$$KE_{max} = V_0e$$

$$= (5.0 \text{ V})(1.6 \times 10^{-19} \text{ C})$$

$$= 8.0 \times 10^{-19} \text{ J } [5.0 \text{ eV}]$$

The graph below represents an experiment that relates the color (i.e., the frequency) of the incident light to the maximum kinetic energy of the photoelectrons.

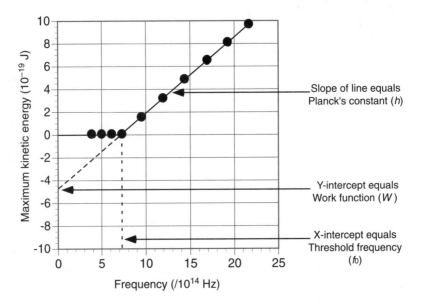

Slope of line equals Planck's constant (h)

Y-intercept equals Work function (W)

X-intercept equals Threshold frequency (f_0)

For any photoemissive material tested, the graph is a straight line whose slope is Planck's constant. There is a frequency, called the *threshold frequency* (f_0) at which the maximum kinetic energy of the photoelectrons is zero. The energy of a photon at this frequency is just sufficient to break the bonds holding the electron to the surface of the material. This energy, known as the *work* function (W), is the y-intercept of the graph. At frequencies *less* than the threshold value, photons cannot produce a photoelectric effect.

Einstein predicted these results before there was any experimental verification. He theorized that light behaved as if it were a collection of particles called *photons*. Einstein summarized his theory by means of his *photoelectric equation*:

PHYSICS CONCEPTS

$$KE_{max} = hf - W$$

When a photon with energy hf strikes a surface, a part of its energy (the work function, W) frees the electron from its bonds. The remainder of the energy ($hf - W$) gives the electron its kinetic energy (KE_{max}).

PROBLEM

A photon with a frequency of 8.0×10^{14} hertz strikes a photoemissive surface whose work function is 1.7×10^{-19} joule. Calculate (a) the maximum kinetic energy of the ejected photoelectrons and (b) the threshold frequency of the surface.

SOLUTION

(a) $KE_{max} = hf - W$

$$= (6.6 \times 10^{-34} \text{ J} \cdot \text{s})(8.0 \times 10^{14} \text{ Hz}) - 1.7 \times 10^{-19} \text{ J}$$

$$= 3.6 \times 10^{-19} \text{ J}$$

(b) The threshold frequency (f_0) is the frequency at which the maximum kinetic energy of the photoelectrons is just zero. Applying the photoelectric equation gives

$$KE_{max} = hf - W$$

$$0 = hf_0 - W$$

$$W = hf_0$$

$$1.7 \times 10^{-19} \text{ J} = (6.6 \times 10^{-34} \text{ J} \cdot \text{s})f_0$$

$$f_0 = 2.6 \times 10^{14} \text{ Hz}$$

15.4   THE COMPTON EFFECT

The photon theory of light was strongly supported by the experiments of Arthur Compton, a U.S. physicist. When Compton bombarded a block of graphite with X-rays of known frequency, he discovered that both electrons and X-rays emerged from the block, as shown in the diagram.

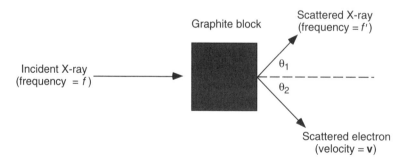

Compton observed that the scattered X-rays had lower frequencies than the incident X-rays, and he recognized that both energy and momentum were conserved in these collisions. Compton used the following relationship for the magnitude of the momentum of a photon:

PHYSICS CONCEPTS

$$p = \frac{E}{c} = \frac{hf}{c} = \frac{h}{\lambda}$$

PROBLEM

Calculate the momentum of an X-ray photon whose wavelength is 1.0×10^{-10} meter.

SOLUTION

$$p = \frac{h}{\lambda} = \frac{6.6 \times 10^{-34} \text{ J·s}}{1.0 \times 10^{-10} \text{ m}} = 6.6 \times 10^{-24} \text{ kg·m/s}$$

This relationship demonstrates that momentum (p), the particle property of light, wavelength (λ), and the wave property of light, are inversely related: High-frequency light (ultraviolet, X-rays, gamma radiation) behaves more like particles and less like waves; low-frequency light (radio waves, microwaves, infrared radiation) behaves more like waves and less like particles.

15.5   MATTER WAVES

In 1924, French physicist Louis De Broglie expressed his belief that nature was symmetrical and that the particle-wave properties of light should be present in ordinary matter as well. He further stated that the same relationship:

PHYSICS CONCEPTS

$$p = mv = \frac{h}{\lambda}$$

---

should hold for both light and matter.

Subsequent experiments demonstrated that particles such as electrons and neutrons produce diffraction patterns similar to those produced by light. In addition, their measured wavelengths agree with the relationship given above. The waves associated with matter, called *matter waves*, do not behave as other waves, however, in that they do not travel in space.

PROBLEM

Calculate the wavelength of a proton whose speed is 5.0×10^6 meters per second.

SOLUTION

$$p = mv = \frac{h}{\lambda}$$

$$\lambda = \frac{h}{mv}$$

$$= \frac{6.6 \times 10^{-34} \text{ J} \cdot \text{s}}{(1.66 \times 10^{-27} \text{ kg})(5.0 \times 10^6 \text{ m/s})} = 8.0 \times 10^{-14} \text{ m}$$

Ordinary-size objects have wavelengths that are much too small ($\sim 10^{-34}$ m) to be measured. Therefore, a batter does not have to worry about the diffraction pattern produced by a baseball!

15.6   MODELS OF THE ATOM

The impact of modern physics is most evident in the development of the atomic model of matter. Though the concept of the atom goes back to ancient Greece and Rome, it was not until the twentieth century that the structure of the atom was reasonably well understood. We use the term *atomic model* to indicate that we are trying to describe the key features of the atom. We really do not know what an atom "looks like" because we have no instruments for its direct observation. A model is like a road map: it gives us information to get from place to place, but it does *not* describe the actual landscape.

Early Models

The first scientific model of the atom was conceived by English chemist and physicist John Dalton at the beginning of the nineteenth century. His model explained the mathematics of chemical combinations, but the internal structure of the atom remained a mystery. At the beginning of the twentieth century, the discoveries of Antoine Henri Becquerel, the Curies, Frédéric Joliot-Curie, Iréne Joliot-Curie, and J.J. Thomson led to the idea that an atom is constructed of positively and negatively charged particles, most notably the electron.

Rutherford Model

British physicist Ernest Rutherford and his two assistants, Hans Geiger and Ernest Marsden, bombarded thin metallic foils, such as gold, with massive positively charged particles known as *alpha particles*. When the tracks of the scattered particles were analyzed, it was observed that:

1. Most of the particles passed through the foils without being deflected.
2. A very small number of particles were deflected through large angles, some approaching 180°.
3. The paths of the deflected particles were hyperbolic in shape.

The diagrams below represent the results of Rutherford's experiments.

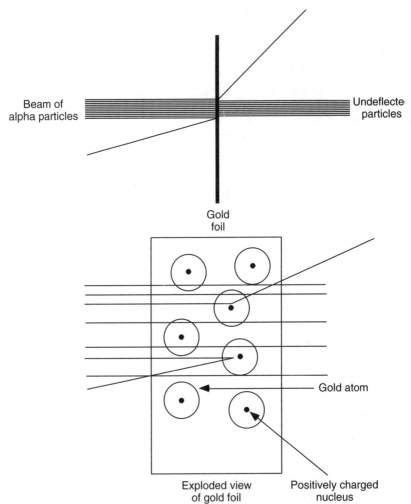

On the basis of these experiments, Rutherford drew the following conclusions:
1. Most of the atom is empty space.
2. Most of the mass of the atom is concentrated in a dense, positively charged *nucleus*. The negative electrons orbit the nucleus much as the planets orbit the Sun.
3. The hyperbolic paths are the result of the (Coulomb's law) repulsion between the alpha particles and the atomic nuclei.

Bohr Model

Although Rutherford's nuclear model was an important advance in our knowledge of atomic structure, it had two serious failings:

1. Orbiting electrons are *accelerating* charges and, as such, should produce electromagnetic radiation. This release of energy should cause any electron's orbit to decrease, leading to the collapse of any and all atoms. However, atoms are generally stable; they do not collapse.
2. When heated or subjected to a high potential difference, atomic gases, such as hydrogen, produce *line emission spectra,* rather than the continuous emission spectrum seen in rainbows. Why these gases emit only certain frequencies of light could not be explained by the Rutherford model.

The diagram below represents a partial line spectrum (in the visible-light region) for atomic hydrogen gas. Note that the wavelengths for the spectral lines are given in nanometers (1 nm = 10^{-9} m) .

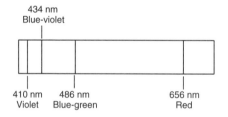

(Partial) Visible Line Spectrum
for Atomic Hydrogen

Danish physicist Neils Bohr was able to resolve the difficulties of the Rutherford model for hydrogen, and he derived a mathematical relationship that allows the wavelengths (or frequencies) of the spectral lines of hydrogen to be calculated precisely. Bohr accepted the idea that atoms have positively charged nuclei and orbiting electrons. However, he proposed that the hydrogen atom possesses distinct orbits. At any instant, the single electron of hydrogen can be associated with one and only one of these orbits. Each orbit gives the hydrogen atom a specific amount of energy. (For this reason, Bohr called the orbits *energy levels.*) Energy levels are identified by integers: 1, 2, 3, . . . , *n*. Today these integers are known as *quantum numbers*.

If the electron is on the first energy level, the atom is said to be in the **ground state**. On any other level, the atom is in an **excited state**. No energy is either emitted or absorbed by the atom when it is in any single state, known as a *stationary state*. Rather, energy is *emitted* (as photons) when the atom goes from a higher energy to a lower energy state, and energy is *absorbed* (as photons) when the atom goes from a lower energy to a higher energy state.

Bohr's relationship for the energy levels in a hydrogen atom is as follows:

$$\boxed{\text{PHYSICS CONCEPTS}}$$

$$E_n = -\frac{13.6 \text{ eV}}{n^2}$$

In the ground state ($n = 1$), the energy of the hydrogen atom is -13.6 eV. Then, for example, in the excited state where $n = 3$, the energy of the hydrogen atom is

$$E_3 = -\frac{13.6 \text{ eV}}{3^2} = -1.51 \text{ eV}$$

If n is considered infinitely large, the energy of the hydrogen atom is taken to be zero because the electron is no longer considered to be associated with the atom, a condition known as **ionization**.

PROBLEM
Calculate the energy of a hydrogen atom when its electron is associated with energy level 2.

SOLUTION

$$E_n = -\frac{13.6 \text{ eV}}{n^2}$$

$$E_2 = -\frac{13.6 \text{ eV}}{2^2} = -3.40 \text{ eV}$$

The diagram below is a graphic representation of the energy levels of the hydrogen atom:

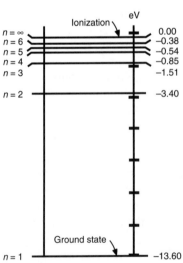

Energy Levels of the Hydrogen Atom

When a hydrogen atom changes from energy state $E_{initial}$ to energy state E_{final}, the amount of energy emitted or absorbed by the atom is determined as follows:

PHYSICS CONCEPTS

$$\Delta E = E_{final} = E_{initial} = \frac{-13.6 \text{ eV}}{n^2_{final}} - \frac{-13.6 \text{ eV}}{n^2_{initial}}$$

which can be simplied to

$$\Delta E = 13.6 \text{ eV} \left(\frac{1}{n^2_{initial}} - \frac{1}{n^2_{final}} \right)$$

If ΔE is a *negative* number, the energy is emitted; if it is positive, the energy is absorbed. The frequency of the photon that is emitted (or absorbed) can be calculated from the familiar relationship $\Delta E = hf$.

PROBLEM
Calculate the energy of the photon that is emitted when a hydrogen atom changes from energy state $n = 3$ to $n = 2$.

SOLUTION
We have already calculated the two energy states: $E_3 = -1.51$ eV; $E_2 = -3.40$ eV.

$$\Delta E = E_{final} - E_{initial} = E_2 - E_3$$
$$= (-3.40 \text{ eV}) - (-1.51 \text{ eV}) = -1.89 \text{ eV}$$

A 1.89–eV photon is emitted by the hydrogen atom in changing from $n = 3$ to $n = 2$. This corresponds to the red line of the hydrogen spectrum shown above.

The visible line spectrum for atomic hydrogen is known as the *Balmer series*. It consists of electron transitions to energy level 2 from higher levels (3, 4, 5, . . .). Transitions down to energy level 1 yield a line spectrum in the ultraviolet region, while transitions to level 3, 4 or 5 yield line spectra in the infrared region.

PROBLEM
How much energy is needed to ionize a hydrogen atom in the ground state?

SOLUTION
We must raise the energy level of the hydrogen atom from level 1 to "infinity."

$$\Delta E = E_\infty - E_1 = 0.0 \text{ eV} - (-13.6 \text{ eV}) = +13.6 \text{ eV}$$

Thus, 13.6 eV must be absorbed by a hydrogen atom in the ground state in order to ionize it. This energy is known as the *ionization energy* (or *ionization potential*) of the hydrogen atom.

When De Broglie proposed the existence of matter waves, as discussed in Section 15.5, he was able to show that the Bohr model of the hydrogen atom could be explained by considering the orbits of the electron as a series of standing waves. Unfortunately, the Bohr model is not successful in explaining atoms that have more than one electron.

The diagram below represents a portion of the line spectrum of mercury. Notice that the spacing of the lines is more complex and less regular than for hydrogen. The Bohr model *cannot* be applied to this kind of atom. Nevertheless, we can still calculate the energy of an electron transition between states because we have been given the energy values (labeled *a* through *j*) associated with them.

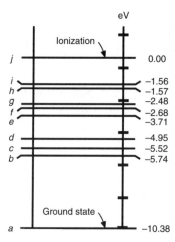

A Few Energy Levels for the Mercury Atom

Cloud Model

The current model of the atom is able to explain the structure of all of the atoms in the Periodic Table. This model is based on a mathematical area of physics known as *quantum mechanics* or *wave mechanics*. Quantum mechanics does not place electrons in specific orbits; rather, it indicates the *probability* that an electron will be in a region of space near the nucleus. The most probable regions of the electron's location define what is known as the **electron cloud**. The energy levels of the Bohr model are subdivided into other levels termed *sublevels* and *orbitals*. Together, these define the energy and structure of a particular atom.

15.7 EINSTEIN'S SPECIAL THEORY OF RELATIVITY

Relative Velocity

Is it possible for a car to travel at 25 meters per second [east], 0 meter per second, 25 meters per second [west], and 5 meters per second [east] *all at the same time?* The answer is yes!

When we measure an object's velocity, we measure it *relative* to another object. For example, a speedometer reading measures the velocity of a car relative to the road. Suppose two cars are traveling at 25 meters per second [east] with respect to the road; the velocity of one car *relative to the other car* will be 0 meter per second.

In the language of the physicist, we must establish a *frame of reference* in order to measure a quantity such as velocity. A frame of reference is a coordinate grid and a set of synchronized clocks that can be used to measure the position and time of an event.

PROBLEM

How can a car traveling at 25 meters per second [east] relative to the road have velocities of (a) 25 meters per second [west] and (b) 5 meters per second [east]?

SOLUTION

(a) Consider that a second car is traveling at 50 m/s [east]. A person in this car will measure the relative velocity of the first car as 25 m/s [west]. In other words, the velocity of the first car is 25 m/s [west] *when the second car is used as a frame of reference.*

(b) If the frame of reference is a second car traveling at 20 m/s [east], the relative velocity of the first car will be 5 m/s [east].

We denote the relative velocity of an object by the symbol $\mathbf{v}_{ab}$, where a is the object whose velocity is being measured and b is the frame of reference. For example, the velocity of a car with respect to the road can be written as $\mathbf{v}_{cr}$.

If the velocity of the car relative to the road is 25 meters per second [east], what is the velocity of the *road relative to the car?* Although this is an odd thought, nothing prevents us from using the car as the frame of reference. Using this frame, we could say that the road is moving *west at 25 meters per second!* In our notation, we write this general relationship as follows:

PHYSICS CONCEPTS

$$\mathbf{v}_{ab} = -\mathbf{v}_{ba}$$

In our problem dealing with two cars and the road, we usually use the road as the frame of reference because we think of the road as being at rest. But is it really at rest? The road is part of the Earth, which is rotating on its axis and revolving around the Sun. In addition, the Sun is dragging the Earth (including the road) through the Milky Way galaxy. Clearly, the road is *not* at rest; it is merely a convenient frame of reference.

Is there anything in the universe that we can say is *absolutely* at rest, a reference point that could be the standard for measuring all motion? The answer is no. All we can do is measure the motion of an object with respect to a frame of reference that we choose arbitrarily.

Inertial Frames of Reference and Newton's Laws of Motion

We will now examine more closely the motion of an object—in this case, a railroad car—in different frames of reference.

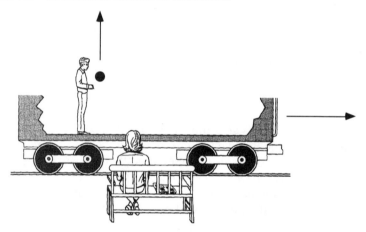

The diagram above illustrates a railroad car traveling at constant velocity with respect to the ground. A person in the car tosses an object up vertically and allows it to return to the ground. Two cameras, one mounted in the car and one mounted alongside the tracks, take time-lapse photographs of the object's motion.

The diagram below shows that the path of the object is a straight line when viewed *from inside the car.*

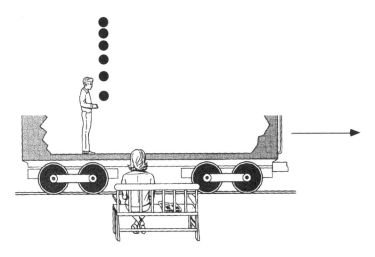

The next diagram shows that the path of the object is a parabola when viewed *from the tracks*. The reason is that the horizontal velocity of the object is zero relative to the car but is not zero relative to the tracks.

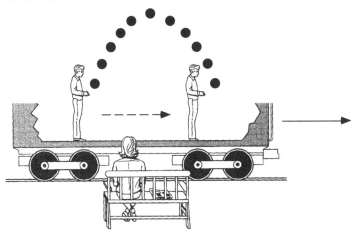

We could use either of the last two diagrams to calculate the gravitational force present on the object by calculating the acceleration and then using Newton's second law of motion. In either case we would obtain the same result.

In fact, we would obtain the same result if we performed the calculations in *any* frame of reference that is traveling at *constant velocity* with respect to the Earth. All such frames of reference are called *inertial frames of reference* because Newton's laws (particularly, Newton's *first* law) are valid in each of these frames. As a result, we can conclude that there is no *preferred* inertial frame of reference; that is, it is impossible to say that any specific frame of reference is absolutely at rest in the universe. The principle of *Gallilean-Newtonian relativity* summarizes this idea:

Electromagnetism and Relativity

In Chapter 11, we learned that the *relative motion* between a coil of wire and a magnet can induce a potential difference. It would appear that the principle of relativity should extend to electromagnetism as well. In actuality, the discovery and existence of electromagnetic waves posed a problem for nineteenth century physicists.

In the nineteenth century, Scottish physicist James Clerk Maxwell developed four equations that unified electricity and magnetism. As a result of this work, he derived a mathematical expression stating that all electromagnetic waves (including light) would travel in space at a speed of approximately 3×10^8 meters per second. His relationship did not specify a frame of reference.

At the time, physicists felt that *every* wave needed a medium in which to travel, light being no exception. They suggested the existence of a massless medium called the *ether,* which was absolutely at rest and could serve as the reference frame for measuring the speed of electromagnetic waves. Two American physicists, Albert Michelson and Edward Morley, tested this hypothesis with a series of very precise experiments but could find no evidence for the existence of such a medium.

Einstein's Solution: The Special Theory of Relativity

Einstein examined both mechanics and electromagnetism and concluded that no state of absolute rest exists. In 1905, he published his special theory of relativity. The two fundamental postulates of this theory are as follows:

The first postulate refers only to *inertial* frames of reference. If an object is in an *accelerating* frame of reference (as on a merry-go-round), the laws of physics will not take the same form as in an inertial frame.

The second postulate explains Maxwell's conclusion that the speed of light is a universal constant, but it violates the "commonsense" notion of relative velocity that we examined earlier in this section. For example, the diagram below illustrates a person on a railroad car, traveling at velocity **v**, emitting a light beam traveling in the same direction at velocity **c**.

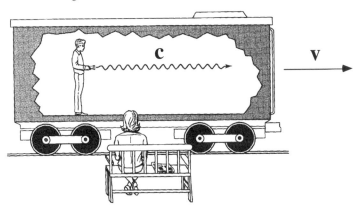

What is the velocity of light relative to the person who is observing the experiment while sitting on the bench outside the car? "Common sense" tells us that velocity ought to be **c + v**. Einstein's second postulate, however, indicates that the velocity of light must be exactly **c** when measured in *any* frame of reference!

In his special theory of relativity, Einstein provides a more general formula for the addition of two velocities, **u** and **v**, to produce a resultant velocity, **w**:

PHYSICS CONCEPTS

$$w = \frac{u + v}{1 + \dfrac{uv}{c^2}}$$

In everyday situations, when both **u** and **v** are very small compared with **c**, the fraction can be ignored in the denominator of Einstein's formula.. Then the sum simplifies to our "commonsense" notion of **w = u + v**.

For the diagram above, we calculate the relative velocity of light by substituting **c** for the term **u** in Einstein's equation:

$$w = \frac{u + v}{1 + \dfrac{uv}{c^2}}$$

$$= \frac{c + v}{1 + \dfrac{cv}{c^2}}$$

$$= \frac{c + v}{1 + \frac{v}{c}} = \frac{c + v}{\frac{c + v}{c}} = c$$

We note that $w = c$, as Einstein stated in his second postulate.

CONSEQUENCES OF EINSTEIN'S POSTULATES

If we accept Einstein's postulates of special relativity, we must also accept their consequences, namely, that time, length, and mass are not absolute, invariable quantities, but instead vary with the reference frames in which they are measured.

The Relativity of Time

Since we normally travel at speeds much slower than the speed of light, we generally accept the Newtonian premise that time is an absolute quantity in every frame of reference. From Einstein's point of view, however, this is not so.

Consider the problem of *simultaneity*. According to Newton, if two events occur at the same time in one frame of reference, these events must occur at the same time in every other frame of reference. The following "thought experiment" is adapted from one originally proposed by Einstein and is illustrated in the diagram below.

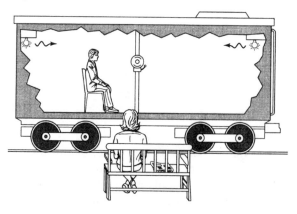

A screen with an attached bell is set up in the center of a moving railroad car so that the bell will ring if two light beams from opposite directions hit the screen simultaneously. At a given instant, two lights mounted inside the car are turned on. Let's analyze this experiment from the point of view of (1) a person seated inside the car at the screen and (2) a person seated on a bench along the tracks.

In case (1), the person inside the car hears the bell ring, meaning that the light beams hit the screen together. Since the two beams traveled the same distance, and the speed of light is a constant, this person concludes that the lights were turned on at the same time, that is, the events were *simultaneous*.

In case (2), the person outside the car also hears the bell, meaning that the light beams hit the screen together. However, since the car is moving toward the beam on the right, this beam must have traveled a shorter distance to reach the screen than the beam on the left. Since the speed of light is always a constant, this person concludes that the beam on the left was turned on *before* the beam on the right, that is, the events were *not* simultaneous!

Diagrams (1) and (2) below illustrate the experiment from both points of view.

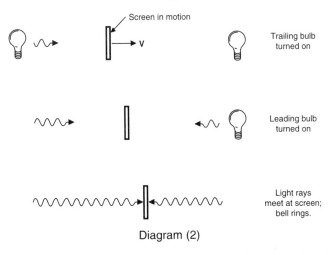

Diagram (1)

Diagram (1) represents the experiment as viewed from inside the car; the light bulbs are turned on simultaneously.

Diagram (2)

Diagram (2) represents the experiment as viewed from outside the car; the light bulb at the left is turned on before the light bulb at the right.

We might ask which observer is correct. The answer is that *both* persons are correct! The relative occurrences of two events depend on the frame of reference.

Time Dilation

Since time is not an absolute quantity, we might ask how the measurement of time differs in different frames of reference. Suppose a person is traveling on a rocket ship at one-half the speed of light (0.500c). The rocket ship has a giant clock that can be read on Earth as well as on the ship. Another consequence of special relativity is that an observer on Earth would *not* agree with an observer on the ship as to the passage of time.

The time intervals on the ship and on Earth are related by this equation:

PHYSICS CONCEPTS

$$\Delta t = \frac{\Delta t'}{\sqrt{1 - \dfrac{v^2}{c^2}}}$$

The term $\Delta t'$ is known as the *proper time*: it is the time interval measured in the frame of reference where the clock is located. In this case, proper time is measured on the rocket ship. The term Δt is the time interval measured in the frame of reference in which the clock is *not* located, the Earth in this case.

According to the equation, Δt will be greater than $\Delta t'$ for all speeds greater than zero, meaning that *time passes more slowly on the rocket ship than it does on Earth*. This phenomenon, known as *time dilation*, has been verified experimentally.

In the example we have been considering, suppose that 1.00 second has elapsed on the rocket ship. How much time has elapsed on Earth? We can apply the time dilation equation as follows:

$$\Delta t = \frac{\Delta t'}{\sqrt{1 - \dfrac{v^2}{c^2}}}$$

$$= \frac{100 \text{ s}}{\sqrt{1 - \dfrac{(0.500c)^2}{c^2}}} = 1.15 \text{ s}$$

Therefore, when an astronaut records the passage of 1.00 second, a person on Earth records the passage of 1.15 seconds.

One verification of time dilation is the existence of *muons* at the surface of the Earth. Muons are unstable elementary particles that are produced high in the Earth's atmosphere (approximately 4800 m above sea level) and have a mean lifetime of 2.2×10^{-6} second. Traveling at 0.99c, muons should be able to travel only approximately 600 meters before they disintegrate; yet they reach the Earth's surface.

The explanation is that the muon's lifetime (2.2×10^{-6} s) is measured in *its own* frame of reference. In the *Earth's* frame of reference, the time interval (found by applying the time dilation equation) is approximately 1.6×10^{-5} second, which is sufficient for the muons to reach the surface of the Earth.

The Twin Paradox

An interesting application of time dilation is given in the following hypothetical situation involving 15-year-old twin sisters: One of the twins journeys on a rocket ship to and from a star located 35 light-years from Earth at a speed very close to the speed of light. Meanwhile, her sister remains on Earth. When the traveling twin returns from her journey, she finds that her Earth-bound sister has grown very old, while she herself has aged only a fraction of this time.

Since there is no preferred frame of reference, why has the sister on Earth aged more than the sister in the rocket ship? This situation is known as the *twin paradox*. The answer lies in the fact that the sister on the rocket ship is *not* in an inertial frame of reference: her trip includes accelerations, and therefore the time relationships based on special relativity are not valid. Only for the sister on Earth do these relationships apply, and she, indeed, ages more rapidly than her twin in the rocket ship.

Length Contraction

Another consequence of special relativity is the fact that the length of an object changes with its motion. Suppose again that a person is traveling on a rocket ship at one-half the speed of light (0.500c). This person measures the length of an object as 1.00 meter. What would the length of the object be when measured by a person on the Earth?

The lengths of the object on the ship and on Earth are related by this equation:

PHYSICS CONCEPTS

$$L = L' \sqrt{1 - \frac{v^2}{c^2}}$$

The term L' is known as the *proper length*: it is the length measured in the frame of reference in which the object is at rest. In this case, proper length is measured on the rocket ship. The term L is the length of the object measured from the Earth's frame of reference.

According to the equation, L will be smaller than L' for all speeds greater than zero, a phenomenon known as *length contraction*. The length of the object as measured on Earth can be calculated by using the equation given above:

$$L = L' \sqrt{1 - \frac{v^2}{c^2}}$$

$$= 1.00 \text{ m} \sqrt{1 - \frac{(0.500c)^2}{c^2}} = 0.867 \text{ m}$$

The shortening of length occurs only along the direction of the object's motion.

Mass and Special Relativity

Einstein also showed that the mass of an object depends on its motion. We will let m_0 be the mass of an object measured by an observer at rest with respect to the object. This quantity is known as the object's *rest mass*. If m_v is the mass of the object measured by an observer moving at a speed v with respect to the object, Einstein's relation takes this form:

PHYSICS CONCEPTS

$$m_v = \frac{m_0}{\sqrt{1 - \frac{v^2}{c^2}}}$$

According to this relationship, as the speed of an object increases, its mass (m_v) also increases. For example, an object with a rest mass of 1.00 kilogram will have a mass of 1.15 kilograms if its speed is 0.500c. This relationship has been demonstrated experimentally with electrons traveling at speeds close to the speed of light.

The mass relationship also shows that the speed of light is the *ultimate limiting speed*: as an object approaches the speed of light, its mass increases without bound, and the energy needed to accelerate it also increases without bound. At the speed of light, the mass of the object is no longer defined. According to special relativity, objects with mass can never travel at the speed of light (although they may come very close to it).

Energy and Special Relativity

According to special relativity, the kinetic energy of an object is given by the following relationship:

PHYSICS CONCEPTS

$$KE = mc^2 - m_0c^2$$

The term m_0c^2 is known as the *rest energy* of the object; this quantity does not depend on the speed of the object. The rest energy demonstrates that mass is simply another form of energy. Since c^2 is a large number, a very small quantity of mass corresponds to a very large amount of energy. This fact has been verified by the existence of devices such as nuclear reactors and nuclear weapons.

The term mc^2 is the total energy of the object; it is the sum of the object's kinetic energy and rest energy. We state, without showing the mathematics, that at speeds that are low compared to the speed of light the kinetic energy of an object reduces to the familiar form:

$$KE = \frac{1}{2} mv^2$$

QUESTIONS

1. The ratio of the energy of a quantum of electromagnetic radiation to its frequency is
   (1) the electrostatic constant   (3) the gravitational constant
   (2) the electron-volt   (4) Planck's constant

2. What is the energy of a photon with a frequency of 5.0×10^{15} hertz?
   (1) 3.3×10^{-18} J   (3) 1.5×10^{24} J
   (2) 2.0×10^{-6} J   (4) 7.5×10^{48} J

3. The energy of a photon varies directly with its
   (1) frequency   (2) wavelength   (3) speed   (4) rest mass

4. According to the quantum theory of light, the energy of light is carried in discrete units called
   (1) alpha particles   (3) photons
   (2) protons   (4) photoelectrons

5. Which graph best represents the energy of a photon as a function of its frequency?

(1)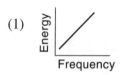

(3)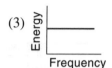

(2)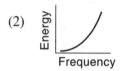

(4)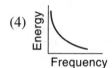

6. Which graph best represents the relationship between the energy of a photon and its wavelength?

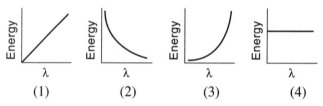

(1)　　　　　(2)　　　　　(3)　　　　　(4)

7. Light demonstrates the characteristics of
   (1) particles, only
   (2) waves, only
   (3) both particles and waves
   (4) neither particles nor waves

8. Which color of light has the greatest energy per photon?
   (1) red　　　(2) green　　　(3) blue　　　(4) violet

9. What is the energy of a photon of blue light whose frequency is 6×10^{14} hertz?
   (1) 4×10^{-6} J　(2) 4×10^{-10} J　(3) 4×10^{-14} J　(4) 4×10^{-19} J

10. The wavelength of photon A is greater than that of photon B. Compared to the energy of photon A, the energy of photon B is
    (1) less　　　(2) greater　　　(3) the same

11. Which phenomenon best supports the particle theory of light?
    (1) photoelectric effect
    (2) diffraction
    (3) interference
    (4) polarization

12. As the intensity of light above the threshold frequency falling on a photoelectric surface increases, the number of photoelectrons emitted
    (1) decreases   (2) increases   (3) remains the same

13. As the frequency of photons incident upon a photoemissive surface is increased, the maximum energy of the photoelectrons
    (1) decreases   (2) increases   (3) remains the same

14. As the intensity of monochromatic light on a photoemissive surface increases, the maximum kinetic energy of the photoelectrons emitted
    (1) decreases   (2) increases   (3) remains the same

15. The threshold frequency of a metal surface is in the violet light region. What type of radiation will cause photoelectrons to be emitted from the metal's surface?
    (1) infrared light   (3) ultraviolet light
    (2) red light   (4) radio waves

16. The threshold frequency for a certain photoelectric surface is 6.5×10^{14} hertz. The work function of the surface is
    (1) 1.2×10^{-48} J   (3) 7.5×10^{-18} J
    (2) 4.3×10^{-19} J   (4) 9.8×10^{47} J

17. A photon with an energy of 9.3 electron-volts strikes a metallic surface. If the most energetic electron emitted is stopped by a retarding potential of 3.1 volts, then the work function of the metal is
    (1) 12.4 eV   (2) 6.2 eV   (3) 3.0 eV   (4) 3.1 eV

18. Photons with energies of 3.9×10^{-19} joule strike a photoemissive surface whose work function is 2.9×10^{-19} joule. The maximum kinetic energy of the ejected photoelectrons is
    (1) 1.0×10^{-19} J   (3) 7.0×10^{-19} J
    (2) 7.5×10^{-20} J   (4) 1.2×10^{-18} J

19. Photons with a frequency of 1.0×10^{20} hertz strike a metal surface. If electrons with a maximum kinetic energy of 3.0×10^{-14} joule are emitted, the work function of the metal is
    (1) 1.0×10^{-14} J   (3) 3.6×10^{-14} J
    (2) 2.2×10^{-14} J   (4) 6.6×10^{-14} J

20. Photons with an energy of 5 electron-volts strike a photoemissive surface causing the emission of 2-electron-volt photoelectrons. If photons with 10. electron-volts of energy strike the same photoemissive surface, what will be the energy of the emitted photoelectrons?
    (1) 5 eV   (2) 2 eV   (3) 7 eV   (4) 8 eV

21. As the photoelectric work function of a substance increases, its threshold frequency
  (1) decreases   (2) increases   (3) remains the same

Base your answers to questions 22 through 24 on the information below.

Light of constant intensity strikes a metal surface. The frequency of the light is increased from 6.0×10^{14} hertz to 9.0×10^{14} hertz. Photoelectrons are emitted by the metal surface when the frequency reaches 8.0×10^{14} hertz.

22. The work function of the metal surface is approximately
  (1) 6.0×10^{-19} J   (3) 5.3×10^{-19} J
  (2) 2.0×10^{-19} J   (4) 4.0×10^{-19} J

23. As the frequency of the incident light increases, the photons striking the metal surface increase in
  (1) number   (2) energy   (3) speed   (4) wavelength

24. If only the intensity of the incident light is increased as the frequency is kept constant, the maximum kinetic energy of the emitted photoelectrons
  (1) decreases   (2) increases   (3) remains the same

Base your answers to questions 25 through 27 on the graph, which represents the relationship between the maximum photoelectron energy (KE_{max}) and the frequency of the incident radiation (f) for four target metals: a, b, c, and d.

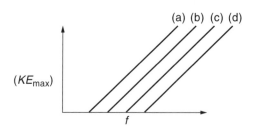

25. Which metal has the highest threshold frequency?
  (1) a (2) b (3) c (4) d

26. The energy of photons producing photoelectric emission from metal b is 4.2 electron-volts. If the maximum KE of the photoelectrons is 2.5 electron-volts, then the work function of the metal is
  (1) 1.7 eV   (2) 3.5 eV   (3) 4.2 eV   (4) 6.7 eV

27. The energy of a photon with a frequency of 6×10^{14} hertz is approximately
  (1) 7×10^{-34} J   (2) 1×10^{-24} J   (3) 4×10^{-19} J   (4) 4×10^{48} J

Base your answers to questions 28 through 31 on the graph below, which represents the maximum kinetic energy of the photoelectrons emitted by a metal surface upon exposure to a beam of light of varying frequency.

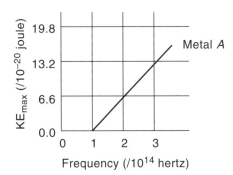

28. The work function for metal A is
(1) 0.0 J
(2) 6.6×10^{-20} J
(3) 6.6×10^{-34} J
(4) 6.6×10^{-48} J

29. The slope of the line for metal A is
(1) 6.6 J · s
(2) 6.6×10^{-6} J · s
(3) 6.6×10^{-20} J · s
(4) 6.6×10^{-34} J · s

30. As the frequency of the light falling on metal A increases above the threshold frequency, the maximum kinetic energy of the photoelectrons
(1) decreases   (2) increases   (3) remains the same

31. Metal B with a work function of 2.0×10^{-19} joule is substituted for metal A. If photons with an energy of 6.0×10^{-19} joule are incident on the surface of metal B, the kinetic energy of the emitted photoelectrons will be
(1) 12.0×10^{-19} J
(2) 2.0×10^{-19} J
(3) $3 0 \times 10^{-19}$ J
(4) 4.0×10^{-19} J

Base your answers to questions 32 through 36 on the graph below, which represents the maximum kinetic energies of photoelectrons for varying frequencies for three different metals.

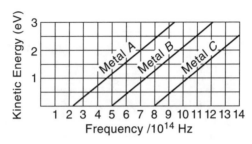

32. The slope of each graph represents
  (1) the work function (3) the threshold frequency
  (2) Planck's constant (4) the kinetic energy

33. If light intensity capable of ejecting photoelectrons from metal C is increased, the
  (1) slope of metal C will increase
  (2) threshold frequency for metal C will increase
  (3) work function of metal C will increase
  (4) number of photoelectrons will increase

34. Which metal has the greatest work function?
  (1) A (2) B (3) C

35. Which metal has the highest threshold frequency?
  (1) A (2) B (3) C

36. The frequency of light incident upon metal A is increased from 3×10^{14} hertz to 5×10^{14} hertz. The kinetic energy of the photoelectrons will
  (1) decrease   (2) increase   (3) remain the same

Base your answers to questions 37 through 39 on the diagram below, which represents photons entering a quartz phototube, causing photoelectrons to be ejected from the negative plate. Each photon of light has an energy of 3.0×10^{-19} joule, and the work function of the photoemissive surface is 2.4×10^{-19} joule.

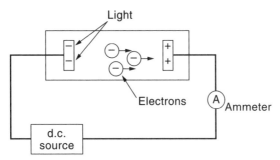

37. What is the frequency of the light entering the phototube?
    (1) 2.2×10^{-15} Hz   (3) 3.6×10^{14} Hz
    (2) 2.0×10^{-52} Hz   (4) 4.5×10^{14} Hz

38. The maximum energy of the ejected photoelectrons is
    (1) 6×10^{-20} J   (3) 3.0×10^{-19} J
    (2) 5.4×10^{-19} J   (4) 4×10^{-20} J

39. If the frequency of the original light were increased, the maximum energy of the emitted photoelectrons would
    (1) decrease   (2) increase   (3) remain the same

Base your answers to questions 40 through 42 on the diagram below, which represents monochromatic light incident upon photoemissive surface A. Each photon has 8.0×10^{-19} joule of energy. B represents the particle emitted when a photon strikes surface A.

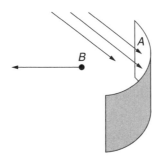

40. What is particle B?
    (1) an alpha particle   (3) a neutron
    (2) an electron   (4) a proton

41. If the work function of metal A is 3.2×10^{-19} joule, the energy of particle B is
(1) 3.0×10^{-19} J   (3) 8.0×10^{-19} J
(2) 4.8×10^{-19} J   (4) 11×10^{-19} J

42. The frequency of the incident light is approximately
(1) 1.2×10^{15} Hz   (3) 3.7×10^{15} Hz
(2) 5.3×10^{15} Hz   (4) 8.3×10^{15} Hz

Base your answers to questions 43 through 47 on the graph below, which represents the relationship between the maximum kinetic energy of the photoelectrons emitted from a photoemissive surface and the frequency of the incident electromagnetic radiation.

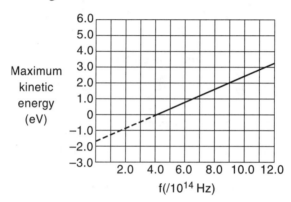

43. What is the threshold frequency of this surface?
(1) 2.6×10^{-4} Hz   (3) 3.0×10^{14} Hz
(2) 2.0×10^{14} Hz   (4) 4.0×10^{14} Hz

44. The number of photoelectrons emitted from the surface may be increased by increasing the
(1) threshold frequency of the surface
(2) work function of the surface
(3) intensity of the incident light
(4) wavelength of the incident light

45. The work function of this photoemissive surface is
(1) 2.6×10^{-19} eV   (3) 4.0×10^{14} eV
(2) 2.6×10^{-19} J   (4) 4.0×10^{14} J

46. An electron at the surface of the photoemissive material absorbs an 8.8×10^{14}-hertz photon. The maximum kinetic energy of the photoelectron is approximately
(1) 1.0 eV   (2) 2.0 eV   (3) 3.0 eV   (4) 4.0 eV

47. As the intensity of the electromagnetic radiation falling on the surface increases, the maximum kinetic energy of the emitted photoelectrons
(1) decreases   (2) increases   (3) remains the same

Base your answers to questions 48 through 52 on the graph below, which represents the relationship between the maximum kinetic energy of emitted photoelectrons and the frequencies of the photons incident upon a photoemissive surface.

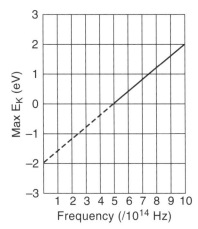

48. What is the frequency in hertz of a photon that would result in the emission of a photoelectron with a maximum kinetic energy of 2.0 eV?
(1) 0   (2) 2.0×10^{14} (3) 1.5×10^{14} (4) 1.0×10^{15}

49. The work function of the photoemissive surface is approximately
(1) 0 eV   (2) 2.0 eV   (3) 1.5 eV   (4) 4.0 eV

50. A photon whose frequency is equal to the threshold frequency strikes the photoemissive surface. What is the maximum kinetic energy of the emitted photoelectron?
(1) 5.0 eV   (2) 2.0 eV   (3) –2.0 eV   (4) 0 eV

51. As the frequency of the incident photons decreases, their momentum
(1) decreases   (2) increases   (3) remains the same

52. The photoemissive surface is replaced with a surface having a smaller work function. Compared to the threshold frequency of the original photoemissive surface, the threshold frequency of the new photoemissive surface is
(1) less   (2) greater   (3) the same

Base your answers to questions 53 through 55 on the graph below, which shows the maximum kinetic energy of the photoelectrons ejected when photons of different frequencies strike a metal surface.

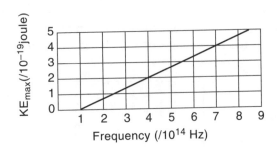

53. Which is the lowest frequency photon that will produce photoelectrons?
   (1) 0 Hz
   (2) 1.0 Hz
   (3) 1.0×10^{14} Hz
   (4) 15×10^{14} Hz

54. Photons with a frequency of 4×10^{14} Hz will produce photoelectrons with a maximum kinetic energy of
   (1) 4.0×10^{14} J
   (2) 1.3 J
   (3) 1.3×10^{-19} J
   (4) 2.0×10^{-19} J

55. Compared to the energy of the bombarding photon, the energy of the emitted photoelectron is
   (1) less   (2) greater   (3) the same

Base your answers to questions 56 through 60 on the information below.

An ultraviolet photon with a wavelength of 3.0×10^{-7} meter strikes a metal surface that has a work function of 2.0 electron volts. A photoelectron is emitted that loses its kinetic energy to the surrounding air. When it comes to rest, it is struck by an X-ray photon that has a frequency of 3.0×10^{18} hertz. Both the X-ray photon and the electron recoil from the collision.

56. The frequency of the ultraviolet photon is
   (1) 1.0×10^{14} Hz
   (2) 2.0×10^{14} Hz
   (3) 1.0×10^{15} Hz
   (4) 2.0×10^{15} Hz

57. If the ultraviolet photon has an energy of 4.2 eV, then the maximum kinetic energy of the photoelectron is
   (1) 1.2 eV   (2) 2.2 eV   (3) 6.2 eV   (4) 7.2 eV

58. What was the energy of the X-ray photon before it struck the electron?
   (1) 1.9×10^{-14} J
   (2) 2.0×10^{-15} J
   (3) 1.9×10^{-16} J
   (4) 2.0×10^{-17} J

59. As a result of the collision between the X-ray photon and the electron, the wavelength of the X-ray photon
(1) decreased   (2) increased   (3) remained the same

60. Compared to the total energy of the X-ray photon-electron system before the collision, the total energy of the system after the collision is
(1) less   (2) greater   (3) the same

61. As the wavelength of a ray of light increases, the momentum of the photons of the light ray will
(1) decrease   (2) increase   (3) remain the same

62. Which graph best represents the relationship between the wavelength and the momentum of a photon?

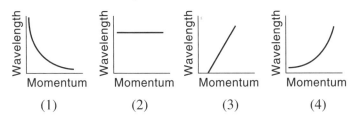

(1)   (2)   (3)   (4)

63. Which occurs when a photon and a free electron collide?
(1) Momentum is conserved.
(2) Only the kinetic energy of the photon is conserved.
(3) The momentum of the photon is increased.
(4) The wavelength of the photon is unchanged.

64. The relationship for the momentum of a photon is $p = \dfrac{hf}{c}$ What is the speed of the photon?
(1) less than 3×10^8 m/s
(2) greater than 3×10^8 m/s
(3) equal to 3×10^8 m/s

65. A gamma photon makes a collision with an electron at rest. During the interaction, the momentum of the photon will
(1) decrease   (2) increase   (3) remain the same

66. If an X-ray photon collides with an electron, the frequency of the photon will
(1) decrease   (2) increase   (3) remain the same

Base your answers to questions 67 through 71 on the information and diagram below.

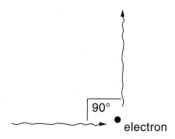

An incident photon with a frequency of 5.0×10^{16} hertz strikes a stationary electron. The scattered photon rebounds at an angle of 90°, and the electron moves away with a kinetic energy of 100. electron-volts.

67. Which vector best represents the direction of motion of the electron after the collision?

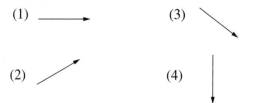

68. What is the energy of the incident photon?
(1) 1.3×10^{-18} J
(3) 5.0×10^{16} J
(2) 5.0×10^{-17} J
(4) 3.3×10^{-17} J

69. If the energy of the incident photon is Y electron-volts, what is the energy of the scattered photon?
(1) $(Y + 100.)$ eV
(3) $(Y \times 100.)$ eV
(2) $(Y - 100.)$ eV
(4) $(Y \div 100.)$ eV

70. Compared to the speed of the incident photon, the speed of the scattered photon is
(1) smaller   (2) greater   (3) the same

71. Compared to the wavelength of the incident photon, the wavelength of the scattered photon is
(1) shorter   (2) longer   (3) the same

72. As the speed of an electron increases, its wavelength
(1) decreases   (2) increases   (3) remains the same

73. All of the following particles are traveling at the same speed. Which has the greatest wavelength?
(1) a proton
(3) a neutron
(2) an alpha particle
(4) an electron

74. Which property of a particle has the greatest influence on its wave nature?
(1) mass
(3) volume
(2) shape
(4) specific heat

75. A mass m moving with a speed v has a wavelength of
(1) $h \times \dfrac{1}{2} mv^2$ (2) $\dfrac{h}{1/2 \; mv^2}$ (3) $h \times mv$ (4) $\dfrac{h}{mv}$

76. What is the wavelength of the matter wave associated with a bird of 1.0-kilogram mass flying at 2.0 meters per second?
(1) 3.3×10^{34} m
(3) 3.3×10^{-34} m
(2) 1.3×10^{-33} m
(4) 8.6×10^{-34} m

77. In the Rutherford scattering experiment, gold foil was bombarded by
(1) alpha particles
(3) electrons
(2) protons
(4) neutrons

78. As an alpha particle approaches the nucleus of an atom, the coulomb force of repulsion between the nucleus and the alpha particle will
(1) decrease   (2) increase   (3) remain the same

79. When alpha particles are directed against a thin metal foil, their deflection paths are
(1) circular   (2) elliptical   (3) parabolic   (4) hyperbolic

80. Evidence that the positive charges of the atom are concentrated in the nucleus was obtained from the
(1) spectra of elements
(2) scattering pattern of alpha particles
(3) ionization potentials of elements
(4) photoelectric threshold energies of different metals

81. In the Rutherford experiment, a beam of alpha particles was directed at a thin gold foil. The deflection pattern of the alpha particles showed that
(1) the electrons of gold atoms have waves
(2) the nuclear volume is a small part of the atomic volume
(3) the energy levels of a gold atom are quantized
(4) gold atoms can emit photons under bombardment

82. According to the Rutherford model of the atom, the volume of an atom is composed mainly of
(1) electrons   (2) protons   (3) neutrons   (4) empty space

83. Which type of force causes the hyperbolic trajectory of alpha particles in Rutherford's scattering experiment?
(1) gravitational   (2) electrostatic   (3) magnetic   (4) nuclear

84. A scattering experiment is performed in which alpha particles from a single source are deflected by the nuclei of various atomic elements. The nuclei causing the greatest amount of alpha-particle scattering are those having the
(1) smallest neutron number   (3) smallest mass number
(2) greatest photon number   (4) greatest atomic number

85. Rutherford observed that most of the alpha particles directed at a metallic foil appear to pass through unhindered, with only a few deflected at large angles. What did he conclude?
(1) Alpha particles behave like waves when they interact with atoms.
(2) Atoms have most of their mass distributed loosely in an electron cloud.
(3) Atoms can easily absorb and reemit alpha particles.
(4) Atoms consist mainly of empty space and have small, dense nuclei.

86. High-speed alpha particles strike a metal foil. Which element, when used in the foil, will tend to scatter the alpha particles through the greatest angles?
(1) platinum, with 78 elementary charges per nucleus
(2) silver, with 47 elementary charges per nucleus
(3) copper, with 29 elementary charges per nucleus
(4) vanadium, with 23 elementary charges per nucleus

Base your answers to questions 87 through 91 on Rutherford's experiments in which alpha particles were allowed to pass into a thin gold foil. All alpha particles had the same speed.

87. The paths of the scattered alpha particles were
(1) hyperbolic   (2) circular   (3) parabolic   (4) elliptical

88. Some of the alpha particles were deflected. The explanation for this phenomenon is that
(1) electrons have a small mass
(2) electrons have a small charge
(3) the gold leaf was only a few atoms thick
(4) the nuclear charge and mass are concentrated in a small volume

89. The alpha particles were scattered because of
   (1) gravitational forces   (3) magnetic forces
   (2) coulomb forces   (4) nuclear forces

90. As the distance between the nuclei of the gold atoms and the paths of the alpha particles increases, the angle of scattering of the alpha particles
   (1) decreases   (2) increases   (3) remains the same

91. If a foil were used whose nuclei had a greater atomic number, the angle of scattering of the alpha particles would
   (1) decrease   (2) increase   (3) remain the same

92. In alpha-particle scattering, the nucleus produces an effect on the scattering angles. This is due primarily to the fact that the nucleus
   (1) has a small total charge
   (2) has a mass close to that of the alpha particles
   (3) exerts coulomb forces
   (4) is widely dispersed throughout the atom

93. Which diagram best represents the path of a positively charged particle as it passes near the nucleus of an atom?

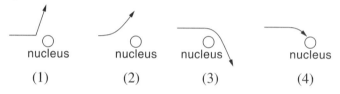

Base your answers to questions 94 through 97 on the following information and the diagram.

In the Rutherford scattering experiments, alpha particles are fired at a thin gold foil. The scattering angle of the alpha particles is θ, and the aiming error or the impact parameter is P.

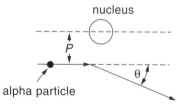

94. The particles with the greatest value of P will
   (1) lose most of their momentum as they collide with the thin gold foil
   (2) pass through the thin gold foil without appreciable deflection
   (3) be scattered uniformly in all directions
   (4) be scattered back toward the source

95. Which factor has the *least* effect on angle θ?
(1) the charge on the target nucleus
(2) the kinetic energy of the alpha particles
(3) the magnitude of *P*
(4) the planetary electrons in the target material

96. What is the value of angle θ when *P* = 0?
(1) 0°   (2) 90°   (3) 180°   (4) 270°

97. If *P* is held constant and the speed of an alpha particle is increased, angle θ will
(1) decrease   (2) increase   (3) remain the same

98. As metal foils of increasingly larger atomic number are bombarded by alpha particles, the number of alpha particles scattered beyond a given angle
(1) decreases   (2) increases   (3) remains the same

99. As excited hydrogen atoms return to the ground state, they emit
(1) electrons   (2) protons   (3) photons   (4) neutrons

100. As an electron orbits a nucleus in the same energy level, the energy of the electron
(1) decreases   (2) increases   (3) remains the same

101. As the radius of an electron orbit in a Bohr atom increases, the magnitude of the energy of the atom
(1) decreases   (2) increases   (3) remains the same

102. Which is a characteristic of both the Bohr and the Rutherford atomic model?
(1) Neutrons exist in the nuclei of all atoms.
(2) Only a limited number of specified orbits is permitted.
(3) The nucleus is concentrated in a small, dense core.
(4) Electron energy level changes are in discrete amounts.

103. When a hydrogen atom changes from one energy level (E_1) to a lower energy level (E_2), the expression for the frequency of the emitted photon is
(1) $\dfrac{\lambda}{E}$ (2) $\dfrac{h\lambda}{E_1 - E_2}$ (3) $\dfrac{E_1 - E_2}{h}$ (4) $(E_1 - E_2)\dfrac{\lambda}{2}$

104. At low pressure, spectra characterized only by bright lines are produced by
(1) transparent liquids   (3) incandescent solids
(2) opaque solids   (4) luminous gases

472

105. The lowest energy state of an atom is called its
(1) ground state
(3) initial energy state
(2) ionized state
(4) final energy state

106. When an excited atom emits a photon, the total energy of the atom
(1) decreases   (2) increases   (3) remains the same

107. When an electron changes from a higher energy state to a lower energy state within an atom, a quantum of energy is
(1) fissioned   (2) fused   (3) emitted   (4) absorbed

108. According to the Bohr model of the atom, an electron in a stable orbit does *not*
(1) have potential energy
(3) undergo acceleration
(2) emit radiation
(4) have kinetic energy

109. Which are emitted as atoms of a given element return to the ground state?
(1) electrons
(3) alpha particles
(2) photons
(4) neutrons

110. In his model of the atom, Bohr assumed that the electrons
(1) are distributed evenly throughout the atom
(2) are located only in the nucleus of the atom
(3) are located only in a limited number of specified orbits
(4) emit energy while in orbit

111. A hydrogen atom undergoes a transition from the $n = 3$ state to the ground state. The total number of different possible photon energies that may be emitted is
(1) 1   (2) 2   (3) 3   (4) 4

112. A photon with an energy of 10.2 electron-volts is absorbed by a hydrogen atom. This may cause the energy state of the hydrogen atom to move from
(1) $n = 1$ to $n = 2$
(3) $n = 1$ to $n = 4$
(2) $n = 1$ to $n = 3$
(4) $n = 1$ to $n = 5$

113. The energy needed to ionize a hydrogen atom in the ground state is
(1) 2.9 eV   (2) 3.2 eV   (3) 13.06 eV   (4) 13.6 eV

114. The minimum energy needed to ionize a hydrogen atom in the $n = 3$ energy state is
(1) 13.6 eV   (2) 12.1 eV   (3) 3.00 eV   (4) 1.51 eV

115. What is the minimum amount of energy required to ionize a hydrogen atom in the $n = 2$ state?
(1) 13.6 eV   (2) 10.2 eV   (3) 3.40 eV   (4) 0 eV

116. An atom changing from an energy state of –0.54 eV to an energy state of –0.85 eV will emit a photon whose energy is
(1) 0.31 eV   (2) 0.54 eV   (3) 0.85 eV   (4) 1.39 eV

117. A hydrogen atom can be raised from the $n = 2$ state to the $n = 3$ state by a photon with an energy of
(1) 1.89 eV   (2) 10.2 eV   (3) 12.1 eV   (4) 22.3 eV

118. A hydrogen atom in the ground state receives 10.2 electron-volts of energy. To which energy level may the atom become excited?
(1) $n = 5$ (2) $n = 2$ (3) $n = 3$ (4) $n = 4$

119. A photon having an energy of 15.5 electron-volts is incident upon a hydrogen atom in the ground state. The photon may be absorbed by the atom and
(1) ionize the atom   (3) excite the atom to $n = 3$
(2) excite the atom to $n = 2$ (4) excite the atom to $n = 4$

120. A hydrogen atom is in the $n = 5$ energy state after having absorbed a 0.97-eV photon. What was the original energy state of the hydrogen atom?
(1) $n = 1$ (2) $n = 2$ (3) $n = 3$ (4) $n = 4$

121. A model of the atom in which the electrons can exist only in specified orbits was suggested by
(1) Bohr   (2) Planck   (3) Einstein   (4) Rutherford

122. In a hydrogen atom the electron makes the following successive transitions:

$$n = 5 \rightarrow n = 4 \rightarrow n = 3 \rightarrow n = 2 \rightarrow n = 1$$

The emitted photon energies for the successive transitions
(1) decrease   (2) increase   (3) remain the same

Base your answers to questions 123 through 125 on the Energy Levels for Hydrogen chart in the Physics Reference Tables.

123. A hydrogen atom changes from the $n = 1$ energy state to the $n = 3$ energy state. This change could be caused by a single photon that has an energy of
(l) 1.5 eV   (2) 10.2 eV   (3) 12.1 eV   (4) 13.6 eV

124. Which photon could be absorbed by a hydrogen atom in the ground state?
(1) a 11.0-eV photon   (3) a 3.4-eV photon
(2) a 10.2-eV photon   (4) a 0.54-eV photon

125. Which energy-level jump would show as a bright line in the visible spectrum of hydrogen?
(1) 1 to 2   (2) 2 to 3   (3) 3 to 2   (4) 4 to 7

Base your answers to questions 126 through 128 on the information below.

A hydrogen atom in its ground state is struck by a photon, and the atom is excited to the $n = 4$ state.

126. The energy of the photon striking the atom is
(1) 0.85 eV   (2) 3.4 eV   (3) 12.75 eV   (4) 13.6 eV

127. When the atom is in the $n = 4$ state, the minimum additional energy needed to ionize the hydrogen atom is
(1) 13.6 eV   (2) 3.4 eV   (3) 1.5 eV   (4) 0.85 eV

128. If the atom changes from the $n = 4$ to the $n = 1$ state by emitting a single photon, the photon must have an energy of
(1) 10.2 eV   (2) 12.1 eV   (3) 12.75 eV   (4) 13.6 eV

129. A hydrogen atom, in the ground state, is bombarded by an 11-electron-volt photon. Which statement best describes the interaction that occurs?
(1) The photon collides elastically and leaves the atom with an energy of 11 electron-volts.
(2) The photon collides inelastically and retains an energy of 0.8 electron-volt.
(3) The photon collides inelastically and disappears.
(4) The atom is completely ionized to a +1 ion, and the photon disappears.

Base your answers to questions 130 through 132 on the information below.

A photon with an energy of 20. electron-volts is completely absorbed by a hydrogen atom in the ground state, ionizing the atom.

130. What is the approximate energy of the incoming photon?
(1) 8.0×10^{-20} J
(3) 3.2×10^{-18} J
(2) 1.6×10^{-19} J
(4) 20. J

131. What is the kinetic energy of the hydrogen electron at the end of the interaction?
(1) 0 eV
(2) 6.4 eV
(3) 13.6 eV
(4) 20. eV

132. Which energy level transition of a hydrogen atom would result in the emission of a spectral line in the Balmer series?
(1) $n = 3$ to $n = 1$
(3) $n = 4$ to $n = 2$
(2) $n = 5$ to $n = 4$
(4) $n = 4$ to $n = 1$

133. A hydrogen atom undergoes a transition from the $n = 4$ state to the $n = 1$ state. The energy of the single photon emitted during the transition is approximately
(1) 13.6 eV
(2) 12.75 eV
(3) 2.55 eV
(4) 0.85 eV

Base your answers to questions 134 through 137 on the information and diagram below and on the Physics Reference Tables.

The diagram represents the model of the Bohr hydrogen atom in the ground state. The speed of the electron is 2.17×10^6 meters per second.

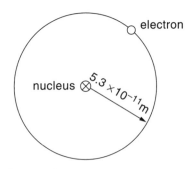

134. What is the kinetic energy of the electron in the ground state?
(1) 4.36×10^{-18} J
(3) 9.87×10^{-24} J
(2) 2.14×10^{-18} J
(4) 0

135. Compared to the electrostatic force between the proton and the electron, the centripetal force on the electron is
(1) one-fourth as much   (3) the same
(2) one-half as much   (4) twice as much

136. The electrostatic force between the electron and the proton is approximately
(1) 4.3×10^{-18} N   (3) 8.2×10^{-8} N
(2) 4.1×10^{-8} N   (4) 5.1×10^{1} N

137. Compared to the electrostatic force between the proton and the electron, the gravitational force between them is
(1) less   (2) greater   (3) the same

Base your answers to questions 138 through 139 on the diagram below, which shows some of the energy levels of a mercury atom. The range of the energies of visible photons is approximately 1.5 eV to 3.0 eV

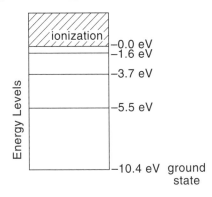

138. Which energy level transition would cause the emission of visible photons?
(1) from –5.5 eV to –10.4 eV   (3) from –3.7 eV to –1.6 eV
(2) from –3.7 eV to –5.5 eV   (4) from –5.5 eV to –1.6 eV

139. The minimum energy required to ionize a mercury atom in the ground state is
(1) 4.9 eV   (2) 6.7 eV   (3) 8.8 eV   (4) 10.4 eV

140. A photon with a wavelength of 6.0×10^{-7} meter would have an energy of
(1) 9.4×10^{-20} J   (3) 6.0×10^{-7} J
(2) 3.3×10^{-19} J   (4) 9.4×10^{18} J

141. A 3.0-electron-volt photon would have a wavelength of approximately
(1) 4×10^{-7} m   (3) 6×10^{-7} m
(2) 5×10^{-7} m   (4) 7×10^{-7} m

142. An inertial frame of reference is defined as
(1) any coordinate system
(2) a coordinate system for which Newton's first law of motion is valid
(3) a coordinate system at rest
(4) a nonrotating coordinate system

143. Which statement best describes the results of the Michelson-Morley experiment?
(1) They established the theory of special relativity.
(2) They linked the speed of light to the speed of all electromagnetic waves.
(3) They demonstrated the existence of photons
(4) They questioned the credibility of the ether hypothesis.

144. Which statement most accurately describes the results of experimental evidence about the nature of light?
(1) Particles that travel faster than the speed of light have been discovered.
(2) A particle with mass can approach, but never reach, the speed of light.
(3) Electrons in an atom travel at the speed of light.
(4) Matter waves travel at the speed of light.

145. As an object's speed increases, the object's length in the direction of its velocity
(1) decreases   (2) increases   (3) remains the same

146. The lifetime of a stationary particle is measured to be 5.0×10^{-5} second in an Earth laboratory. When moving at 0.80 times the speed of light, the lifetime of the particle measured in the same laboratory will be
(1) 8.3×10^{-5} s   (3) 3.0×10^{-5} s
(2) 6.4×10^{-5} s   (4) 3.9×10^{-5} s

Base your answers to questions 147 and 148 on the diagram below, which represents two spaceships, each approaching the other at a speed of 0.60 times the speed of light (c).

0.60 c 0.60 c

| 1 | | 2 |

147. The speed of ship 1 relative to that of ship 2 is
    (1) 0.60c (2) 0.88c (3) c (4) 1.2c

148. If ship 2 flashes a light beam toward ship 1, ship 1 will measure the speed of the beam as
    (1) 2.2c (2) 1.2c (3) c   (4) 0.60c

NUCLEAR ENERGY

Chapter Sixteen

KEY IDEAS

The main components of the atomic nucleus are protons and neutrons, also called nucleons. The atomic number of an atom is the number of protons its nucleus contains; the mass number is the sum of the numbers of protons and neutrons in the nucleus.

Isotopes are atoms with the same atomic numbers but differing mass numbers. The unit of atomic mass is based on the isotope carbon–12. One atomic mass unit is equivalent to 931 million electron-volts of energy. The mass of a nucleus is always less than the mass of its individual nucleons. The energy equivalent of this lost mass is related to the stability of the nucleus. Inside the nucleus, the existence of short-range forces contribute to this stability.

To track nuclear reactions, a number of detection devices such as the Geiger counter are in use. A nuclear reaction involves changes in one or more atomic nuclei. Nuclear equations represent nuclear reactions. A nuclear equation is balanced when the atomic and mass numbers agree on both sides of the arrow.

Radioactive decay is a series of naturally occurring nuclear reactions, which occur in order to increase the stability of the resulting nuclei. In alpha decay, alpha particles are ejected from the nucleus. In beta decay and positron emission, negative and positive beta particles, respectively, are ejected from their parent nuclei. In electron capture, a nucleus absorbs one of the inner electrons surrounding it. In gamma decay, a nucleus ejects a gamma photon, thereby lowering the nuclear energy.

Radioactive decay occurs according to strict mathematical laws. As a result, the half life of a radioactive substance, that is, the amount of time required to reduce a sample of the substance to one-half its initial value, is constant under all conditions.

Nuclear reactions can by induced by bombarding nuclei with subatomic particles such as alpha particles. The energies of the bombarding particles needed to produce a reaction are achieved by introducing the particles into a particle accelerator such as a cyclotron. Only charged particles may be accelerated in this way.

When a neutron bombards certain heavy nuclei, a fission reaction results in which the nucleus splits into two nuclei of similar size; this fission is accompanied by the release of large quantities of energy. The process, known as *nuclear fission*, produces more neutrons, which can be used in turn to bombard other atoms, creating a chain reaction. Fission reactors use this principle and harness the energy created to generate electricity.

Energy may also be produced by forcing light nuclei to combine, a process known as *nuclear fusion*. Stars use fusion reactions to produce their immense energies. At present, physicists are experimenting with ways to control fusion and to build fusion reactors as energy sources.

It is now known that the proton and the neutron are not fundamental particles. The existence of these nucleons and other nuclear particles has been explained by the presence of six varieties of particles called *quarks*. The electric charge on these particles is either –1/3 or +2/3 of the elementary charge.

KEY OBJECTIVES

At the conclusion of this chapter you will be able to:
- Define the term *nucleon,* and distinguish between the two nucleons.
- Interpret the parts of nuclear symbol, and define the terms *atomic number, mass number,* and *isotope.*
- Define the terms *mass defect* and *binding energy*, and explain how they contribute to the stability of the nucleus.
- Explain how nuclear forces differ from gravitational and electromagnetic forces.
- Describe how particle detectors are used in the study of nuclear physics.
- Balance a nuclear equation.
- List examples of nuclear reactions that occur naturally.
- Interpret the uranium-238 decay series graph.
- Define the term *half-life*, and solve problems involving half-lives.
- Explain how particle accelerators are used to induce nuclear reactions.
- Describe various types of induced nuclear reactions.
- Describe the process of nuclear fission and the reactions by which it produces vast quantities of energy.
- Describe the components of a fission reactor and the purposes for which each component is used.
- Describe the process of nuclear fusion, and compare it with nuclear fission.
- Describe how quarks are involved in nuclear structure.

16.1 INTRODUCTION

After British physicist Ernest Rutherford proposed his nuclear model in the early part of the twentieth century, physicists began to question whether the nucleus had a structure of its own. At present, there is still no complete answer to this question. By the mid-1930s, however, a simple nuclear model was in place and we will examine that model in this chapter.

16.2 NUCLEONS

The components of the nucleus are called **nucleons** and are described by a number of properties, including electric charge. Nuclear charge is usually measured in terms of the elementary charge (e) rather than the coulomb.

Two of the principal nucleons are the *proton* and the *neutron*. The proton has a charge of $+1e$, and the neutron is uncharged.

16.3 NUCLEAR SYMBOLS

All atomic nuclei (also called *nuclides*)—and their component nucleons— may be represented by the same general symbol:

$$^A_Z X$$

The letter X represents the letter(s) used to identify the particle; the letter Z, representing the **atomic number**, indicates the number of elementary charges present (assumed to be positive unless a negative sign is written); and the letter A, representing the **mass number**, is equal to the sum of neutrons and protons present.

The number of neutrons (N) present in an atomic nucleus is given by this expression:

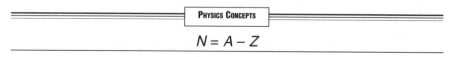

PHYSICS CONCEPTS

$$N = A - Z$$

Using this representation, we can write the symbols for the proton, neutron, and electron, respectively, as follows:

$$^1_1 H \text{ (proton)} \quad ^1_0 n \text{ (neutron)} \quad ^0_{-1} e \text{ (electron)}$$

The symbol for the proton is a result of the fact that a proton is the nucleus of the simplest hydrogen atom. While the electron is not normally considered a nuclear particle, there are occasions when it is produced in the nucleus.

PROBLEM

Identify the nucleons present in the atomic nucleus whose symbol is $_{10}^{22}$Ne.

SOLUTION

The atomic number of neon (Ne) is 10; therefore, the nucleus contains 10 protons.

The mass number of this nucleus is 22; therefore, the nucleus contains 22 protons *and* neutrons.

The number of neutrons is found by subtracting the atomic number from the mass number:

$$22 \text{ protons and neutrons} - 10 \text{ protons} = 12 \text{ neutrons.}$$

PROBLEM

Calculate the number of protons and neutrons in these nuclei: (a) $_1^2$H and (b) $_1^3$H.

SOLUTION

(a) $_1^2$H contains 1 proton and 1 neutron.

(b) $_1^3$H contains 1 proton and 2 neutrons.

16.4   ISOTOPES

All the nuclei in a sample of a given element contain the same number of protons. They may, however, contain different numbers of neutrons. Nuclei that have the same atomic number but have different mass numbers are called **isotopes**. For example, $_1^1$H, $_1^2$H, and $_1^3$H are all isotopes of the element hydrogen. Sometimes an isotope is written without its atomic number (^{22}Ne) or with the name of its element and its mass number (hydrogen–2).

16.5   NUCLEAR MASSES

Since nuclear particles have very small masses, they are usually measured in terms of the *atomic mass unit* (u) rather than the kilogram. The proton and the neutron each have an approximate mass of 1 atomic mass unit, although the neutron is slightly more massive than the proton. The basis of the nuclear-mass scale is the isotope carbon-12, and an *atom* of this isotope is assigned an exact mass of 12 atomic mass units. One atomic mass unit is approximately equal to 1.66×10^{-27} kilogram.

Equivalents of Nuclear Masses

Nuclear masses may also be expressed in terms of their *energy equivalents.* Using Einstein's famous mass-energy relationship ($E = mc^2$), it can be shown (see the next problem) that 1 atomic mass unit of mass is equivalent to 931 mega-electron-volts of energy.

PROBLEM
Calculate the energy equivalent (in MeV) of 1 atomic mass unit of mass.

SOLUTION
We need the following relationships in order to solve the problem:

$$1 \text{ eV} = 1.60 \times 10^{-19} \text{ J}$$

$$1 \text{ MeV} = 10^6 \text{ eV}$$

$$1 \text{ u} = 1.66 \times 10^{-27} \text{ kg}$$

$$c = 3.00 \times 10^8 \text{ m/s}$$

First, we calculate the energy equivalent of 1 u in *joules,* using $E = mc^2$:

$$E = (1.66 \times 10^{-27} \text{ kg})(3.00 \times 10^8 \text{ m/s})^2 = 1.49 \times 10^{-10} \text{ J}$$

Next, we convert this number to electron-volts and then to mega-electron-volts:

$$1.49 \times 10^{-10} \text{ J}\left(\frac{1 \text{ eV}}{1.60 \times 10^{-19} \text{ J}}\right) = 9.31 \times 10^8 \text{ eV}$$

$$9.31 \times 10^8 \text{ eV}\left(\frac{1 \text{ MeV}}{10^6 \text{ eV}}\right) = 931 \text{ MeV}$$

The following table lists the masses and energy equivalents of some nuclear particles:

Particle	Mass (u)	Energy Equivalent (MeV)
Electron	0.0005486	0.5110
Proton	1.007276	938.3
Hydrogen (1 atom)	1.007825	938.8
Neutron	1.008665	939.6

Mass Defect and Binding Energy

A nucleus such as $^{56}_{26}$Fe contains 26 (positive) protons concentrated in an extremely small space. We might suppose that the repulsion of the positive charges ought to tear the nucleus apart. However, this nucleus is quite stable. To explain the reason, we need to compare two quantities: the mass of the nucleus itself and the total mass of its nucleons.

PROBLEM
Compare the mass of a $^{56}_{26}$Fe nucleus (mass = 55.9206 u) with the total mass of its nucleons.

SOLUTION
Using the preceding table and the fact that this nucleus contains 26 protons and 30 neutrons (why?), we can calculate the total mass of the nucleons:

Mass of 26 protons	= (26)(1.007276 u)	= 26.1892 u
Mass of 30 neutrons	= (30)(1.008665 u)	= 30.2600 u
Total mass of nucleons		= 56.4492 u

Subtracting the two values (56.4992 u – 55.9206 u), we see that the mass of the nucleus is 0.5286 u *less* than the mass of its nucleons!

The mass difference that we just calculated is known as the **mass defect** of the nucleus. Its energy equivalent is approximately 492 mega-electron-volts (How can we calculate this quantity?) and is called the *binding energy* of the nucleus. The **binding energy** of a nucleus is the amount of energy that must be added in order to separate the nucleus into its component nucleons.

16.6 AVERAGE BINDING ENERGY PER NUCLEON

One way of estimating the stability of a nucleus is by referring to a quantity known as the average *binding energy per nucleon*. It is calculated by dividing the total binding energy of the nucleus by the number of nucleons present (i.e., the mass number). In the problem in Section 16.5, the binding energy of 492 mega-electron-volts is divided by 56 nucleons to yield 8.79 mega-electron-volts per nucleon. In general, the larger the binding energy per nucleon, the more stable is the nucleus.

The graph below illustrates how the binding energy per nucleon varies with the number of nucleons in a nucleus. The nuclei located toward the center of the graph (e.g., ^{56}Fe) have larger values than the nuclei located at either end (e.g., ^{2}H and ^{238}U). In Section 16.10, we will see how some of the less stable nuclei can be used in the production of energy.

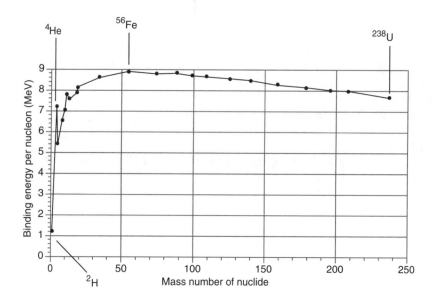

16.7 NUCLEAR FORCES

The stability of a nucleus is tied to the existence of two *nuclear forces*. These forces, called the *strong* and *weak interactions*, are much more powerful at the very small distances present within the nucleus than are gravitational or electromagnetic forces. At larger distances, however, the strong and weak interactions lose their effectiveness, and for this reason they are called *short-range forces*.

16.8 NUCLEAR REACTIONS

Detection Devices

In order to study nuclear reactions and their products, scientists have invented a number of *detection devices*. The principle underlying all these devices is that the particles produced in nuclear reactions leave their "fingerprints" as they pass through the detector.

In a *Geiger counter*, an electric current is produced that yields an audible click or activates an electronic counter when a particle is encountered; in a *scintillation counter*, the particles produce flashes of light. In a *cloud chamber*, a supercooled gas (i.e., a gas below the boiling point) is condensed around the moving particles, leaving tracks that look like thin streaks of cloudlike material; a *bubble chamber* uses the opposite effect: a superheated liquid (i.e., a liquid above the boiling point) is vaporized around the moving particles, leaving streaks of gas bubbles. However, devices as simple as photographic plates have been used to detect the products of nuclear reactions.

Nuclear Equations

A nuclear reaction is a change that occurs within or among atomic nuclei and is represented by a *nuclear equation*, such as the following:

$$^{15}_{7}N + ^{1}_{1}H \rightarrow ^{12}_{6}C + ^{4}_{2}He$$

If we examine this equation carefully, we note that the sum of the atomic numbers on the left side (7 + 1) equals the sum of the atomic numbers on the right side (6 + 2). This equality demonstrates the fact that electric charge must be conserved in a nuclear reaction. Similarly, the sum of the mass numbers on the left side of the equation (15 + 1) equals the sum of the mass numbers on the right side (12 + 4).

This nuclear equation is considered to be *balanced* because both charge and mass number are conserved. All of the nuclear equations we write will be balanced equations.

Radioactive Decay

Less stable nuclei may break down spontaneously by a number of processes known collectively as *radioactive decay*. When a nucleus undergoes radioactive decay, it does so in order to become more stable. One aspect of the strong interaction mentioned in Section 16.7 is that the stability of a nucleus depends on the relative numbers of neutrons and protons present.

The graph shows how the *neutron-proton* (*N/P*) ratio varies in *stable* nuclei.

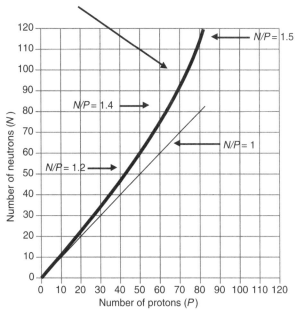

As the number of protons increases, the N/P ratio rises from 1 to 1.5. The larger number of neutrons reduces the repulsion among the positively charged protons. The graph does not continue beyond 83 protons because there are no stable isotopes with atomic numbers greater than 83.

Radioactive decay is an attempt to "correct" the ratio of neutrons to protons. However, stability may not occur immediately; a series of decay reactions may be required before a stable nucleus is finally produced.

ALPHA DECAY

The following nuclear reaction:

$$^{238}_{92}U \rightarrow {}^{234}_{90}Th + {}^{4}_{2}He$$

is an example of *alpha decay*. The uranium-238 nucleus (the *parent nucleus*) breaks down to produce a thorium-234 nucleus (the *daughter nucleus*) and a helium-4 nucleus, also known as an *alpha particle*. A symbol for the alpha particle is α.

Whenever alpha decay occurs, the atomic number of the daughter nucleus (as compared with its parent) is *decreased* by 2 and its mass number is *decreased* by 4. Many heavier radioactive nuclei, especially those with atomic numbers greater than 83, undergo alpha decay as a way of reducing the number of protons and neutrons present.

PROBLEM
The nuclide $^{222}_{86}Rn$ undergoes alpha decay and forms an isotope of polonium (Po). Write the nuclear equation for this process.

SOLUTION
Our problem is to find the atomic and mass numbers of the Po nuclide. We write the alpha decay as follows:

$$^{222}_{86}Rn \rightarrow {}^{A}_{Z}Po + {}^{4}_{2}He$$

Since the atomic and mass numbers must be equal on both sides of the equation, the atomic number of polonium must be 84 and the mass number of this nuclide must be 218. Therefore, we can now write the complete equation:

$$^{222}_{86}Rn \rightarrow {}^{218}_{84}Po + {}^{4}_{2}He$$

BETA (–) DECAY

Certain nuclei undergo radioactive decay and produce an *electron,* also known as a *beta (–) particle*, in the reaction. A symbol for the beta (–) particle is β^-.

488

Beta decay is governed by the weak interaction mentioned in Section 16.7. The following equation illustrates the process of beta (−) decay:

$$^{214}_{82}\text{Pb} \rightarrow \, ^{214}_{83}\text{Bi} + \, ^{0}_{-1}\text{e}$$

In beta (−) decay, the atomic number of the daughter is increased by 1 while its mass number remains unchanged. Beta (−) decay occurs in nuclei whose *N/P* ratios lie *above* the band of stability illustrated in the graph on page 487.

Actually, another subatomic particle, called an *antineutrino* is also produced in beta (−) decay. Even though charge and mass number are conserved, an antineutrino must be produced in order to conserve momentum and energy as well.

The antineutrino is an example of an *antiparticle*. Every subatomic particle is associated with its own unique antiparticle. Particles and their antiparticles have the same mass, but their electric charges are *opposite* in sign. If a particle and its antiparticle are brought into contact, all of the mass is transformed entirely into electromagnetic energy in the form of two gamma-ray photons.

PROBLEM

The nuclide $^{14}_{6}\text{C}$ undergoes beta (−) decay and forms an isotope of nitrogen (N). Write the nuclear equation for this process.

SOLUTION

Our problem is to find the atomic and mass numbers of the N nuclide. We will write the beta (−) decay as follows:

$$^{14}_{6}\text{C} \rightarrow \, ^{A}_{Z}\text{N} + \, ^{0}_{-1}\text{e}$$

Since the atomic and mass numbers must be equal on both sides of the equation, the atomic number of nitrogen must be 7 and the mass number of this nuclide must be 14. Therefore, we can now write the complete equation:

$$^{14}_{6}\text{C} \rightarrow \, ^{14}_{7}\text{N} + \, ^{0}_{-1}\text{e}$$

Note that the *N/P* ratio of the parent nucleus (^{14}C) is 1.33 (8/6), while the *N/P* ratio of the daughter nucleus (^{14}N) is 1.00 (7/7). Therefore, beta (−) decay is a process that *decreases* the *N/P* ratio.

POSITRON DECAY

The name *positron* is a combination of the word parts positive electron. The positron is the *antiparticle* of the electron. The symbol for the positron is $^{0}_{+1}\text{e}$; another symbol for the positron is β+.

The following equation is an example of positron decay:

$$_{10}^{19}\text{NE} \rightarrow {}_{9}^{19}\text{F} + {}_{+1}^{0}\text{e}$$

In positron decay, the atomic number of the daughter nucleus is *decreased* by 1 while its mass number remains *unchanged*. An additional particle (called the *neutrino* and symbolized as v) is also produced in positron decay in order to conserve energy and momentum.

If we examine the *N/P* ratios of the parent and daughter nuclei in the equation given above, we see that positron decay *increases* the *N/P* ratio.

ELECTRON CAPTURE

This type of reaction occurs when a nucleus captures one of the inner electrons orbiting the atom. In this reaction, a neutrino is also produced. The following equation is an example of electron capture:

$$_{4}^{7}\text{Be} + {}_{-1}^{0}\text{e} \rightarrow {}_{3}^{7}\text{Li}$$

Electron capture also serves to *increase* the *N/P* ratio of nuclei.

GAMMA DECAY

Like the electrons in an atom, the nucleus contains energy levels. Occasionally, a nucleus will enter an excited state, known as a *metastable* state. We symbolize a metastable nucleus by adding an *m* to its mass number (e.g., 99m). Eventually, the nucleus will return to its normal state and then a high-energy gamma-ray photon will be emitted.

The symbol for the gamma-ray photon is γ. The following equation is an example of gamma decay:

$$_{43}^{99m}\text{Tc} \rightarrow {}_{43}^{99}\text{Tc} + \gamma$$

URANIUM ($_{92}^{238}$U) DECAY SERIES

As noted above, radioactive decay occurs in order to produce stable nuclei. Sometimes, this process requires more than one step. The unstable nuclide $_{92}^{238}$U decays to the stable nuclide $_{82}^{206}$Pb in a series of steps involving both alpha and beta decay.

The graph indicates how the decay of $_{92}^{238}$U occurs. Atomic numbers are plotted along the *x*-axis, and mass numbers along the *y*-axis. Each diagonal line represents an alpha decay; each horizontal line, a beta decay. The daughter nucleus at each step (atomic number and mass number) is indicated by a dot (•).

490

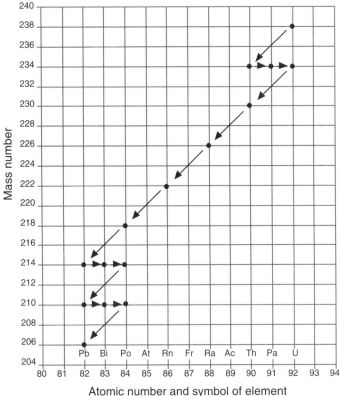

PROBLEM

(a) How many alpha and beta decays are there in the $^{238}_{92}$U decay series?

(b) How does the nuclide $^{214}_{82}$Pb decay?

SOLUTION

(a) Referring to the graph, we count eight diagonal lines and six horizontal lines. Therefore, the series has eight alpha decays and six beta decays.

(b) We locate the dot corresponding to $^{214}_{82}$Pb on the graph and we follow the (horizontal) arrow to the next dot, which corresponds to $^{214}_{83}$Bi. Therefore, $^{214}_{82}$Pb decays to $^{214}_{83}$Bi by beta decay.

16.9 RADIOACTIVE DECAY AND HALF-LIFE

The rate of decay of a radioactive isotope is measured in terms of a quantity called the **half-life**. The half-life is defined as the time in which a sample of the isotope decays to one-half of its original mass.

For example, the half-life of iodine-131 is 8 days. If we have a 16-milligram sample of this isotope, its decay over a period of time occurs as follows:

$$16 \text{ mg} \xrightarrow{8\,d} 8 \text{ mg} \xrightarrow{8\,d} 4 \text{ mg} \xrightarrow{8\,d} 2 \text{ mg} \ldots$$

After 24 days of decay, 2 milligrams of iodine-131 remains unchanged. The other 14 milligrams has not simply disappeared; it has been transformed into other substances.

The half-life is constant: it cannot be altered by changes in pressure, temperature or chemical activity. Moreover, the isotope will *never* decay to 0 milligram; it will simply come closer and closer to this value.

The graph shows how iodine-131 decays over a period of time.

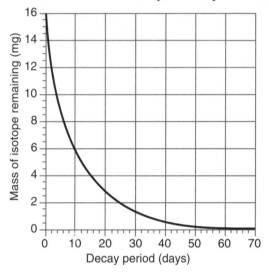

PROBLEM
According to the graph shown above, approximately how much of the isotope iodine-131 remains after 20 days of decay?

SOLUTION
By inspecting the graph, we see that approximately 3 mg of the isotope remains.

A simple mathematical relationship allows us to calculate the mass of an isotope remaining after a given number of half-lives have elapsed:

PHYSICS CONCEPTS

$$m_f = \frac{m_i}{2^n}$$

In this equation, m_i is the initial mass of the isotope, n is the number of half-life periods of decay, and m_f is the mass of the isotope that remains when the decay period has ended.

PROBLEM

An 800-milligram sample of iodine-131 (half-life = 8 d) decays for 40 days. How much isotope remains at the end of this time period?

SOLUTION

To calculate the number of half-life periods, we divide the decay period (40 d) by the half-life (8 d) to obtain five half-life periods. This is the value of the exponent n in the equation given above. The initial mass (m_i) is 800 mg. Therefore,

$$m_f = \frac{m_i}{2^n}$$

$$= \frac{800 \text{ mg}}{2^5} = \frac{800 \text{ mg}}{32} = 25 \text{ mg}$$

16.10 INDUCED NUCLEAR REACTIONS

All of the nuclear reactions we have studied so far have been *natural* processes. We now turn our attention to nuclear changes that have been produced artificially or *induced*. To induce a nuclear reaction, a target nucleus is bombarded with a nuclear particle.

Particle Accelerators

A *particle accelerator* is a device that uses electric and magnetic fields to provide a bombarding nuclear particle with sufficient kinetic energy to induce the desired nuclear reaction. As an analogy, consider a bullet fired at a wall at a speed of 10 miles per hour. At this slow speed, the kinetic energy of the bullet would have hardly any effect on the wall. If the bullet were fired at 600 miles per hour, however, its effect on the wall would be devestating!

Examples of modern particle accelerators include the *Van de Graaff accelerator*, the *linear accelerator*, the *cyclotron*, the *synchrotron,* and the *large electron-positron (LEP) collider*. These devices can supply bombarding particles with kinetic energies ranging from 10^3 to 10^{12} electron-volts.

Artificial Transmutation

The first induced nuclear reactions used alpha particles (because of their large masses) as the bombarding particles. In 1919, Rutherford bombarded nitrogen-14 nuclei with alpha particles and noted that protons were emitted. In this reaction:

$$^{14}_{7}\text{N} + ^{4}_{2}\text{He} \rightarrow ^{1}_{1}\text{H} + ^{17}_{8}\text{O}$$

nitrogen-14 was artificially changed or *transmuted* into oxygen-17. In 1932, English physicist Sir James Chadwick bombarded beryllium-9 and identified a stream of uncharged particles that we now call *neutrons*. The reaction is shown below:

$$^{9}_{4}\text{Be} + ^{4}_{2}\text{He} \rightarrow ^{1}_{0}\text{n} + ^{12}_{6}\text{C}$$

In 1934, French physicists Frédéric Joliot-Curie and Irène Joliot-Curie bombarded aluminum-27 and produced the first artificially radioactive isotope, phosphorus-30:

$$^{27}_{13}\text{Al} + ^{4}_{2}\text{He} \rightarrow ^{1}_{0}\text{n} + ^{30}_{15}\text{P}$$

The phosphorus-30 undergoes positron decay (see pages 489–490) and forms silicon-30.

Nuclear Fission and Fusion

In the bombardment reactions studied so far, only minor changes to the target nuclei occurred. We will now discuss processes that cause major nuclear changes.

Referring to the graph of binding energy per nucleon on page 486, we see that the nuclides located in the left and right portions of the graph have lower values than the nuclides located toward the center. The implication is that very light and very heavy nuclei ought to be less stable than nuclei having moderate masses. We might guess that the heavy elements could split, and the lighter elements could join, to form nuclei whose masses would lie toward the middle of the binding-energy curve.

NUCLEAR FISSION

In the 1930s, Enrico Fermi, an Italian-American physicist, suggested that neutrons be used in bombardment reactions because they are uncharged particles and therefore would not be repelled by the target nuclei. In 1938, German physicists Otto Hahn and Fritz Strassman bombarded uranium atoms with neutrons and discovered that some of the uranium atoms split into roughly two equal fragments. This reaction, known as *nuclear fission*, also produced more neutrons and a large amount of energy.

One Type of Fission Reaction

There are many types of fission reactions; one type is shown below:

$$^{235}_{92}\text{U} + ^{1}_{0}\text{n} \rightarrow \left[^{236}_{92}\text{U} \right] \rightarrow ^{141}_{56}\text{Ba} + ^{92}_{36}\text{Kr} + 3^{1}_{0}\text{n} + \text{Energy}$$

The actual amount of energy released is approximately 200 mega-electron-volts per fission. This is an enormous quantity for a nuclear reaction, and it is produced because the mass of the reactants (uranium-235 and the neutron) is considerably greater than the mass of the products (barium-141, krypton-92, and the three neutrons).

Fission Reactors

A number of physicists recognized that the neutrons released in nuclear fission could be used, in turn, to cause other uranium-235 nuclei to fission, producing a *chain reaction*. If this chain could be controlled, the fission reactions could provide a continuous source of energy. This controlled chain reaction is the principle behind the *fission reactor*. (Uncontrolled chain reactions are produced in fission weapons such as the atomic bomb.)

Fission reactors are used in the production of electrical energy, as research tools and as a means of producing other radioactive isotopes. The diagram illustrates a typical fission reactor used to produce electricity.

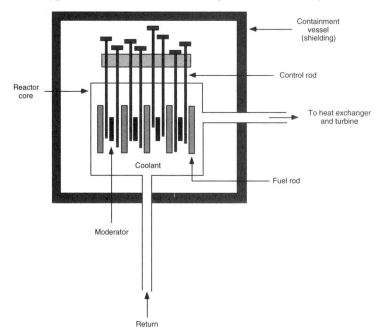

The primary system of a fission reactor has the following components:

- The *containment vessel* (concrete and steel) provides shielding for the reactor.
- The *fuel rods*, located in the *core*, serve as sources of energy. Uranium-233 and uranium-235 are used as fuels.
- The *moderator*, also located in the core, slows the neutrons so that they will

be absorbed by the fuel nuclei. These slow neutrons have kinetic energies that are close to the kinetic energies of air molecules at room temperature. For this reason, they are known as *thermal neutrons*. Moderators are usually composed of water (containing either hydrogen-1 or hydrogen-2), graphite, or beryllium.

- The *control rods* in the core regulate the rate of fission by absorbing neutrons. Control rods are usually made of cadmium or boron.
- The *coolant* (water or liquid sodium) removes thermal energy from the core.
- The *heat exchanger* receives the thermal energy and produces steam for the generation of electrical energy by the secondary system (turbine) of the reactor.

The most common isotope of uranium, uranium-238, comprises about 99% of the naturally occurring element but does *not* undergo fission. The isotope uranium-235 does undergo fission but it comprises less than 1% of the naturally occurring element. For this reason, it is necessary to *enrich* uranium in order to use it as a fissionable fuel.

When uranium-238 absorbs a neutron, it is converted into uranium-239, which then decays in two steps to form plutonium-239:

$$^{238}_{92}U + ^{1}_{0}n \rightarrow ^{239}_{92}U$$

$$^{239}_{92}U \rightarrow ^{239}_{93}Np + ^{0}_{-1}e$$

$$^{239}_{93}Np \rightarrow ^{239}_{94}Pu + ^{0}_{-1}e$$

The $^{239}_{94}Pu$ isotope of plutonium *does* fission, and it has been suggested that *breeder reactors* be constructed. These reactors would be able to produce their own fuel. No breeder reactors have been operated in the United States, however, because of the serious health and environmental hazards attributable to plutonium and also because of the possibility that the plutonium might be used to produce nuclear weapons.

A number of problems are associated with any fission reactor. For example, the heat energy produced by the reactor contributes to *thermal pollution*. There is also the serious problem of *radioactive waste disposal*. Solid and liquid wastes are placed in corrosion-resistant containers and stored in isolated underground areas. Wastes with low levels of radioactivity may be diluted until they are considered harmless and then released into the environment. Gaseous wastes, such as krypton-85, nitrogen-16, and radon-222, are stored and allowed to decay until it is considered safe to release them into the atmosphere.

NUCLEAR FUSION

The joining of light nuclei to form heavier, more stable nuclei is known as *nuclear fusion* and is the process by which stars, including our own Sun, produce their energy. In stars, fusion is a series of reactions that depend on the temperature of the particular star. The *net* fusion reaction in our Sun is as follows:

$$4{}_1^1H \rightarrow {}_2^4He + 2{}_{+1}^0e + 2\gamma + 2\nu + 26.7 \text{ MeV}$$

where γ represents a gamma photon and ν represents a neutrino.

For a fusion reaction to occur, very high temperatures are needed to give the positively charged nuclei the kinetic energy they need to overcome their mutual repulsion. For this reason, all fusion devices are referred to as *thermonuclear devices*.

For a given mass of fuel, nuclear fusion yields more energy than nuclear fission. For this reason, work is being done to develop fusion reactors as a means of producing power. The three fusion reactions most likely to succeed in a reactor are as follows:

$$ {}_1^2H + {}_1^2H \rightarrow {}_1^3H + {}_1^1H + 4.03 \text{ MeV} $$

$$ {}_1^2H + {}_1^2H \rightarrow {}_2^3He + {}_0^1n + 3.27 \text{ MeV} $$

$$ {}_1^2H + {}_1^3H \rightarrow {}_2^4He + {}_0^1n + 17.59 \text{ MeV} $$

At this time, no successful fusion reactor has been constructed. The problems to be overcome include the production of a high *ignition* temperature, the packing of nuclei into a space small enough to allow a sufficient number of collisions, the control of the fusion reaction once it has begun, and the development of materials that can withstand the high operating temperatures and radiation levels of the reactor.

16.11 FUNDAMENTAL PARTICLES AND INTERACTIONS

The model of nuclear structure continues to evolve. Over the years, as physicists probed the nucleus, a host of particles were discovered whose functions were largely unknown. In an attempt to explain the existence of these particles, a number of theories were developed.

One of the most successful models is known as the *standard model*, which assumes that four fundamental interactions exist in the universe: electromagnetic, weak, strong, and gravitational. The standard model describes the behavior of the first three of these interactions. It also assumes that there are particles more fundamental than protons and neutrons, known as *quarks*. There are six varieties of quarks and six varieties of antimatter quarks, whimsically named *up, down, charm, strange, top,* and *bottom*. The property with which the names of quarks are associated is known as *flavor*. (Who says physicists don't have senses of humor?) The electric charges of the six quarks are all less than one elementary charge, as shown in the table.

Flavor	Charge
Up (*u*)	+2/3
Down (*d*)	−1/3
Charm (*c*)	+2/3
Strange (*s*)	−1/3
Top (*t*)	+2/3
Bottom (*b*)	−1/3

According to the standard model scheme, the proton has the structure *uud* and the neutron has the structure *udd*.

The binding energies of the quarks in protons and neutrons are so high that these quarks cannot be isolated as separate particles. Therefore, their existence has been demonstrated only by indirect means.

Other nuclear particles have also been classified according to the standard model scheme. Whether these particles represent the ultimate structure of matter or whether there are even smaller subunits continues to be the subject of research.

QUESTIONS

1. Neutral atoms always have equal numbers of
 (1) protons and neutrons (3) protons and electrons
 (2) electrons and neutrons (4) protons and positrons

2. What is the relationship between the atomic number Z, the mass number A, and the number of neutrons N in a nucleus?
 (1) $A = Z + N$ (3) $A = N/Z$
 (2) $A = Z - N$ (4) $A = NZ$

3. The ratio of the magnitude of charge on an electron to the magnitude of charge on a proton is
 (1) $1 : 2$ (3) $1 : 6.25 \times 10^{18}$
 (2) $1 : 1$ (4) $1 : 1840$

4. What is the mass number of an atom with 9 protons, 11 neutrons, and 9 electrons?
 (1) 9 (2) 18 (3) 20 (4) 29

5. An atom consists of 9 protons, 9 electrons, and 10 neutrons. The number of nucleons in this atom is
 (1) 0 (2) 9 (3) 19 (4) 28

6. A neutral atom could be composed of
(1) 4 electrons, 5 protons, 6 neutrons
(2) 5 electrons, 5 protons, 6 neutrons
(3) 6 electrons, 3 protons, 6 neutrons
(4) 0 electrons, 5 protons, 5 neutrons

7. Isotopes of the same element have the same number of
(1) neutrons and protons, only
(2) neutrons and electrons, only
(3) protons and electrons, only
(4) electrons, protons, and neutrons

8. If the number of neutrons in an atom increases, the atomic number of the atom
(1) decreases   (2) increases   (3) remains the same

9. As the mass number of an isotope increases, its atomic number
(1) decreases   (2) increases   (3) remains the same

10. As the number of protons in a nucleus increases, its atomic number
(1) decreases   (2) increases   (3) remains the same

11. A lithium nucleus contains 3 protons and 4 neutrons. What is the atomic number of the nucleus?
(1) 1   (2) 7   (3) 3   (4) 4

12. What is the number of neutrons in the nucleus of $^{222}_{86}$Rn?
(1) 86   (2) 136   (3) 222   (4) 308

13. A neutral atom has 24 neutrons and 20 protons. The number of electrons in the atom is
(1) 24   (2) 20   (3) 44   (4) 4

14. Which atom is an isotope of $^{238}_{92}$U?
(1) $^{238}_{91}$X   (2) $^{235}_{92}$X   (3) $^{238}_{93}$X   (4) $^{235}_{93}$X

15. What type of particle has a charge of 1.6×10^{-19} coulomb and a rest mass of 1.67×10^{-27} kilogram?
(1) a proton   (3) a neutron
(2) an electron   (4) an alpha particle

16. An atomic mass unit (u) is approximately equal to the mass of
(1) an alpha particle   (3) a photon
(2) an electron   (4) a proton

17. The mass of a nucleus is less than the total mass of its nucleons. This fact indicates that some of the mass has been converted to
(1) radioactivity
(3) binding energy
(2) photoelectric effect
(4) thermal energy

18. What is the energy equivalent of a mass of 1 kilogram?
(1) 9×10^{16} J
(2) 9×10^{13} J
(3) 9×10^{10} J
(4) 9×10^{7} J

19. How much energy would be produced if 1.0×10^{-3} kilogram of matter was entirely converted to energy?
(1) 9.0×10^{13} J
(3) 9.0×10^{16} J
(2) 3.0×10^{16} J
(4) 3.0×10^{19} J

20. As the binding energy of a nucleus increases, the energy required to separate the nucleus into nucleons
(1) decreases
(2) increases
(3) remains the same

21. In the reaction $Q + {}^{2}_{1}H \rightarrow {}^{1}_{1}H + {}^{1}_{0}n$, Q represents the energy needed to separate the neutron from the deuterium nucleus.

Mass of deuterium (${}^{2}_{1}H$) = 2.0141 u

Mass of hydrogen (${}^{1}_{1}H$) = 1.0078 u

Mass of neutron (${}^{1}_{0}n$) = 1.0087 u

What is the value of the mass equivalent of Q?
(1) 0.0009 u
(2) 0.0024 u
(3) 2.0165 u
(4) 4.0306 u

22. When ${}^{238}_{92}U$ decays to ${}^{228}_{90}Th$ plus ${}^{4}_{2}He$, how much energy is released? [Refer to the data below for mass values.]

${}^{238}_{92}U = 232.0372$ u

${}^{228}_{90}Th = 228.0287$ u

${}^{4}_{2}He = 4.0026$ u

(1) 0 MeV
(2) 5.5 MeV
(3) 58 MeV
(4) 230 MeV

23. Which graph best represents the relationship between energy and mass in the mass-energy equation?

E | E | E | E
m (1) | m (2) | m (3) | m (4)

24. When 4 hydrogen nuclei combine into a helium nucleus, the mass difference is 0.02871 atomic mass units. The radiated energy is approximately

(1) 3.08×10^{-10} eV
(3) 3.25×10^8 eV
(2) 2.67×10^{14} eV
(4) 2.67×10^7 eV

25. For a particular nuclear decay, the mass of the products is 0.01 atomic mass unit less than the original nucleus. The total energy released during this decay is

(1) 1.07×10^{-4} Mev
(3) 9.31 Mev
(2) 1.07 Mev
(4) 9.31×10^6 Mev

26. If the mass of one proton is totally converted into energy, it will yield a total energy of

(1) 5.1×10^{-19} J
(3) 9.3×10^8 J
(2) 1.5×10^{-10} J
(4) 9.0×10^{16} J

27. The nuclear force that binds nucleons together in the atom is relatively

(1) strong and of long range
(3) weak and of short range
(2) strong and of short range
(4) weak and of long range

28. What is the purpose of a cloud chamber?

(1) to accelerate particles
(3) to fuse particles
(2) to split particles
(4) to detect particles

29. Which device could be used to determine whether a substance is radioactive?

(1) a Van de Graaff generator
(3) a cyclotron
(2) a Geiger counter
(4) a linear accelerator

30. Which would produce a track in a cloud chamber?

(1) a neutron
(3) a gamma photon
(2) an electron
(4) a hydrogen atom

31. The synchrotron and cyclotron are examples of

(1) high-energy particles
(3) beta-emitting nuclei
(2) radiation detectors
(4) particle accelerators

32. Which particle *cannot* be accelerated in a particle accelerator?

(1) a neutron
(3) an electron
(2) a proton
(4) an alpha particle

33. Which radioactive decay products may be deflected by a magnetic field?
    (1) alpha particles, only
    (2) gamma rays, only
    (3) gamma rays and beta particles
    (4) alpha particles and beta particles

34. Which group of particles can *all* be accelerated by a cyclotron?
    (1) alpha particles, electrons, and neutrons
    (2) electrons, neutrons, and protons
    (3) neutrons, protons, and alpha particles
    (4) protons, alpha particles, and electrons

35. In the reaction $^9_4Be + ^4_2He \rightarrow ^{12}_6C + X$, particle X is
    (1) an electron   (2) a neutron   (3) a positron   (4) a proton

36. When a gamma ray is emitted by a nucleus, the atomic number of the nucleus
    (1) decreases   (2) increases   (3) remains the same

37. When $^{238}_{92}U$ emits an alpha particle, the resulting nucleus will be
    (1) $^{230}_{90}Th$ (2) $^{234}_{90}Th$ (3) $^{234}_{91}Pa$ (4) $^{234}_{92}U$

38. After the decay of a radioactive sample, helium gas was one of the products. This fact suggests that the nuclear disintegrations consisted of
    (1) alpha decays   (3) gamma emissions
    (2) positron emissions   (4) beta decays

39. If, after beta decay, a nucleus is $^{234}_{91}Pa$, what was the nucleus just before the release of the beta particle?
    (1) $^{234}_{90}Th$ (2) $^{235}_{91}Pa$ (3) $^{238}_{91}Pa$ (4) $^{234}_{92}U$

40. Given the equation $^{27}_{13}Al + ^4_2He \rightarrow ^{30}_{15}P + X$. The correct symbol for X is
    (1) $^0_{+1}e$ (2) $^0_{-1}e$ (3) ^4_2He (4) ^1_0n

41. When lead $^{214}_{82}Pb$ emits a beta (–) particle, the resultant nucleus will be
    (1) $^{214}_{81}Tl$ (2) $^{213}_{82}Pb$ (3) $^{214}_{83}Bi$ (4) $^{214}_{84}Po$

42. In the equation $^{239}_{92}U \rightarrow ^{239}_{93}Np + X$, particle X is
    (1) a proton   (3) an alpha particle
    (2) a neutron   (4) a beta (–) particle

43. When a nucleus captures an electron, the atomic number of the nucleus
    (1) decreases   (2) increases   (3) remains the same

44. When a radioactive nucleus emits a beta particle, the mass number of the nucleus will
(1) decrease   (2) increase   (3) remain the same

45. Which atom has a stable nucleus?
(1) ^{234}Th   (2) ^{218}Po   (3) ^{210}Pb   (4) ^{206}Pb

46. The diagram below represents an inverted test tube over a sample of a radioactive material. Helium has collected in the test tube.

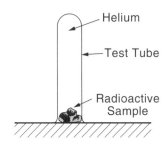

The presence of helium indicates that the sample is most probably undergoing the process of
(1) alpha decay   (3) neutron decay
(2) beta decay   (4) gamma emission

47. What kind of nuclear reaction is represented by the equation below?

$$^{14}_{7}N + ^{4}_{2}He \rightarrow ^{17}_{8}O + ^{1}_{1}H$$

(1) alpha decay   (3) artificial transmutation
(2) beta decay   (4) nuclear fission

48. In the reaction $^{24}_{11}Na \rightarrow ^{24}_{12}Mg + X$, what does X represent?
(1) an alpha particle   (3) a neutron
(2) a beta (–) particle   (4) a positron

Base your answers to questions 49 and 50 on the information below.

A pure sample of radon (Rn) gas is sealed inside a glass tube. The half-life of radon is 4 days.

49. If the pressure inside the tube were decreased to one-half, the half-life of the radon would be
(1) unchanged   (3) decreased to one-half
(2) doubled   (4) decreased to one-quarter

50. Twenty days after the radon was sealed inside the tube, the fraction of radon gas remaining would be

(1) $\dfrac{1}{2}$ (2) $\dfrac{1}{8}$ (3) $\dfrac{1}{16}$ (4) $\dfrac{1}{32}$

51. Which isotope is *not* in the uranium disintegration series?

(1) $^{238}_{92}U$ (2) $^{234}_{92}U$ (3) $^{235}_{90}Pa$ (4) $^{226}_{88}Ra$

52. In which reaction does X represent a beta particle?

(1) $^{234}_{92}U \rightarrow ^{230}_{90}Th + X$ (3) $^{226}_{88}Ra \rightarrow ^{222}_{86}Rn + X$

(2) $^{214}_{84}Pa \rightarrow ^{210}_{82}Pb + X$ (4) $^{214}_{82}Pb \rightarrow ^{214}_{83}Bi + X$

Base your answers to questions 53 through 54 on the information below. The following equations represent a two-stage nuclear reaction.

$$^{27}_{13}Al + ^{4}_{2}He \rightarrow ^{30}_{15}P + X$$

$$^{30}_{15}P \rightarrow ^{30}_{14}Si + Y + energy$$

53. Which nucleus in the two equations has the greatest number of neutrons?

(1) $^{27}_{13}Al$ (2) $^{4}_{2}He$ (3) $^{30}_{15}P$ (4) $^{30}_{14}Si$

54. What is particle X?

(1) a positron   (2) an electron   (3) a proton   (4) a neutron

55. Particle Y represents

(1) $^{1}_{0}n$ (2) $^{1}_{1}H$ (3) $^{0}_{+1}e$ (4) $^{0}_{-1}e$

56. If the temperature of a radioactive isotope is doubled, its half-life will be

(1) doubled   (2) quadrupled   (3) halved   (4) the same

57. The half-life of an isotope is 14 days. How many days will it take 8 grams of this isotope to decay to 1 gram?

(1) 14   (2) 21   (3) 28   (4) 42

58. As the original mass of a radioactive substance decreases, its half-life

(1) decreases   (2) increases   (3) remains the same

Base your answers to questions 59 through 63 on the graph below, which represents the disintegration of a sample of a radioactive element. At time $t = 0$ the sample has a mass of 4.0 kilograms.

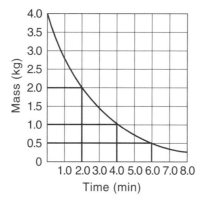

59. What mass of the material remains at 4.0 minutes?
    (1) 1 kg   (2) 2 kg   (3) 0 kg   (4) 4 kg

60. What is the half-life of the isotope?
    (1) 1.0 min   (2) 2.0 min   (3) 3.0 min   (4) 4.0 min

61. How many half-lives of the isotope occurred during 8.0 minutes?
    (1) 1   (2) 2   (3) 8   (4) 4

62. How long did it take for the mass of the sample to reach 0.25 kilogram?
    (1) 1 min   (2) 5 min   (3) 3 min   (4) 8 min

63. If the mass of this material had been 8.0 kilograms at time $t = 0$, its half-life would have been
    (1) less   (2) greater   (3) the same

64. The half-life of $^{223}_{88}$Ra is 11.4 days. If M kilograms of this radium isotope are present initially, how much remains at the end of 57 days?
    (1) $\dfrac{1}{2} M$ (2) $\dfrac{1}{4} M$ (3) $\dfrac{1}{5} M$ (4) $\dfrac{1}{32} M$

65. If the half-life of $^{234}_{90}$Th is 24 days, the amount of a 12-gram sample remaining after 96 days is
    (1) 1 g   (2) 0.75 g   (3) 6 g   (4) 1.5 g

66. After 2 hours, 1/16 of the initial amount of a radioactive isotope remains undecayed. What is the half-life of the isotope?
    (1) 15 min   (2) 30 min   (3) 45 min   (4) 60 min

67. A substance has a half-life of 300,000 years. In how many years would only 25 kilograms of the original 100-kilogram sample remain?
   (1) 75,000   (2) 300,000   (3) 600,000   (4) 1,200,000

68. An atom of U-235 splits into two nearly equal parts. This is an example of
   (1) alpha decay   (2) beta decay   (3) fusion   (4) fission

69. Neutrons are used in some nuclear reactions as bombarding particles because they are
   (1) positively charged and are repelled by the nucleus
   (2) uncharged and are not repelled by the nucleus
   (3) negatively charged and are attracted by the nucleus
   (4) uncharged and have negligible mass

70. In a nuclear reactor, control rods are used to
   (1) slow down neutrons   (3) absorb neutrons
   (2) speed up neutrons   (4) produce neutrons

71. The function of a moderator in a nuclear reactor is to
   (1) slow down neutrons   (3) produce more neutrons
   (2) speed up neutrons   (4) absorb neutrons

72. The uranium isotope $^{238}_{92}U$ is used to produce
   (1) shielding   (3) control rods
   (2) fissionable plutonium   (4) heavy water

73. The splitting apart of a heavier nucleus to form lighter nuclei is called
   (1) fusion   (3) positron emission
   (2) fission   (4) beta emission

74. The fission process in a nuclear reactor is controlled by regulating the number of available
   (1) electrons   (2) neutrons   (3) protons   (4) positrons

75. Which nuclear reaction is represented by the equation below?

$$^{235}_{92}U + ^{1}_{0}n \rightarrow ^{141}_{56}Ba + ^{92}_{36}Kr + 3^{1}_{0}n + Q$$

   (1) fission   (2) fusion   (3) ionization   (4) diffusion

76. During nuclear fission, energy is released as a result of the
   (1) splitting of heavy nuclei   (3) combining of light nuclei
   (2) combining of heavy nuclei   (4) splitting of light nuclei

77. Which reaction is an example of nuclear fusion?

(1) $^{226}_{88}\text{Ra} \rightarrow {}^{222}_{86}\text{Rn} + {}^{4}_{2}\text{He} + Q$

(2) $^{214}_{83}\text{Bi} \rightarrow {}^{214}_{84}\text{Po} + {}^{0}_{-1}\text{e} + Q$

(3) $^{235}_{92}\text{U} + {}^{1}_{0}\text{n} \rightarrow {}^{141}_{56}\text{Ba} + {}^{92}_{36}\text{Kr} + 3{}^{1}_{0}\text{n} + Q$

(4) $^{3}_{1}\text{H} + {}^{1}_{1}\text{H} \rightarrow {}^{4}_{2}\text{He} + Q$

78. In the fusion reaction $^{3}_{1}\text{H} + {}^{2}_{1}\text{H} \rightarrow {}^{4}_{2}\text{He} + X +$ energy, X is
(1) a proton (3) an alpha particle
(2) a neutron (4) a beta particle

Base your answers to questions 79 through 80 on the nuclear equation below, which represents the induced transmutation of an element.

$$^{27}_{13}\text{Al} + {}^{4}_{2}\text{He} \rightarrow {}^{A}_{Z}\text{X} + {}^{1}_{0}\text{n} + \text{energy}$$

79. The symbol $^{4}_{2}\text{He}$ represents
(1) a hydrogen nucleus (3) a beta particle
(2) a thermal neutron (4) an alpha particle

80. The atomic number of nucleus X is
(1) 13 (2) 15 (3) 30 (4) 31

81. The mass number of nucleus X is
(1) 13 (2) 15 (3) 30 (4) 31

Base your answers to questions 82 through 86 on the equation of a nuclear reaction and information below.

$$^{3}_{1}\text{H} + {}^{1}_{1}\text{H} \rightarrow {}^{4}_{2}\text{He} + Q$$

The mass of each of the nuclei is:

$$^{1}_{1}\text{H} = 1.00813 \text{ u}$$

$$^{3}_{1}\text{H} = 3.01695 \text{ u}$$

$$^{4}_{2}\text{He} = 4.00388 \text{ u}$$

82. In the equation, $^{3}_{1}\text{H}$ and $^{1}_{1}\text{H}$ are
(1) nucleons (3) isotopes
(2) alpha particles (4) quanta

507

83. The total number of nucleons in ^4_2He is
  (1) 1   (2) 2   (3) 3   (4) 4

84. Energy is produced in the reaction. The mass equivalent of this energy is approximately
  (1) 1 u   (2) 2 u   (3) 0.01 u   (4) 0.02 u

85. The equation represents
  (1) fusion   (2) fission   (3) alpha decay (4) beta decay

86. In the equation, the alpha particle is represented by the symbol
  (1) ^3_1H (2) ^1_1H (3) ^4_2He (4) Q

Base your answers to questions 87 through 89 on the Physics Reference Tables.

87. Which change is the result of the loss of two negative beta particles?
  (1) U to Th   (2) Th to U   (3) Pb to Bi   (4) Bi to Pb

88. Which pair of isotopes is included in the uranium disintegration series?

  (1) $^{222}_{84}\text{Po}$ and $^{218}_{84}\text{Po}$ (3) $^{214}_{83}\text{Bi}$ and $^{210}_{83}\text{Bi}$

  (2) $^{226}_{88}\text{Ra}$ and $^{222}_{88}\text{Ra}$ (4) $^{239}_{92}\text{U}$ and $^{238}_{92}\text{U}$

89. Which particle is emitted as $^{234}_{90}\text{Th}$ changes to $^{234}_{91}\text{Pa}$?
  (1) a neutron   (3) a proton
  (2) an alpha particle   (4) a negative beta particle

Base your answers to questions 90 through 95 on the following equation.

$$^{29}_{15}\text{A} + {}^1_0\text{n} \rightarrow {}^0_{-1}\text{D} + {}^w_t\text{E}$$

90. What type of particle is represented by D?
  (1) an electron   (3) an alpha particle
  (2) a positron   (4) a gamma ray

91. For particle E, the value of w is
  (1) 26   (2) 29   (3) 30   (4) 31

92. As particle E emits gamma radiation, its atomic number will
  (1) decrease by 4   (3) remain the same
  (2) increase by 1   (4) decrease by 2

93. For particle E, the value of t is
  (1) 13   (2) 14   (3) 15   (4) 16

94. In 5.0 years, 16 grams of element A decays to 4.0 grams. What is the half-life of element A?
(1) 0.80 year   (2) 1.25 years   (3) 2.5 years   (4) 5.0 years

95. How many neutrons are in the nucleus of $^{29}_{15}$A ?
(1) 14   (2) 15   (3) 29   (4) 44

Base your answers to questions 96 through 99 on the information below.

An atom of $^{238}_{92}$U absorbs a neutron as indicated in this equation:

$$^{238}_{92}U + ^{1}_{0}n \rightarrow Y$$

96. The atomic number of element Y is
(1) 92   (2) 91   (3) 90   (4) 89

97. The mass number of Y is
(1) 240   (2) 239   (3) 238   (4) 237

98. How many neutrons does $^{238}_{92}$U have in its nucleus?
(1) 146   (2) 147   (3) 237   (4) 238

99. If the $^{238}_{92}$U atom decays before the neutron can be absorbed, then according to the uranium disintegration series, the $^{238}_{92}$U atom will emit
(1) a neutron   (3) a beta particle
(2) a positron   (4) an alpha particle

Base your answers to questions 100 through 105 on the list of nuclear equations below and the Uranium Disintegration Series chart in the Physics Reference Tables.

(1) $^{230}_{90}Th \rightarrow ^{214}_{82}Pb + x\,^{4}_{2}He$

(2) $^{14}_{7}N + ^{4}_{2}He \rightarrow ^{17}_{8}O + Y$

(3) $^{235}_{92}U + ^{1}_{0}n \rightarrow ^{90}_{38}Sr + ^{140}_{54}Xe + 6^{1}_{0}n$

(4) $^{3}_{1}H + ^{1}_{1}H \rightarrow ^{4}_{2}He + energy$

100. Which equation represents a fission reaction?
(1) 1   (2) 2   (3) 3   (4) 4

101. Which equation represents a part of the uranium disintegration series?
(1) 1   (2) 2   (3) 3   (4) 4

102. In equation 1, what number of ^4_2He particles is represented by the coeffcient *x*?
    (1) 1   (2) 2   (3) 3   (4) 4

103. In equation 2, Y represents
    (1) a neutron   (3) a proton
    (2) an electron   (4) an alpha particle

104. The number of neutrons in an atom of $^{235}_{92}\text{U}$ is
    (1) 92   (2) 143   (3) 235   (4) 327

105. In equation 4, compared to the nuclear mass of ^4_2He, the sum of the nuclear masses of ^3_1H and ^1_1H is
    (1) less   (2) greater   (3) the same

Base your answers to questions 106 through 109 on the information below.

In the equation $^{221}_{87}\text{Fr} \rightarrow \text{X} + \gamma + Q$, the letter X represents the nucleus produced by the reaction, γ represents a gamma photon, and Q represents additional energy released in the reaction.

106. Which nucleus is represented by X?
    (1) $^{217}_{85}\text{X}$ (2) $^{221}_{87}\text{X}$ (3) $^{220}_{87}\text{X}$ (4) $^{217}_{85}\text{X}$

107. The rest mass of the gamma-ray photon is approximately
    (1) one atomic mass unit   (3) the mass of a neutron
    (2) the mass of a proton   (4) zero

108. If energy Q equals 9.9×10^{-13} joule, the mass equivalent of this energy is
    (1) 0 kg   (3) 1.1×10^{-29} kg
    (2) 9.1×10^{-31} kg (4) 3.3×10^{-21} kg

109. The sample of $^{221}_{87}\text{Fr}$ (half-life = 4.8 min) will decay to one-fourth of its original amount in
    (1) 4.8 min   (2) 9.6 min   (3) 14.4 min   (4) 19.2 min

Base your answers to questions 110 through 113 on the following nuclear reactions.

(1) $^{14}_{7}N + ^{4}_{2}He \rightarrow ^{17}_{8}O + ^{1}_{1}H$

(2) $^{24}_{11}Na \rightarrow ^{24}_{12}Mg + ^{0}_{-1}e$

(3) $^{226}_{88}Ra \rightarrow ^{222}_{86}Rn + ^{4}_{2}He$

(4) $^{235}_{92}U + ^{1}_{0}n \rightarrow ^{90}_{38}Sr + ^{143}_{54}Xe + y^{1}_{0}n + Q$

(5) $^{1}_{1}H + ^{2}_{1}H \rightarrow ^{3}_{2}He + Q$

110. Which reaction represents beta decay?
  (1) 1   (2) 2   (3) 3   (4) 5

111. Which reaction represents transmutation in the uranium disintegration series?
  (1) 1   (2) 2   (3) 3   (4) 5

112. In reaction 4, the number represented by y is
  (1) 1   (2) 2   (3) 3   (4) 6

113. Which reaction is a fusion reaction?
  (1) 5   (2) 2   (3) 3   (4) 4

Base your answers to questions 114 through 118 on the information below:

When tellurium is bombarded with protons, the following reaction occurs:

$$^{1}_{1}H + ^{130}_{52}Te \rightarrow ^{130}_{53}I + Y$$

The iodine produced is radioactive, has a half-life of 12.6 hours, and decays according to the reaction:

$$^{130}_{53}I \rightarrow ^{130}_{54}Xe + ^{0}_{-1}e$$

114. The number of neutrons in a nucleus of $^{130}_{52}Te$ is

  (1) 52   (2) 78   (3) 130   (4) 182

115. Particle Y is
  (1) a proton   (2) a neutron   (3) a positron   (4) an electron

116. The first equation represents an example of
  (1) induced transmutation   (3) nuclear fission
  (2) natural radioactivity   (4) nuclear fusion

117. A 2.4×10^{-5} kilogram sample of $^{130}_{53}\text{I}$ is left to decay. After 37.8 hours, the amount of iodine present will be
(1) 8.0×10^{-6} kg
(3) 3.0×10^{-6} kg
(2) 2.0×10^{-5} kg
(4) 4.0×10^{-6} kg

118. When $^{130}_{53}\text{I}$ decays, it emits
(1) a proton
(3) a beta particle
(2) an alpha particle
(4) a neutron

119. One atomic mass unit is defined as 1/12 of the mass of an isotope of the element
(1) hydrogen   (2) oxygen   (3) uranium   (4) carbon

Base your answers to questions 120 through 124 on the information below:

A photon strikes a stationary deuterium nucleus, $^{2}_{1}\text{H}$, producing two separated, stationary nucleons. The masses of the particles are as follows:

$$\text{proton} = 1.0076 \text{ u}$$
$$\text{neutron} = 1.0090 \text{ u}$$
$$^{2}_{1}\text{H} = 2.0142 \text{ u}$$

120. If Q represents energy in atomic mass units, then the equation for this nuclear reaction is
(1) $Q + {}^{2}_{1}\text{H} \rightarrow {}^{1}_{1}\text{H} + {}^{1}_{0}\text{n}$
(3) $^{1}_{1}\text{H} + {}^{1}_{1}\text{H} \rightarrow {}^{2}_{1}\text{H} + Q$
(2) $Q + {}^{2}_{1}\text{H} \rightarrow {}^{1}_{1}\text{H} + {}^{1}_{1}\text{H}$
(4) $^{1}_{1}\text{H} + {}^{1}_{0}\text{n} \rightarrow {}^{2}_{1}\text{H} + Q$

121. For this nuclear reaction to occur, the minimum photon energy needed is
(1) 0 u   (2) 0.0024 u   (3) 2.0142 u   (4) 2.0166 u

122. What is the binding energy of the deuterium nucleus?
(1) 2.6×10^{-6} MeV
(3) 2.2 MeV
(2) 2.2×10^{-3} MeV
(4) 1.9×10^{3} MeV

123. As the photon energy is increased, the mass defect of the deuterium nucleus will
(1) decrease   (2) increase   (3) remain the same

124. As the binding energy per nucleon of a nucleus increases, the stability of the nucleus
(1) decreases   (2) increases   (3) remains the same

Glossary

absolute index of refraction The ratio of the speed of light in a vacuum to the speed of light in a medium.

absolute temperature The temperature as measured on the Kelvin scale; a measure of the average kinetic energy of the molecules of a body.

absolute zero The temperature at which the internal energy of an object is at a minimum (0 K or $-273°C$).

absorption spectrum A series of dark spectral lines or bands formed by the absorption of specific wavelengths of light by atoms or molecules.

acceleration The time rate of change in velocity. The SI unit is meters per second^2.

acceptor material A substance used to dope a semiconductor in order to increase the number of positive holes in the semiconductor.

accuracy The agreement of a measured value with an accepted standard.

alpha decay A natural radioactive process that results in the emission of an alpha particle from a nuclide.

alpha particle A helium nucleus; a particle consisting of two protons and two neutrons.

alternating current An electric current that varies in magnitude and alternates in direction.

ammeter A device used to measure electric current. It is constructed by placing a low-resistance shunt across the coil of a galvanometer.

ampere (A) The SI unit of electric current, equivalent to the unit coulomb per second.

amplitude The maximum displacement in periodic phenomena such as wave motion, pendulum motion, and spring oscillation.

angle of incidence The angle made by the incident wave with the surface of a medium; the angle made by the incident ray with the normal to the surface of the medium.

angle of reflection The angle made by the reflected wave with the surface of a medium; the angle made by the reflected ray with the normal to the surface of the medium.

angle of refraction The angle made by the refracted wave with the surface of a medium; the angle made by the refracted ray with the normal to the surface of the medium.

anode The positive terminal of a DC source of potential difference.

antimatter One or more atoms composed entirely of antiparticles.

antinode The point or locus of points on an interference pattern (such as a standing wave or double slit pattern) that results in maximum constructive interference.

antiparticle The counterpart of a subatomic particle. An antiparticle has the same mass as its companion particle, but its electric charge is opposite in sign.

atomic mass unit (u) A unit of mass defined as one-twelfth the mass of an atom of carbon-12.

atomic number The number of protons in the nucleus of an atom. The atomic number defines the element.

avalanche The breakdown of a semiconductor diode caused by placing an excessively large potential difference across it when it is reverse biased.

back emf The potential difference that develops in a circuit that opposes the potential difference of the source. A back emf arises as a consequence of Lenz's law.

Balmer series The visible-ultraviolet line spectrum of atomic hydrogen. It is the result of electrons falling from higher levels to the $n = 2$ state.

base The thin middle portion of a bipolar transistor.

battery A combination of two or more electric cells.

beta (−) decay A natural radioactive process that results in the emission of a beta (−) particle from a nuclide.

beta (−) particle An electron formed in the nucleus by the disintegration of a neutron.

beta (+) decay A natural radioactive process that results in the emission of a beta (+) particle from a nuclide.

beta (+) particle A positron, the antiparticle of the electron, formed in the nucleus by the disintegration of a proton.

binding energy The energy equivalent of the mass defect of a nucleus.

boiling The condition in which the liquid and gaseous phases of matter are in equilibrium. A synonym for boiling is *vaporization*.

Boyle's law The volume of an ideal gas is inversely proportional to the pressure at constant temperature.

breeder reactor A fission reactor that produces additional fissionable fuel as a result of neutron bombardment of uranium-238.

cathode The negative terminal of a DC source of potential difference.

cathode ray tube A device for visualizing an electron beam. It consists of an evacuated tube with a source of electrons at one end and a fluorescent screen at the other end. The electron beam is controlled by electric and magnetic fields.

celsius scale (°C) The temperature scale that fixes the (atmospheric) freezing point of water at 0° and the boiling point of water at 100°.

center of curvature In spherical mirrors and lenses, the point on the principal axis that is located a distance of one radius from the center of the mirror or lens.

centripetal acceleration The acceleration that is directed along the radius and toward the center of a curved path in which an object is moving.

centripetal force The force that causes centripetal acceleration. It is responsible for changing an object's direction, not its speed.

chain reaction In nuclear fission, a continuing series of reactions that results from bombardment by neutrons that are, themselves, fission products.

Charles's law The volume occupied by an ideal gas is directly proportional to the Kelvin temperature at constant pressure.

chromatic aberration A lens defect in which different colors of light are focused at different points.

circuit A closed loop formed by a source of potential difference connected to one or more resistances.

cloud chamber A device that tracks the path of a charged particle by condensing an undercooled gas into a liquid.

coefficient of kinetic friction The ratio of the force of kinetic friction on an object to the normal force on it.

coherent light A series of light waves that have a fixed phase relationship; the type of light produced by a laser. Lasers produce beams of monochromatic coherent light.

collector The outer layer in a bipolar transistor, which receives the majority of charge carriers.

commutator A split ring that is connected to the armature coil of a DC motor or generator, causing the current to reverse direction with each half-turn.

component One of the two or more vectors into which a given vector may be resolved.

concave lens A lens that is thinner in the middle than at the edges.

concave mirror A curved mirror whose reflecting surface is the inner surface of the curve.

concurrent forces Two or more forces acting at the same point.

conduction band The energy band that contains the electrons that are able to carry an electric current.

conductivity The reciprocal of a material's resistivity. The SI unit of conductivity is the mho, which is equivalent to the $ohm^{-1} \cdot meter^{-1}$.

conductor A material that allows electrons to flow through it freely. Metals such as copper and silver are conductors.

constructive interference The combination of two in-phase wave disturbances to produce a single wave disturbance whose amplitude is the sum of the amplitudes of the individual disturbances.

control rod A device used to regulate the rate of a nuclear chain reaction by absorbing neutrons.

converging lens A lens that focuses its transmitted light to a point. Generally, convex lenses are converging lenses.

converging mirror A mirror that focuses its reflected light to a point. Generally, concave mirrors are converging mirrors.

convex lens A lens that is thicker in the middle than at the edges.

convex mirror A curved mirror whose reflecting surface is the outer surface of the curve.

coolant A circulating fluid that removes and transfers the heat energy generated by devices such as fission reactors.

core The part of a fission reactor that is the focus of the fission reaction.

coulomb (C) The SI unit of electric charge, approximately equal to 6.25×10^{18} elementary charges.

Coulomb's law The electrostatic force between two point charges is directly proportional to the product of the charges and inversely proportional to the square of the distance between the charges.

critical angle The angle of incidence for which the corresponding angle of refraction is $90°$.

critical mass The minimum amount of a fissionable nuclide needed to sustain a chain reaction.

cycle One complete repetition of the pattern in any periodic phenomenon.

cyclotron One of a number of devices that accelerate charged nuclear particles by using electric and magnetic fields.

de broglie wavelength The wavelength of a matter wave.

destructive interference The combination of two out-of-phase wave disturbances to produce a single wave disturbance whose amplitude is the difference of the amplitudes of the individual disturbances.

deuterium An isotope of hydrogen that contains one proton and one neutron in its nucleus.

diode A device that permits charge to flow in one direction only. A semiconductor diode consists of a P-type semiconductor joined to an N-type semiconductor.

diffraction The bending of a wave around a barrier.

diffuse reflection The reflection of parallel light rays by irregular surfaces.

direct current An electric current that flows in one direction only.

dispersion The separation of polychromatic light into its individual colors.

dispersive medium A medium in which the speed of a wave depends on its frequency.

displacement A change of position in a specific direction.

diverging lens A lens that spreads its transmitted light outward. Generally, concave lenses are diverging lenses.

diverging mirror A mirror that spreads its reflected light outward. Generally, convex mirrors are diverging mirrors.

donor material A substance used to dope a semiconductor in order to increase the number of electrons in the semiconductor.

doping The process of inserting small amounts of certain impurities into a semiconductor in order to increase the number of electrons or positive holes that it contains.

Doppler effect An apparent change in frequency that results when a wave source and an observer are in relative motion with respect to each other.

efficiency The degree to which a device such as a transformer or machine compares with an ideal device. Efficiency is usually expressed as a percentage.

Einstein's postulates of special relativity
(1) The laws of physics are valid in all inertial frames of reference.
(2) The speed of light has the same value in all frames of reference.

elastic potential energy The energy stored in a spring when it is compressed or stretched.

electric current The time rate of flow of charged particles. The SI unit of electric current is the ampere (A).

electric field The region of space around a charged object that affects other charges.

electric field barrier The region in a semiconductor, established by the combination of holes and electrons, that prevents further migration of charge carriers across the P-N junction.

electric field intensity The ratio of the force that an electric field exerts on a charge to the magnitude of the charge.

electric motor A device that converts electrical energy into mechanical energy.

electric potential The total work done by an electric field in bringing 1 coulomb of positive charge from infinity to a specific point. The potential is a positive number if the charge is repelled by the field and a negative number if the charge is attracted by the field. At infinity, the potential is taken to be zero. Electric potential is measured in volts.

electromagnet A solenoid whose magnetic field is intensified by the insertion of certain ferromagnetic materials.

electromagnetic induction The process by which the magnetic field and the mechanical energy are used to generate a potential difference.

electromagnetic radiation The propagation of electromagnetic waves in space.

electromagnetic spectrum The entire range of electromagnetic waves from the lowest to the highest frequencies.

electromagnetic wave A periodic wave, consisting of mutually perpendicular electric and magnetic fields, that is radiated away from the vicinity of an accelerating charge.

electromotive force The potential difference produced as a result of the conversion of other forms of energy into electrical energy.

electron A fundamental, negatively charged, subatomic particle.

electron capture A radioactive process in which a nucleus absorbs one of an atom's innermost electrons.

electron cloud In quantum theory, the region of space where an electron is most likely to be found.

electron-volt (eV) A unit of energy equal to the work needed to move an elementary charge across a potential difference of 1 volt.

electroscope A device used to detect the presence of electric charges.

elementary charge The magnitude of charge present on a proton or an electron. An elementary charge is approximately equal to 1.6×10^{-19} coulomb.

emission spectrum A series of bright spectral lines or bands formed by the emission of certain wavelengths of light by excited atoms falling to lower energy states.

emitter One of the outer layers of a bipolar transistor, which supplies the charge carriers to the rest of the transistor.

energy A quantity related to work.

entropy A measure of the disorder or randomness present in a system.

equilibrant A single balancing force that maintains the static equilibrium of an object.

equivalent resistance A single resistance that can be substituted for a group of resistances in series or in parallel.

ether A hypothetical medium whose existence was proposed as the carrier of all electromagnetic waves.

excited state A condition in which the energy of an atom is greater than its lowest energy state.

extrinsic semiconductor Silicon or germanium that has been made semiconductive by doping.

farad (F) The SI unit of capacitance, equivalent to the unit coulomb per volt.

ferromagnetic Referring to a material, such as iron, that has the ability to strengthen greatly the magnetic field of a current-carrying coil.

field lines A series of lines used to represent the magnitude and direction of a field.

first law of thermodynamics A change in the internal energy of a system is equal to the difference of the heat energy absorbed by the system and the work done by it. It is a statement of the law of conservation of energy.

fission The process of splitting a heavy nucleus, such as uranium-235, into lighter fragments. Fission is accompanied by the release of large quantities of energy.

flux density The number of magnetic field lines per unit area. The flux density is one way of measuring the strength of a magnetic field.

focal length The distance between the center of a lens or mirror and its focal point.

focal point The point of a lens or mirror where incoming parallel light rays meet. The focal point is also known as the *principal focus*.

force A push or a pull on an object. If the force is unbalanced, an acceleration will result.

forward bias A potential difference applied across a *P-N* junction in a direction that facilitates both electron and hole flow across the junction.

frame of reference A coordinate grid and a set of synchronized clocks that can be used to determine the position and time of an event.

free fall A motion in the Earth's gravitational field without regard to air resistance.

frequency The number of repetitions produced per unit time by periodic phenomena.

friction The force present as the result of contact between two surfaces. The direction of a frictional force is opposite to the direction of motion.

fuel rods Rods packed with nuclear fuel pellets and placed in the core of a fission reactor, which is the source of energy from the fission reaction.

fusion (1) The process of uniting lighter nuclei, such as deuterium, into a heavier nucleus. Fusion is accompanied by the release of large quantities of energy. (2) In the study of heat and thermodynamics, a synonym for *melting*.

Gallilean-Newtonian relativity principle The laws of mechanics are valid in all inertial frames of reference.

galvanometer A device, consisting of a coil-shaped wire placed between the opposite poles of a permanent magnet, that is used to detect small amounts of electric current.

gamma radiation Very high energy photons of electromagnetic radiation. Gamma photons have the highest frequencies in the electromagnetic spectrum.

geiger counter A device that detects charged nuclear particles.

generator A device that uses a magnetic field and mechanical energy to induce a source of electromotive force.

geosynchronous orbit An orbit in which the period of a satellite is equal to the period of the Earth's rotation (approximately 24 hours). A satellite in a geosynchronous orbit remains continually in the same position above the Earth's surface.

gravitational field The region of space around a mass that affects other masses.

gravitational field intensity The ratio of the force that a gravitational field exerts on a mass to the magnitude of the mass, numerically equal to the acceleration due to gravity.

gravitational force The universal attraction between two masses.

gravitational potential energy The energy that an object acquires as a result of the work done in moving the object against a gravitational field.

ground An extremely large source or reservoir of electrons, which can supply or accept electrons as the need arises.

ground state The lowest energy state of an atom.

half-life The time needed reduce the number of radioactive nuclei present in a sample to one-half its initial value.

heat energy The energy that is transferred from warmer objects to cooler ones because of a temperature difference between them.

heat of fusion The quantity of heat energy needed to convert a unit mass of a solid entirely into a liquid at its melting temperature.

heat of vaporization The quantity of heat energy needed to convert a unit mass of a liquid entirely into a gas at its boiling temperature.

hertz (Hz) The SI unit of frequency, equivalent to the unit second^{-1}.

hole An electron vacancy in a semiconductor. In an electric field, holes behave like positive charges.

ideal gas A simplified model of a gas in which the molecules have mass, but no volume, and do not exert forces except during collisions. Real gases may approximate ideal behavior at low pressure and high temperature.

image An optical reproduction of an object.

impulse The product of the net force acting on an object and the time during which the force acts. The impulse delivered to an object is equal to its change in momentum. The direction of the impulse is the direction of the force. The SI unit of impulse is the newton · second, which is equivalent to the kilogram · meter per second.

incident ray A ray of a wave impinging on a surface.

incident wave A wave impinging on a surface.

induced current An electric current that is the result of an induced electromotive force.

induced emf A potential difference created when a magnetic field is interrupted over a time period.

induction (1) A method of charging a neutral object by using a charged object and a ground. The induced charge is always opposite to the charge on the charged object. (2) See *induced current* and *induced emf.*

inertia The property of matter that resists changes in motion. Mass is the quantitative measure of inertia.

inertial frame of reference A frame of reference in which Newton's first law of motion is valid.

instantaneous velocity The ratio of displacement to time at any given instant; the slope of a line tangent to a displacement-time graph at any given point.

insulator A material that is a very poor conductor because it has few conduction electrons. Wood and glass are examples of insulators.

integrated circuit A miniaturized circuit containing an array of semiconductor devices.

interference pattern Regions of constructive and destructive interference that are present in a medium as a result of the combination of two or more waves.

internal energy The total kinetic and potential energy associated with the atoms and molecules of an object.

ionization energy See *ionization potential*.

ionization potential The quantity of energy needed to remove a single electron from an atom or ion.

isotopes Atoms with identical atomic numbers but different mass numbers. Two isotopes of the same element have identical numbers of protons but different numbers of neutrons.

joule (J) The SI unit of work and energy, equivalent to the unit newton · meter.

k-capture Electron capture.

Kelvin (K) The SI unit of temperature.

Kelvin scale (K) The absolute temperature scale. The single fixed point on the Kelvin scale is the triple point of water, which is set at 273.16K.

Kepler's first law The orbits of all the planets are elliptical, with the Sun placed at one focus of the ellipse.

Kepler's second law A line from the Sun to a planet sweeps out equal areas in equal periods of time.

Kepler's third law The ratio of the cube of the mean radius of a planet's orbit to the square of its period of revolution about the Sun is the same for all planetary bodies in the solar system.

kilogram (kg) The SI unit of mass; a fundamental unit.

kinetic energy The energy that an object possesses because of its motion.

kinetic-molecular theory of gases A simplified model of gas behavior based on the properties of an ideal gas.

laser An acronym for light amplification by the stimulated emission of radiation. A laser is a device that emits extremely intense, monochromatic, coherent light.

length contraction A consequence of Einstein's theory of special relativity. The length measured in a moving frame of reference is shorter than the length measured in a stationary frame of reference (the *proper length*).

linear accelerator A device that accelerates charged particles continuously in an electric field along a linear path.

longitudinal wave A wave in which the disturbance is parallel to the direction of the wave's motion. Sound waves are longitudinal.

magnet Any material that aligns itself, when free to do so, in an approximate north-south direction. Magnets exert forces on one another and on charged particles in motion.

magnetic field The region of space around a magnet or charge in motion that exerts a force on magnets or other moving charges.

magnetic induction The force that a magnetic field exerts on a 1-meter-long wire in the field when the wire carries a current of 1 ampere. The unit of magnetic induction is the tesla.

magnification The ratio of image size to object size.

mass (1) The measure of an object's ability to obey Newton's second law of motion. (2) The measure of an object's ability to obey Newton's law of universal gravitation. The SI unit of mass is the kilogram.

mass defect The mass lost by a nucleus when it is assembled from its nucleons. (See also *binding energy*.)

mass number The sum of the number of protons and neutrons in a nucleus; the number of nucleons the nucleus contains.

mass spectrometer A device that uses a magnetic field to separate charged particles of differing masses.

matter waves According to quantum theory, the waves associated with moving particles.

mechanical advantage In a simple machine, the ratio of the input force (*resistance*) to the output force (*effort*).

medium A material through which a disturbance, such as a wave, travels.

moderators Materials used in fission reactors to slow neutrons so that they can be absorbed by the fuel nuclei.

momentum The product of mass and velocity. The direction of an object's momentum is the direction of its velocity. The SI unit of momentum is the kilogram·meter per second.

natural frequency A specific frequency with which an elastic body may vibrate if disturbed.

net force The unbalanced force present on an object; the accelerating force.

neutrino A subatomic particle with no charge and questionable mass. It and its antiparticle are products of beta-decay reactions.

Newton (N) The SI unit of force, equivalent to the unit kilogram · meter per second^2.

Newton's first law of motion Objects remain in a state of uniform motion unless acted upon by an unbalanced force.

Newton's law of universal gravitation Any two bodies in the universe are attracted to each other with a force that is directly proportional to their masses and inversely proportional to the square of the distance between them.

Newton's second law of motion The unbalanced force on an object is equal to the product of its mass and acceleration.

Newton's third law of motion If object *A* exerts a force on object *B*, then object *B* exerts an equal and opposite force on object *A*.

node The point or locus of points on an interference pattern, such as a standing wave or double-slit pattern, that results in total destructive interference.

nondispersive medium A medium in which the speed of a wave does *not* depend on its frequency.

normal A line perpendicular to a surface.

normal force The force that keeps two surfaces in contact. If an object is on a *horizontal* surface, the normal force on the object is equal to its weight.

nuclear force The attractive, short-range force responsible for binding protons and neutrons in the nucleus.

nucleon A proton or a neutron.

nucleus The dense, positively charged core of an atom.

nuclide An atomic nucleus.

ohm (Ω) The SI unit of electrical resistance, equivalent to the unit volt per ampere.

Ohm's law A relationship in which the ratio of the potential difference across certain conductors to the current in them is constant at constant temperature.

parallel circuit An electric circuit with more than one current path.

particle accelerator A device used to accelerate charged nuclear particles.

pascal (Pa) The SI unit of pressure, equivalent to the unit newton per meter^2.

period The time for one complete repetition of a periodic phenomenon. The SI unit of period is the second.

periodic wave A regularly repeating series of waves.

phase (1) A form in which matter can exist, including liquid, solid, gas, and plasma. (2) In wave motion, the points on the wave that have specific time and space relationships.

photoelectric effect A phenomenon in which light causes electrons to be ejected from certain materials. (See also *photoemissive*.)

photoelectrons Electrons that have been emitted as a result of the photoelectric effect.

photoemissive Referring to materials whose surfaces can eject electrons on exposure to light.

photon The fundamental particle of electromagnetic radiation.

Planck's constant A universal constant (h) relating the energy of a photon to its frequency; its approximate value is 6.62×10^{-34} joule · second.

P-N junction The narrow region in a semiconductor diode in the vicinity of the boundary between the P-type and N-type materials.

point charge A charge with negligible physical dimensions.

polarization A process that produces transverse waves that vibrate in only one plane. Polarization is limited to transverse waves: light can be polarized; sound cannot.

polychromatic Referring to light waves of different colors (frequencies).

positron See *Beta (+) particle*.

potential difference The ratio of the work required to move a test charge between two points in an electric field to the magnitude of the test charge. The unit of potential difference is the volt.

potential energy The energy that a system has because of its relative position or condition.

power The time rate at which work is done or energy is expended. The SI unit of power is the watt, which is equivalent to the unit joule per second

precision The limit of the ability of a measuring device to reproduce a measurement.

pressure The force on a surface per unit area. The SI unit of pressure is the pascal.

primary coil The wire coil of a transformer that is connected to a source of alternating current.

principal axis A line passing through the center of curvature and the center of a curved mirror or lens.

principal quantum number The integer that defines the main energy level of an atom.

proton A positively charged subatomic particle with a charge equal in magnitude to that of the electron.

pulse A nonperiodic disturbance in a medium.

quantum A discrete quantity of energy.

quarks The particles of which protons, neutrons, and certain other subatomic particles are composed. Quarks carry a charge of either one-third or two-thirds of an elementary charge and come in six "flavors": top, bottom, up, down, charm, and strange.

radioactive decay A spontaneous change in the nucleus of an atom.

radioactivity Changes in the nucleus of an atom that produce the emission of subatomic particles or photons.

radius of curvature The distance from the center of curvature to the curved surface of a curved mirror or lens.

ray A straight line indicating the direction of travel of a wave.

real image An image created by the actual convergence of light waves. Real images from single mirrors and single lenses are inverted and can be projected on a screen.

rectifier A device (usually a diode) that converts alternating current into pulsed direct current.

refraction The change in the direction of a wave when it passes obliquely from one medium to another in which it moves at different speed.

regular reflection The reflection of parallel light rays incident on a smooth plane surface.

resistance The opposition of a material to the flow of electrons through it; the ratio of potential difference to current.

resistivity A quantity that allows the resistance of substances to be compared. Numerically, it is the resistance of a 1-meter conductor with a cross-sectional area of 1 square meter. The SI unit of resistivity is the ohm·meter.

resistor A device that supplies resistance to a circuit.

resolution The process of determining the magnitude and direction of the components of a vector.

resonance The spontaneous vibration of an object at a frequency equal to that of the wave that initiates the resonant vibration.

rest energy A quantity derived from the special theory of relativity; the energy equivalent of mass.

resultant A vector sum.

reverse bias A potential difference applied across a *P-N* junction in a direction that inhibits both electron and hole flow across the junction.

satellite A body that revolves around a larger body as a result of a gravitational force.

scalar quantity A quantity, such as mass or work, that has magnitude but not direction.

scintillation counter A device that detects nuclear particles emitted from atoms by converting their kinetic energy into photons of light.

secondary coil In a transformer, the coil that is not connected to the primary source of potential difference.

second law of thermodynamics Heat energy will not flow spontaneously from a cooler body to a warmer body unless work is done. Alternatively, it is impossible to convert heat energy entirely into work.

semiconductor A metalloid whose conductivity can be increased either by raising its temperature or by adding certain impurities. Silicon and germanium are examples of semiconductors.

series circuit A circuit with only one current path.

shunt A low-resistance device for diverting electric current, used to convert a galvanometer into an ammeter.

significant digits The digits that are part of any measurement.

simultaneity The occurrence of two or more events at the same time. According to special relativity, events that are simultaneous in one frame of reference need not be simultaneous in another.

solenoid A coil of wire wound as a helix. When a current is passed through the solenoid, it becomes an electromagnet.

specific heat The amount of heat energy absorbed or liberated by a unit mass of a substance as it changes its temperature by one unit.

speed The time rate of change of distance; the magnitude of velocity. The SI unit of speed is the meter per second.

spherical aberration The failure of mirrors and lenses with spherical surfaces to bring light to a single focus.

spherical lens A lens in which at least one of the refracting surfaces is a portion of a sphere.

spherical mirror A mirror having a reflecting surface that is a portion of a sphere.

spring constant The ratio of the force required to stretch or compress a spring to the magnitude of the stretch or compression.

standard pressure One atmosphere; approximately 1.01×10^5 pascals.

standard temperature Approximately 273 kelvins.

standing wave A wave pattern created by the continual interference of an incident wave with its reflected counterpart. The standing wave does not travel, but oscillates about an equilibrium position.

static equilibrium The condition of a body when a net force of zero is acting on it.

stationary state A condition in which an atom is neither absorbing nor releasing energy.

step-down transformer A transformer in which the potential difference across the secondary coil is less than that across the primary coil.

step-up transformer A transformer in which the potential difference across the secondary coil is greater than that across the primary coil.

superconductor A material with no electrical resistance.

temperature The "hotness" of an object, measured with respect to a chosen standard.

tesla (T) The SI unit of flux density, equal to the units weber per square meter and newton per ampere · meter.

thermal equilibrium The point at which materials in contact reach the same temperature.

thermal neutrons Neutrons able to initiate fission reactions because they have speeds low enough to permit their absorption by fuel nuclei.

thermionic emission The emission of electrons from substances such as metallic filaments when these substances are heated.

thermodynamics The study of the interrelationships among heat energy, work, and other forms of energy.

third law of thermodynamics It is impossible to reach absolute zero in a finite number of operations.

threshold frequency The lowest frequency at which photons striking a specific surface can produce a photoelectric effect.

time dilation A consequence of the theory of special relativity. The time interval measured in a moving frame of reference is longer than the time interval measured in a stationary frame of reference (the *proper time*).

torque A force, applied perpendicularly to a designated line, that tends to produce rotational motion.

total internal reflection The reflection of a wave inside a relatively dense medium produced when the angle of the wave with the boundary exceeds the critical angle.

total mechanical energy The sum of the potential and kinetic energies of a mechanical system.

transformer A device that uses a changing magnetic field to increase or decrease the potential difference between a primary and a secondary circuit.

transistor (bipolar) A semiconducting device in which one type of semiconductor is sandwiched between two semiconductors of the opposite type. Transistors serve as amplification devices.

transmutation The change of one radioactive nuclide into another, either by decay or by bombardment.

transverse wave A wave in which the disturbance is perpendicular to the direction of the wave's motion. Light waves are transverse.

uniform In the study of motion, a term that is equivalent to *constant*.

uranium disintegration series The sequence of alpha and beta decays by which uranium-238 is transformed into lead-206.

valence band The energy band that contains electrons from the outermost energy levels of a group of atoms in a crystal.

vaporization See *boiling*.

vector A representation of a vector quantity; an arrow in which the length represents the magnitude of the quantity and the arrowhead points in the direction of its orientation.

vector quantity A quantity, such as force or velocity, that has both magnitude and direction.

velocity The time rate of change of displacement.

virtual focus The point where incident light waves appear to originate after they are refracted by a diverging lens or reflected by a diverging mirror.

virtual image An image formed by projecting diverging light behind a mirror or a lens.

volt (V) The SI unit of potential difference, equivalent to the unit joule per coulomb.

voltage Another term for *potential difference*.

voltmeter A device used to measure potential difference and constructed by placing a large resistor in series with the coil of a galvanometer.

watt (W) The SI unit of power, equivalent to the unit joule per second.

wave A series of periodic oscillations of a particle or a field both in time and in space.

wave front All points on a wave that are in phase with each other.

wavelength The length of one complete wave cycle.

weber (Wb) The SI unit of magnetic flux, equivalent to the unit joule per ampere.

weight The gravitational force present on an object.

work The product of the force on an object and its displacement. The SI unit of work is the joule.

work function The minimum radiant energy required to remove an electron from a photoemissive surface.

Answers to Questions for Review

CHAPTER ONE

1.	2	4.	3	7.	3	10.	3	13. 2
2.	3	5.	1	8.	3	11.	3	14. 4
3.	2	6.	1	9.	1	12.	2	

CHAPTER TWO

1.	3	12.	3	23.	2	34.	3	45.	2	56.	3
2.	1	13.	2	24.	2	35.	2	46.	1	57.	4
3.	4	14.	3	25.	1	36.	3	47.	4	58.	3
4.	3	15.	1	26.	2	37.	2	48.	1	59.	2
5.	3	16.	1	27.	4	38.	3	49.	1	60.	2
6.	2	17.	4	28.	4	39.	2	50.	3	61.	4
7.	2	18.	4	29.	1	40.	2	51.	1	62.	2
8.	2	19.	4	30.	4	41.	4	52.	2	63.	1
9.	1	20.	4	31.	3	42.	3	53.	1	64.	1
10.	3	21.	2	32.	2	43.	1	54.	1	65.	4
11.	1	22.	2	33.	2	44.	3	55.	2	66.	2

CHAPTER THREE

1.	1	9.	1	17.	2	25.	1	33.	1	41.	2
2.	4	10.	1	18.	2	26.	4	34.	3	42.	3
3.	1	11.	3	19.	3	27.	2	35.	3	43.	1
4.	2	12.	2	20.	4	28.	2	36.	1	
5.	4	13.	2	21.	4	29.	2	37.	1	
6.	3	14.	1	22.	1	30.	3	38.	4	
7.	3	15.	3	23.	1	31.	4	39.	3	
8.	2	16.	3	24.	4	32.	1	40.	4	

CHAPTER FOUR

1.	4	11.	3	21.	2	31.	1	41.	1	51.	3
2.	2	12.	2	22.	3	32.	3	42.	1	52.	1
3.	3	13.	3	23.	4	33.	2	43.	4	53.	1
4.	2	14.	1	24.	2	34.	2	44.	1	54.	4
5.	2	15.	2	25.	2	35.	2	45.	2	55.	3
6.	1	16.	3	26.	1	36.	1	46.	2	56.	4
7.	2	17.	1	27.	4	37.	3	47.	1	57.	4
8.	2	18.	4	28.	4	38.	3	48.	2	58.	1
9.	1	19.	3	29.	2	39.	2	49.	3		
10.	3	20.	4	30.	4	40.	2	50.	2		

CHAPTER FIVE

1.	4	14.	3	27.	2	40.	1	53.	1	66.	3
2.	1	15.	4	28.	1	41.	3	54.	3	67.	4
3.	4	16.	3	29.	4	42.	1	55.	4	68.	3
4.	4	17.	4	30.	4	43.	2	56.	2	69.	1
5.	2	18.	1	31.	2	44.	2	57.	3	70.	3
6.	4	19.	1	32.	2	45.	3	58.	2	71.	1
7.	3	20.	4	33.	3	46.	4	59.	4	72.	2
8.	4	21.	4	34.	4	47.	2	60.	1	73.	3
9.	2	22.	2	35.	3	48.	4	61.	2	74.	4
10.	1	23.	4	36.	3	49.	1	62.	4	75.	3
11.	4	24.	3	37.	1	50.	3	63.	1	76.	3
12.	4	25.	3	38.	4	51.	2	64.	3	77.	2
13.	1	26.	2	39.	3	52.	2	65.	3		

CHAPTER SIX

1.	4	9.	3	17.	3	25.	2	33.	4	41.	1
2.	3	10.	1	18.	2	26.	3	34.	2	42.	3
3.	2	11.	2	19.	4	27.	4	35.	2	43.	4
4.	2	12.	4	20.	1	28.	2	36.	3	44.	2
5.	1	13.	4	21.	2	29.	3	37.	2	45.	4
6.	2	14.	1	22.	4	30.	4	38.	3	46.	4
7.	4	15.	1	23.	3	31.	4	39.	3		
8.	3	16.	2	24.	4	32.	3	40.	3		

CHAPTER SEVEN

1. 3	15. 2	29. 3	43. 2	57. 2	71. 1
2. 4	16. 2	30. 2	44. 3	58. 1	72. 1
3. 3	17. 2	31. 2	45. 3	59. 3	73. 4
4. 3	18. 1	32. 4	46. 2	60. 3	74. 2
5. 1	19. 2	33. 2	47. 1	61. 3	75. 2
6. 3	20. 2	34. 3	48. 3	62. 2	76. 4
7. 3	21. 1	35. 2	49. 1	63. 3	77. 4
8. 1	22. 1	36. 3	50. 2	64. 3	78. 4
9. 3	23. 1	37. 2	51. 3	65. 3	79. 1
10. 3	24. 3	38. 3	52. 2	66. 1	80. 4
11. 2	25. 4	39. 2	53. 4	67. 4	81. 1
12. 2	26. 3	40. 3	54. 2	68. 1	
13. 1	27. 2	41. 2	55. 4	69. 3	
14. 1	28. 2	42. 3	56. 3	70. 1	

CHAPTER EIGHT

1. 1	14. 4	27. 4	40. 3	53. 2	66. 2
2. 4	15. 1	28. 1	41. 2	54. 2	67. 4
3. 2	16. 4	29. 3	42. 1	55. 1	68. 4
4. 3	17. 2	30. 2	43. 4	56. 4	69. 2
5. 3	18. 4	31. 3	44. 1	57. 2	70. 4
6. 2	19. 2	32. 2	45. 4	58. 3	71. 1
7. 3	20. 1	33. 3	46. 3	59. 2	72. 4
8. 3	21. 3	34. 4	47. 4	60. 3	73. 1
9. 4	22. 4	35. 2	48. 2	61. 1	74. 4
10. 1	23. 3	36. 1	49. 1	62. 1	75. 2
11. 1	24. 2	37. 2	50. 2	63. 3	76. 4
12. 3	25. 3	38. 2	51. 2	64. 4	77. 4
13. 1	26. 2	39. 2	52. 3	65. 2	

CHAPTER NINE

1. **4**	23. **1**	45. **1**	67. **4**	89. **4**	111. **2**
2. **4**	24. **4**	46. **3**	68. **2**	90. **2**	112. **3**
3. **2**	25. **4**	47. **4**	69. **1**	91. **1**	113. **2**
4. **2**	26. **1**	48. **4**	70. **3**	92. **4**	114. **2**
5. **2**	27. **2**	49. **1**	71. **2**	93. **4**	115. **4**
6. **3**	28. **2**	50. **3**	72. **1**	94. **2**	116. **3**
7. **2**	29. **3**	51. **1**	73. **1**	95. **3**	117. **4**
8. **3**	30. **3**	52. **3**	74. **1**	96. **2**	118. **1**
9. **2**	31. **1**	53. **2**	75. **2**	97. **4**	119. **1**
10. **1**	32. **3**	54. **4**	76. **2**	98. **3**	120. **2**
11. **3**	33. **4**	55. **2**	77. **1**	99. **3**	121. **4**
12. **3**	34. **2**	56. **3**	78. **2**	100. **2**	122. **3**
13. **3**	35. **1**	57. **4**	79. **4**	101. **3**	123. **2**
14. **1**	36. **2**	58. **3**	80. **2**	102. **4**	124. **3**
15. **4**	37. **1**	59. **2**	81. **2**	103. **2**	125. **2**
16. **1**	38. **1**	60. **1**	82. **4**	104. **1**	126. **3**
17. **2**	39. **4**	61. **1**	83. **3**	105. **1**	127. **2**
18. **3**	40. **4**	62. **4**	84. **4**	106. **3**	
19. **3**	41. **3**	63. **1**	85. **2**	107. **3**	
20. **2**	42. **1**	64. **3**	86. **3**	108. **4**	
21. **4**	43. **1**	65. **2**	87. **3**	109. **3**	
22. **1**	44. **2**	66. **4**	88. **3**	110. **1**	

CHAPTER TEN

1. **1**	18. **4**	35. **1**	52. **4**	69. **3**	86. **3**
2. **4**	19. **1**	36. **1**	53. **1**	70. **3**	87. **3**
3. **4**	20. **2**	37. **2**	54. **4**	71. **4**	88. **1**
4. **2**	21. **3**	38. **3**	55. **3**	72. **4**	89. **4**
5. **4**	22. **4**	39. **2**	56. **2**	73. **2**	90. **4**
6. **2**	23. **4**	40. **2**	57. **3**	74. **4**	91. **4**
7. **2**	24. **1**	41. **4**	58. **3**	75. **1**	92. **1**
8. **4**	25. **4**	42. **2**	59. **1**	76. **3**	93. **2**
9. **4**	26. **2**	43. **3**	60. **2**	77. **1**	94. **3**
10. **3**	27. **3**	44. **4**	61. **2**	78. **2**	95. **2**
11. **4**	28. **1**	45. **1**	62. **3**	79. **2**	96. **2**
12. **2**	29. **4**	46. **3**	63. **4**	80. **4**	97. **3**
13. **3**	30. **3**	47. **1**	64. **1**	81. **2**	98. **4**
14. **3**	31. **3**	48. **2**	65. **4**	82. **1**	99. **2**
15. **4**	32. **3**	49. **1**	66. **3**	83. **2**	100. **4**
16. **2**	33. **4**	50. **2**	67. **3**	84. **3**	101. **1**
17. **3**	34. **2**	51. **4**	68. **1**	85. **1**	

CHAPTER ELEVEN

1.	1	18.	2	35.	1	52.	1	69.	3	86.	3
2.	4	19.	3	36.	1	53.	3	70.	2	87.	1
3.	2	20.	1	37.	3	54.	1	71.	2	88.	4
4.	2	21.	1	38.	2	55.	3	72.	1	89.	2
5.	1	22.	2	39.	4	56.	2	73.	1	90.	3
6.	2	23.	2	40.	2	57.	2	74.	4	91.	2
7.	1	24.	2	41.	1	58.	4	75.	1	92.	1
8.	1	25.	3	42.	2	59.	2	76.	3	93.	4
9.	1	26.	2	43.	2	60.	2	77.	2	94.	2
10.	3	27.	3	44.	1	61.	2	78.	4	95.	2
11.	3	28.	3	45.	2	62.	3	79.	4	96.	1
12.	4	29.	3	46.	4	63.	2	80.	3	97.	1
13.	3	30.	4	47.	4	64.	4	81.	4	98.	1
14.	2	31.	3	48.	1	65.	2	82.	2	99.	2
15.	4	32.	1	49.	2	66.	3	83.	3	100.	1
16.	1	33.	1	50.	4	67.	1	84.	4	101.	3
17.	1	34.	3	51.	2	68.	4	85.	1	102.	2

CHAPTER TWELVE

1.	4	21.	2	41.	2	61.	4	81.	1	101.	4
2.	3	22.	1	42.	1	62.	4	82.	2	102.	2
3.	1	23.	1	43.	1	63.	1	83.	4	103.	4
4.	3	24.	2	44.	2	64.	3	84.	2	104.	2
5.	3	25.	4	45.	4	65.	4	85.	2	105.	4
6.	4	26.	2	46.	3	66.	3	86.	1	106.	2
7.	2	27.	3	47.	2	67.	4	87.	3	107.	2
8.	1	28.	3	48.	1	68.	2	88.	2	108.	4
9.	2	29.	2	49.	4	69.	4	89.	3	109.	3
10.	3	30.	4	50.	1	70.	1	90.	4	110.	1
11.	3	31.	4	51.	4	71.	3	91.	2	111.	1
12.	1	32.	1	52.	2	72.	2	92.	1	112.	1
13.	4	33.	1	53.	3	73.	3	93.	1	113.	3
14.	1	34.	4	54.	4	74.	2	94.	4	114.	4
15.	2	35.	3	55.	1	75.	4	95.	2	115.	1
16.	2	36.	2	56.	3	76.	4	96.	1	116.	4
17.	2	37.	4	57.	1	77.	1	97.	1	117.	2
18.	1	38.	2	58.	3	78.	2	98.	1	118.	4
19.	2	39.	3	59.	4	79.	1	99.	2	119.	4
20.	3	40.	2	60.	4	80.	1	100.	4		

CHAPTER THIRTEEN

1. **4**	36. **3**	71. **3**	106. **2**	141. **1**	176. **1**
2. **1**	37. **4**	72. **3**	107. **4**	142. **1**	177. **1**
3. **1**	38. **2**	73. **1**	108. **4**	143. **3**	178. **2**
4. **1**	39. **3**	74. **3**	109. **4**	144. **4**	179. **3**
5. **1**	40. **1**	75. **2**	110. **2**	145. **3**	180. **4**
6. **3**	41. **2**	76. **1**	111. **2**	146. **4**	181. **1**
7. **3**	42. **3**	77. **1**	112. **1**	147. **1**	182. **3**
8. **1**	43. **3**	78. **1**	113. **4**	148. **3**	183. **1**
9. **3**	44. **2**	79. **1**	114. **2**	149. **3**	184. **1**
10. **4**	45. **2**	80. **2**	115. **1**	150. **3**	185. **2**
11. **1**	46. **3**	81. **3**	116. **3**	151. **3**	186. **2**
12. **3**	47. **4**	82. **2**	117. **4**	152. **4**	187. **1**
13. **1**	48. **2**	83. **3**	118. **2**	153. **2**	188. **4**
14. **2**	49. **4**	84. **4**	119. **2**	154. **1**	189. **4**
15. **2**	50. **2**	85. **1**	120. **2**	155. **2**	190. **3**
16. **3**	51. **4**	86. **4**	121. **1**	156. **1**	191. **4**
17. **1**	52. **2**	87. **3**	122. **4**	157. **1**	192. **3**
18. **3**	53. **1**	88. **4**	123. **4**	158. **1**	193. **1**
19. **3**	54. **1**	89. **1**	124. **1**	159. **3**	194. **3**
20. **1**	55. **2**	90. **1**	125. **4**	160. **4**	195. **4**
21. **4**	56. **2**	91. **3**	126. **2**	161. **1**	196. **2**
22. **2**	57. **1**	92. **1**	127. **3**	162. **1**	197. **1**
23. **4**	58. **3**	93. **4**	128. **2**	163. **4**	198. **4**
24. **4**	59. **2**	94. **3**	129. **1**	164. **1**	199. **3**
25. **4**	60. **2**	95. **1**	130. **3**	165. **2**	200. **3**
26. **3**	61. **2**	96. **3**	131. **1**	166. **4**	201. **2**
27. **3**	62. **1**	97. **2**	132. **2**	167. **2**	202. **1**
28. **4**	63. **3**	98. **4**	133. **1**	168. **3**	203. **3**
29. **1**	64. **1**	99. **1**	134. **2**	169. **2**	204. **1**
30. **4**	65. **2**	100. **3**	135. **3**	170. **2**	205. **2**
31. **1**	66. **2**	101. **4**	136. **1**	171. **1**	206. **2**
32. **4**	67. **1**	102. **4**	137. **1**	172. **1**	207. **3**
33. **3**	68. **2**	103. **1**	138. **3**	173. **3**	208. **4**
34. **1**	69. **2**	104. **1**	139. **2**	174. **1**	
35. **3**	70. **1**	105. **1**	140. **2**	175. **2**	

CHAPTER FOURTEEN

1.	2	13.	3	25.	3	37.	3	49.	4	61.	1
2.	1	14.	2	26.	4	38.	1	50.	4	62.	1
3.	1	15.	1	27.	1	39.	3	51.	2	63.	1
4.	1	16.	4	28.	2	40.	1	52.	1	64.	2
5.	4	17.	1	29.	2	41.	3	53.	3	65.	3
6.	2	18.	1	30.	1	42.	2	54.	2	66.	1
7.	3	19.	3	31.	2	43.	4	55.	3	67.	2
8.	3	20.	1	32.	2	44.	1	56.	3	68.	1
9.	2	21.	3	33.	4	45.	1	57.	2	69.	1
10.	3	22.	3	34.	2	46.	4	58.	2	70.	1
11.	2	23.	4	35.	3	47.	4	59.	4	71.	2
12.	4	24.	1	36.	2	48.	1	60.	3		

CHAPTER FIFTEEN

1.	4	26.	1	51.	2	76.	3	101.	2	126.	3
2.	1	27.	3	52.	1	77.	1	102.	3	127.	4
3.	1	28.	2	53.	3	78.	2	103.	3	128.	3
4.	3	29.	4	54.	4	79.	4	104.	4	129.	3
5.	1	30.	2	55.	1	80.	2	105.	1	130.	3
6.	2	31.	4	56.	3	81.	2	106.	1	131.	2
7.	3	32.	2	57.	2	82.	4	107.	3	132.	3
8.	4	33.	4	58.	2	83.	2	108.	2	133.	2
9.	4	34.	3	59.	2	84.	4	109.	2	134.	2
10.	2	35.	3	60.	3	85.	4	110.	3	135.	3
11.	1	36.	2	61.	1	86.	1	111.	3	136.	3
12.	2	37.	4	62.	1	87.	1	112.	1	137.	1
13.	2	38.	1	63.	1	88.	4	113.	4	138.	2
14.	3	39.	2	64.	3	89.	2	114.	4	139.	4
15.	3	40.	2	65.	1	90.	1	115.	3	140.	2
16.	2	41.	2	66.	1	91.	2	116.	1	141.	1
17.	2	42.	1	67.	3	92.	3	117.	1	142.	2
18.	1	43.	4	68.	4	93.	2	118.	2	143.	4
19.	3	44.	3	69.	2	94.	2	119.	1	144.	2
20.	3	45.	2	70.	3	95.	4	120.	3	145.	1
21.	2	46.	2	71.	2	96.	3	121.	1	146.	1
22.	3	47.	3	72.	1	97.	1	122.	2	147.	2
23.	2	48.	4	73.	4	98.	2	123.	3	148.	3
24.	3	49.	2	74.	1	99.	3	124.	2		
25.	4	50.	4	75.	4	100.	3	125.	3		

CHAPTER SIXTEEN

1.	3	22.	2	43.	1	64.	4	85.	1
2.	1	23.	1	44.	3	65.	2	86.	3
3.	2	24.	4	45.	4	66.	2	87.	2
4.	3	25.	3	46.	1	67.	3	88.	3
5.	3	26.	2	47.	3	68.	4	89.	4
6.	2	27.	2	48.	2	69.	2	90.	1
7.	3	28.	4	49.	1	70.	3	91.	3
8.	3	29.	2	50.	4	71.	1	92.	3
9.	3	30.	2	51.	3	72.	2	93.	4
10.	2	31.	4	52.	4	73.	2	94.	3
11.	3	32.	1	53.	4	74.	2	95.	1
12.	2	33.	4	54.	4	75.	1	96.	1
13.	2	34.	4	55.	3	76.	1	97.	2
14.	2	35.	2	56.	4	77.	4	98.	1
15.	1	36.	3	57.	4	78.	2	99.	4
16.	4	37.	2	58.	3	79.	4	100.	3
17.	3	38.	1	59.	1	80.	2	101.	1
18.	1	39.	1	60.	2	81.	3	102.	4
19.	1	40.	4	61.	4	82.	3	103.	3
20.	2	41.	3	62.	4	83.	4	104.	2
21.	2	42.	4	63.	3	84.	4	105.	2

106.	2
107.	4
108.	3
109.	2
110.	2
111.	3
112.	3
113.	1
114.	2
115.	2
116.	1
117.	3
118.	3
119.	4
120.	1
121.	2
122.	3
123.	3
124.	2

APPENDIX 1 _____

New York State Reference Tables for Physics

If you take the New York State Regents Examination in Physics, you will be provided with a set of reference tables to aid you in answering the questions on the examination. For students not taking this examination, the tables will provide a convenient reference for answering the questions and problems presented in this book. Each of these tables, along with a brief description of it, is given in this appendix.

List of Physical Constants

The most important physical constants and (where appropriate) their symbols are given in this table. Each constant is given to two significant digits along with its units.

For example, if you were asked to calculate the gravitational force between the Moon and the Earth, you would use this table to find the masses of the Moon and the Earth and the mean Earth-Moon distance.

Absolute Indices of Refraction

The *absolute index of refraction* is defined as the ratio of the speed of light in a vacuum to the speed of light in a medium. (Monochromatic yellow-orange light of 5.9×10^{-7} was used to compute these indices.)

The larger the index of refraction, the slower light travels in the medium. Therefore, according to this table, light travels slowest in diamond and fastest in air. In a vacuum, the index of refraction would be exactly 1.

This table is useful in solving Snell's law and critical angle problems, as well as in comparing the speed of light in different media.

LIST OF PHYSICAL CONSTANTS

Name	Symbol	Value(s)
Gravitational constant	G	6.7×10^{-11} N·m^2/kg^2
Acceleration due to gravity (up to 16 km altitude)	g	9.8 m/s^2
Speed of light in a vacuum	c	3.0×10^8 m/s
Speed of sound at STP		3.3×10^2 m/s
Mass-energy relationship		1 u (amu) = 9.3×10^2 MeV
Mass of the Earth		6.0×10^{24} kg
Mass of the Moon		7.4×10^{22} kg
Mean radius of the Earth		6.4×10^6 m
Mean radius of the Moon		1.7×10^6 m
Mean distance from Earth to Moon		3.8×10^8 m
Electrostatic constant	k	9.0×10^9 N·m^2/C^2
Charge of the electron (1 elementary charge)		1.6×10^{-19} C
One coulomb	C	6.3×10^{18} elementary charges
Electronvolt	eV	1.6×10^{-19} J
Planck's constant	h	6.6×10^{-34} J·s
Rest mass of the electron	m_e	9.1×10^{-31} kg
Rest mass of the proton	m_p	1.7×10^{-27} kg
Rest mass of the neutron	m_n	1.7×10^{-27} kg

ABSOLUTE INDICES OF REFRACTION

$(\lambda = 5.9 \times 10^{-7}$ m$)$

Air	1.00
Alcohol	1.36
Canada Balsam	1.53
Corn Oil	1.47
Diamond	2.42
Glass, Crown	1.52
Glass, Flint	1.61
Glycerol	1.47
Lucite	1.50
Quartz, Fused	1.46
Water	1.33

WAVELENGTHS OF LIGHT IN A VACUUM

Violet	$4.0 - 4.2 \times 10^{-7}$ m
Blue	$4.2 - 4.9 \times 10^{-7}$ m
Green	$4.9 - 5.7 \times 10^{-7}$ m
Yellow	$5.7 - 5.9 \times 10^{-7}$ m
Orange	$5.9 - 6.5 \times 10^{-7}$ m
Red	$6.5 - 7.0 \times 10^{-7}$ m

Wavelengths of Light in a Vacuum

The range of wavelengths of *visible* light is presented in this table. Note that each color corresponds to a range and that adjacent colors, such as blue and green, blend into one another.

The table would be useful in calculating the frequencies of various colors of light in a vacuum ($c = f\lambda$), or in determining the color, given the frequency of the light.

Heat Constants

Various constants and their units are provided for a number of substances.

The *(average) specific heat* is defined as the quantity of heat energy (in kilojoules) needed to change the temperature of 1 kilogram of substance 1 Celsius degree. The smaller the specific heat, the more the temperature of the substance will change in response to the addition or removal of heat energy. Note that the specific heat is given for a particular *phase* and that water, ice, and steam have three distinct specific heats.

The *melting and boiling points* are the temperatures at which the solid-liquid and liquid-gas phase transitions occur.

The *heat of fusion* and *heat of vaporization* are defined as the quantities of heat energy (in kilojoules) needed to melt or boil 1 kilogram of substance. Note that the heat of vaporization is always larger than the heat of fusion because conversion of a liquid into a gas requires more energy in order to free the liquid molecules of their bonds.

HEAT CONSTANTS

	Specific Heat (average) (kJ/kg·C°)	Melting Point (°C)	Boiling Point (°C)	Heat of Fusion (kJ/kg)	Heat of Vaporization (kJ/kg)
Alcohol (ethyl)	2.43 (liq.)	−117	79	109	855
Aluminum	0.90 (sol.)	660	2467	396	10500
Ammonia	4.71 (liq.)	−78	−33	332	1370
Copper	0.39 (sol.)	1083	2567	205	4790
Iron	0.45 (sol.)	1535	2750	267	6290
Lead	0.13 (sol.)	328	1740	25	866
Mercury	0.14 (liq.)	−39	357	11	295
Platinum	0.13 (sol.)	1772	3827	101	229
Silver	0.24 (sol.)	962	2212	105	2370
Tungsten	0.13 (sol.)	3410	5660	192	4350
Water { ice	2.05 (sol.)	0	—	334	—
water	4.19 (liq.)	—	100	—	2260
steam	2.01 (gas)	—	—	—	—
Zinc	0.39 (sol.)	420	907	113	1770

Energy Level Diagrams for Mercury and Hydrogen

Each energy state is represented by a horizontal line. Its energy value (in electron-volts) is given at the right. Note that these values are *negative* numbers that increase to a maximum value of *zero*. The lowest energy state is called the *ground state*; every other state is an *excited state*. At a value of 0 electron-volt, the electron is no longer associated with the atom, a condition known as *ionization*.

At the left is the label corresponding to the energy state. For hydrogen, these states are integers known as *principal quantum numbers*. For mercury, the atom is much more complex and the various states are represented as letters.

To calculate the energy involved in a particular transition, one *subtracts* the final energy value from the initial value. If the difference is negative, energy is released by the atom as a photon; if the difference is positive, the energy is absorbed by the atom.

ENERGY LEVEL DIAGRAMS FOR MERCURY AND HYDROGEN

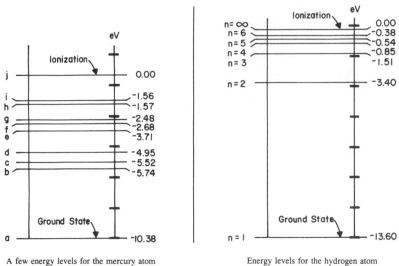

A few energy levels for the mercury atom   Energy levels for the hydrogen atom

Values of Trigonometric Functions

The sines and cosines of the angles between 1° and 90° are given to four digits. If you don't have a calculator (shudder!), you can use these values to calculate vector components and solve problems involving Snell's law (see Chapter 13).

Uranium Disintegration Series

This diagram is a graphical representation of the steps involved in the decay of uranium-238 to the stable nuclide lead-206.

The atomic numbers and symbols of the nuclides are given across the top of the diagram; the mass numbers, along the right side.

Each nuclide involved is represented by a dot (•). The arrow connecting two adjacent nuclides represents a decay reaction. A horizontal arrow represents a beta (–) decay; a diagonal arrow, an alpha decay.

VALUES OF TRIGONOMETRIC FUNCTIONS

Angle	Sine	Cosine	Angle	Sine	Cosine
1°	.0175	.9998	46°	.7193	.6947
2°	.0349	.9994	47°	.7314	.6820
3°	.0523	.9986	48°	.7431	.6691
4°	.0698	.9976	49°	.7547	.6561
5°	.0872	.9962	50°	.7660	.6428
6°	.1045	.9945	51°	.7771	.6293
7°	.1219	.9925	52°	.7880	.6157
8°	.1392	.9903	53°	.7986	.6018
9°	.1564	.9877	54°	.8090	.5878
10°	.1736	.9848	55°	.8192	.5736
11°	.1908	.9816	56°	.8290	.5592
12°	.2079	.9781	57°	.8387	.5446
13°	.2250	.9744	58°	.8480	.5299
14°	.2419	.9703	59°	.8572	.5150
15°	.2588	.9659	60°	.8660	.5000
16°	.2756	.9613	61°	.8746	.4848
17°	.2924	.9563	62°	.8829	.4695
18°	.3090	.9511	63°	.8910	.4540
19°	.3256	.9455	64°	.8988	.4384
20°	.3420	.9397	65°	.9063	.4226
21°	.3584	.9336	66°	.9135	.4067
22°	.3746	.9272	67°	.9205	.3907
23°	.3907	.9205	68°	.9272	.3746
24°	.4067	.9135	69°	.9336	.3584
25°	.4226	.9063	70°	.9397	.3420
26°	.4384	.8988	71°	.9455	.3256
27°	.4540	.8910	72°	.9511	.3090
28°	.4695	.8829	73°	.9563	.2924
29°	.4848	.8746	74°	.9613	.2756
30°	.5000	.8660	75°	.9659	.2588
31°	.5150	.8572	76°	.9703	.2419
32°	.5299	.8480	77°	.9744	.2250
33°	.5446	.8387	78°	.9781	.2079
34°	.5592	.8290	79°	.9816	.1908
35°	.5736	.8192	80°	.9848	.1736
36°	.5878	.8090	81°	.9877	.1564
37°	.6018	.7986	82°	.9903	.1392
38°	.6157	.7880	83°	.9925	.1219
39°	.6293	.7771	84°	.9945	.1045
40°	.6428	.7660	85°	.9962	.0872
41°	.6561	.7547	86°	.9976	.0698
42°	.6691	.7431	87°	.9986	.0523
43°	.6820	.7314	88°	.9994	.0349
44°	.6947	.7193	89°	.9998	.0175
45°	.7071	.7071	90°	1.0000	.0000

URANIUM DISINTEGRATION SERIES

Atomic Number and Chemical Symbol

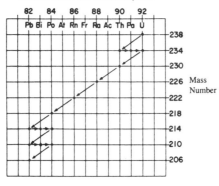

APPENDIX 2 _____

Summary of Equations for New York State Regents Examination in Physics

A summary of the equations used in the New York State Regents physics course is provided in this appendix.

The equations are categorized by areas:

Part I	Part II
Mechanics	Internal energy
Energy	Electromagnetic applications
Electricity and magnetism	Motion in a plane
Wave phenomena	Geometric optics
Modern physics	Nuclear energy

Notes:
The relevant symbols are provided at the right of each set of equations. Your teacher or textbook may use different symbols in connection with these equations.

No equations are provided for the Part II unit on solid-state physics.

Some of the equations used in this book do not appear on these summary tables; you will not be required to use these equations on the New York State Regents Examination.

SUMMARY OF EQUATIONS

MECHANICS

$$\bar{v} = \frac{\Delta s}{\Delta t}$$

$$\bar{v} = \frac{v_f + v_i}{2}$$

$$\bar{a} = \frac{\Delta v}{\Delta t}$$

$$\Delta s = v_i \Delta t + \frac{1}{2}a(\Delta t)^2$$

$$v_f^2 = v_i^2 + 2a\Delta s$$

$$F = ma$$

$$w = mg$$

$$F = \frac{Gm_1 m_2}{r^2}$$

$$p = mv$$

$$J = F\Delta t$$

$$F\Delta t = m\Delta v$$

a = acceleration
r = distance between centers
F = force
g = acceleration due to gravity
G = universal gravitation constant
J = impulse
m = mass
p = momentum
Δs = displacement
t = time
v = velocity
w = weight

ENERGY

$$W = F\Delta s$$

$$P = \frac{W}{\Delta t} = \frac{F\Delta s}{\Delta t} = F\bar{v}$$

$$\Delta PE = mg\Delta h$$

$$KE = \frac{1}{2}mv^2$$

$$F = kx$$

$$PE_s = \frac{1}{2}kx^2$$

F = force
g = acceleration due to gravity
h = height
k = spring constant
KE = kinetic energy
m = mass
P = power
PE = potential energy
PE_s = potential energy stored in a spring
Δs = displacement
t = time
v = velocity
W = work
x = change in spring length from the equilibrium position

ELECTRICITY AND MAGNETISM

$$F = \frac{kq_1 q_2}{r^2}$$

$$E = \frac{F}{q}$$

$$V = \frac{W}{q}$$

$$E = \frac{V}{d}$$

$$I = \frac{\Delta q}{\Delta t}$$

$$R = \frac{V}{I}$$

$$P = VI = I^2 R = \frac{V^2}{R}$$

$$W = Pt = VIt = I^2 Rt$$

d = separation of parallel plates
r = distance between centers
E = electric field intensity
F = force
I = current
k = electrostatic constant
ℓ = length of conductor
P = power
q = charge
R = resistance
t = time
V = electric potential difference
W = energy

Series Circuits:

$$I_t = I_1 = I_2 = I_3 = \ldots$$

$$V_t = V_1 + V_2 + V_3 + \ldots$$

$$R_t = R_1 + R_2 + R_3 + \ldots$$

Parallel Circuits:

$$I_t = I_1 + I_2 + I_3 + \ldots$$

$$V_t = V_1 = V_2 = V_3 = \ldots$$

$$\frac{1}{R_t} = \frac{1}{R_1} + \frac{1}{R_2} + \frac{1}{R_3} + \ldots$$

INTERNAL ENERGY

$$Q = mc\Delta T_c$$

$$Q_f = mH_f$$

$$Q_v = mH_v$$

c = specific heat
H_f = heat of fusion
H_v = heat of vaporization
m = mass
Q = amount of heat
T_c = Celsius temperature

WAVE PHENOMENA

$$T = \frac{1}{f}$$

$$v = f\lambda$$

$$n = \frac{c}{v}$$

$$\sin \theta_c = \frac{1}{n}$$

$$n_1 \sin \theta_1 = n_2 \sin \theta_2$$

$$n_1 v_1 = n_2 v_2$$

$$\frac{\lambda}{d} = \frac{x}{L}$$

c = speed of light in a vacuum
d = distance between slits
f = frequency
L = distance from slit to screen
n = index of absolute refraction
T = period
v = speed
x = distance from central maximum to first-order maximum
λ = wavelength
θ = angle
θ_c = critical angle of incidence relative to air

ELECTROMAGNETIC APPLICATIONS

$$F = qvB$$

$$\frac{N_p}{N_s} = \frac{V_p}{V_s}$$

$$V_p I_p = V_s I_s \text{ (ideal)}$$

% Efficiency =

$$\frac{V_s I_s}{V_p I_p} \times 100$$

$$V = B\ell v$$

B = flux density
F = force
I_p = current in primary coil
I_s = current in secondary coil
N_p = number of turns of primary coil
N_s = number of turns of secondary coil
q = charge
v = velocity
V_p = voltage of primary coil
V_s = voltage of secondary coil
ℓ = length of conductor
V = electric potential difference

MODERN PHYSICS

$$W_o = hf_o$$

$$E_{photon} = hf$$

$$KE_{max} = hf - W_o$$

$$p = \frac{h}{\lambda}$$

$$E_{photon} = E_i - E_f$$

c = speed of light in a vacuum
E = energy
f = frequency
f_o = threshold frequency
h = Planck's constant
KE = kinetic energy
p = momentum
W_o = work function
λ = wavelength

MOTION IN A PLANE

$$v_{iy} = v_i \sin \theta$$

$$v_{ix} = v_i \cos \theta$$

$$a_c = \frac{v^2}{r}$$

$$F_c = \frac{mv^2}{r}$$

a_c = centripetal acceleration
F_c = centripetal force
m = mass
r = radius
v = velocity
θ = angle

GEOMETRIC OPTICS

$$\frac{1}{d_o} + \frac{1}{d_i} = \frac{1}{f}$$

$$\frac{S_o}{S_i} = \frac{d_o}{d_i}$$

d_i = image distance
d_o = object distance
f = focal length
S_i = image size
S_o = object size

NUCLEAR ENERGY

$$E = mc^2$$

$$m_f = \frac{m_i}{2^n}$$

c = speed of light in a vacuum
E = energy
m = mass
n = number of half-lives

APPENDIX 3 _____

Answering Free-Response Questions

A *free-response question* is an examination question that requires the test taker to do more than to choose among several responses or to fill in a blank. You may need to perform numerical calculations, draw and interpret graphs, and provide extended written responses to a question or problem.

Part III of the New York State Regents Examination in Physics contains free-response questions. This appendix is designed to provide you with a number of general guidelines for answering them.

Solving Problems Involving Numerical Calculations

To receive full credit you must:

- Provide the appropriate equation(s).
- Substitute values and units into the equation(s).
- Display the answer, with appropriate units.
- If the answer is a vector quantity, include its direction.

Although SI units are used on the Regents examination, you are expected to have some familiarity with other metric units such as the gram and the kilometer.

You will *not* be penalized if your answer has an incorrect number of significant digits. However, it is always good practice to pay attention to this detail.

You should write as legibly as possible. Teachers are human, and nothing irks them more than trying to decipher a careless, messy scrawl. It is also a good idea to identify your answer clearly, either by placing it in a box or by writing the word "answer" next to it.

A final word: If you provide the correct answer but do not show any work, you will not receive any credit for the problem!

The following is a sample problem and its model solution.

PROBLEM
A 5.0-kilogram object has a velocity of 10. meters per second [east]. Calculate the momentum of this object.

SOLUTION

$$\mathbf{p} = m\mathbf{v}$$

$$\mathbf{p} = (5.0 \text{ kg})(10. \text{ m/s [east]})$$

$$\boxed{\mathbf{p} = 50. \text{ kg·m/s [east]}}$$

Graphing Experimental Data

To receive full credit you must:

- Label both axes with the appropriate variables and units.
- Divide the axes so that the data ranges fill the graph as nearly as possible.
- Plot all data points accurately.
- Draw a best-fit line carefully with a straightedge. The line should pass through the origin *only if the data warrant it.*
- If a part of the question requires that the slope be calculated, calculate the slope *from the line,* not from individual data points.

Generally a graph should have a title, and the *independent variable* is usually drawn along the *x*-axis. However, you will not be penalized on the Regents examination if you do not follow these conventions.

The following is a sample problem and its model solution.

PROBLEM
The weights of various masses, measured on Planet *X*, are given in the table below.

Mass (kg)	Weight (N)
15	21
20.	32
25	35
30.	48
35	56

(1) Draw a graph that illustrates these data.
(2) Use the *graph* to calculate the acceleration due to gravity on Planet *X*.

SOLUTION

(1) The graph shown below incorporates the essential items that were listed in the table.

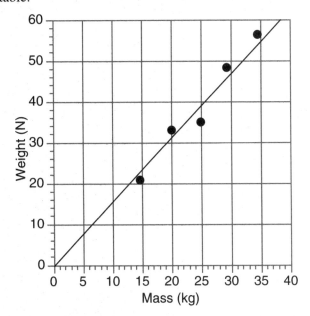

(2) Since the magnitude of the gravitational acceleration can be determined by calculating the ratio of weight to mass ($g = W/m$), we can calculate the value of g from the slope of the graph.

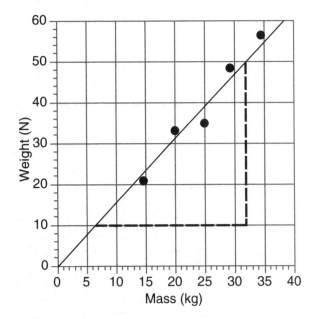

We choose two points on the line; we do not use the data points themselves:

$$g = \frac{\Delta W}{\Delta m} = \frac{50.\ N - 10.\ N}{32\ kg - 6\ kg} = 1.5\ \frac{N}{kg} = 1.5\ m/s^2$$

Drawing Diagrams

To receive full credit you must:

- Draw your diagrams neatly, and label them clearly.
- Draw vectors *to scale* and *in the correct direction.* If you are given a scale, you must draw your vectors to that scale.
- Bring a straightedge and a protractor with you so that you can draw neat, accurate diagrams.

Writing a Free-Response Answer

To receive full credit you must:

- Use complete, clear sentences that make sense to the reader.
- Use correct physics in your explanations.

A sample question and acceptable and unacceptable answers are given below.

QUESTION
If the data from two different photoemissive materials are graphed, which characteristic of the two graphs will be the same?

ACCEPTABLE ANSWERS
- The characteristic that will be the same is the slope of the line.
- It is the slope.
- The two lines have the same slope.

The following answers are unacceptable:

- The slope. (Incomplete sentence)
- Both lines will have the same y-intercept. (Incorrect physics)

APPENDIX 4 _____

The New York State Regents Examination in Physics

The Regents examination in physics is divided into three parts.

Part I consists of 55 multiple-choice questions that count for 65 points. All students must answer every question in this part of the examination.

The questions for Part I are drawn from the first five units of the New York State physics syllabus:

Unit One—Mechanics
Unit Two—Energy
Unit Three—Electricity and Magnetism
Unit Four—Wave Phenomena
Unit Five—Modern Physics

Part II consists of six groups that contain ten multiple-choice questions each. A student must choose only *two* of these groups and answer *every* question in each of the two groups he or she has chosen. Part II counts for 20 points.

The questions for Part II are drawn from the last six units of the New York State physics syllabus:

Unit Six—Motion in a Plane
Unit Seven—Internal Energy
Unit Eight—Electromagnetic Applications
Unit Nine—Geometric Optics
Unit Ten—Solid-State Physics
Unit Eleven—Nuclear Energy

Part III consists of three free-response questions that count for 15 points. All students must answer every question in this part of the examination.

Questions for Part III are drawn from the first five units of the New York State physics syllabus.

You should refer to the detailed outline that follows for a complete listing of the topics covered in each unit of the New York State physics syllabus. Refer to the page numbers that follow each topic to find where that topic is reviewed in the text.

Regents Topic Outline

Examination June 2000

Physics

PART I

Answer all 55 questions in this part. [65]

Directions (1–55): For *each* statement or question, select the word or expression that, of those given, best completes the statement or answers the question. Record the answers to those questions in the spaces provided.

1 The map below shows the route traveled by a school bus.

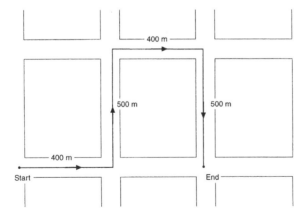

What is the magnitude of the total displacement of the school bus from the start to the end of its trip?

(1) 400 m   (3) 800 m

(2) 500 m   (4) 1,800 m

1 $\dfrac{3}{}$

$V = \dfrac{d}{t}$

Which pair of graphs represent the same motion?

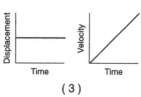

(3)

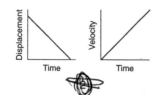

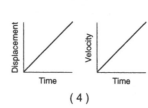

(4)

2 $\dfrac{4}{}$

3 A runner starts from rest and accelerates uniformly to a speed of 8.0 meters per second in 4.0 seconds. The magnitude of the acceleration of the runner is

(1) 0.50 m/s²   (3) 9.8 m/s²

(2) 2.0 m/s²   (4) 32 m/s²

3 $\dfrac{2}{}$

$V_1 = 0$

$V_f = 8 \, m/s$

$t = 4$

$a = \dfrac{8 - 0}{4}$

552

4 A cart moving across a level surface accelerates uniformly at 1.0 meter per second² for 2.0 seconds. What additional information is required to determine the distance traveled by the cart during this 2.0-second interval?

$a = 1 \, m/s^2$

$t = 2 \, sec$

$d = ?$

   1 coefficient of friction between the cart and the surface
   2 mass of the cart
   3 net force acting on the cart
   4 initial velocity of the cart

4 __4__

5 In the diagram below, a force, F, is applied to the handle of a lawnmower inclined at angle θ to the ground.

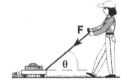

$F = F \cos \theta$

The magnitude of the horizontal component of force F depends on

   1 the magnitude of force F, only
   2 the measure of angle θ, only
   3 both the magnitude of force F and the measure of angle θ
   4 neither the magnitude of force F nor the measure of angle θ

5 __3__

$S \uparrow \downarrow 2$

6 Equilibrium exists in a system where three forces are acting concurrently on an object. If the system includes a 5.0-newton force due north and a 2.0-newton force due south, the third force must be

(1) 7.0 N south   (3) 3.0 N south
(2) 7.0 N north   (4) 3.0 N north

6 __3__

7 A ball is thrown straight up with a speed of 12 meters per second near the surface of Earth. What is the maximum height reached by the ball? [Neglect air friction.]

$V = 12 m/s$

$Vf = ?$

(1) 15 m   (3) 1.2 m
(2) 7.3 m   (4) 0.37 m

7 __2__

$a = -9.81$

$0 = 144 + 2(9.81)(d)$

$d = (12)($

-144

8 Which object weighs approximately 1 newton?

1 dime   3 physics student
2 paper clip   4 golf ball

8 __4__

9 Which terms represent a vector quantity and its respective unit?

1 weight — kilogram
2 mass — kilogram
3 force — newton
4 momentum — newton

9 __3__

10 The vector below represents the resultant of two forces acting concurrently on an object at point *P*.

Which pair of vectors best represents two concurrent forces that combine to produce this resultant force vector?

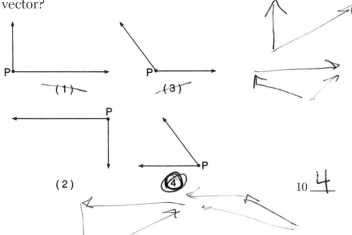

(1)

(2)

(3)

(4)

10 __4__

11 Compared to 8 kilograms of feathers, 6 kilograms of lead has

1 less mass and less inertia
2 less mass and more inertia
3 more mass and less inertia
4 more mass and more inertia

11 __1__

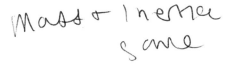

Mass or Inertia
same

12 Two forces are applied to a 2.0-kilogram block on a frictionless horizontal surface, as shown in the diagram below.

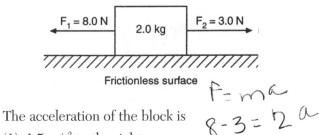

Frictionless surface

The acceleration of the block is

$F = ma$

$8 - 3 = 2\, a$

(1) 1.5 m/s² to the right
(2) 2.5 m/s² to the left
(3) 2.5 m/s² to the right
(4) 4.0 m/s² to the left

12 _2_

$60 = 15\, a$

$4 = a$

13 A 15-kilogram mass weighs 60. newtons on planet X. The mass is allowed to fall freely from rest near the surface of the planet. After falling for 6.0 seconds, the acceleration of the mass is

$V_i = 0$

$t = 6$

(1) 0.25 m/s²   (3) 24 m/s²
(2) 10. m/s²   (4) 4.0 m/s²

13 _4_

14 Sand is often placed on an icy road because the sand

1 decreases the coefficient of friction between the tires of a car and the road
2 increases the coefficient of friction between the tires of a car and the road
3 decreases the gravitational force on a car
4 increases the normal force of a car on the road

14 _2_

$(2)(6) = (3)(v_1)$

15 A 2.0-kilogram cart moving due east at 6.0 meters per second collides with a 3.0-kilogram cart moving due west. The carts stick together and come to rest after the collision. What was the initial speed of the 3.0-kilogram cart?

(1) 1.0 m/s   (3) 9.0 m/s
(2) 6.0 m/s   (4) 4.0 m/s

15 __4__

16 What is the momentum of a 1,200-kilogram car traveling at 15 meters per second due east?

(1) 80. kg•m/s due east $P = m v$
(2) 80. kg•m/s due west $(1200)(15)$
(3) 1.8 × 10⁴ kg•m/s due east
(4) 1.8 × 10⁴ kg•m/s due west

16 __3__

17 A 2,400-kilogram car is traveling at a speed of 20. meters per second. Compared to the magnitude of the force required to stop the car in 12 seconds, the magnitude of the force required to stop the car in 6.0 seconds is

$Ft = \Delta p$

$F(12) = (2400)(20)$

4000

8000

1 half as great   3 the same
2 twice as great   4 four times as great

17 __2__

18 A student applies a 20.-newton force to move a crate at a constant speed of 4.0 meters per second across a rough floor. How much work is done by the student on the crate in 6.0 seconds?

$20 = m(9.8)$

(1) 80. J   (3) 240 J
(2) 120 J   (4) 480 J

18 __4__

average velocity

a

$F \bar{v}$ $W = Fd$

(20)

$(20)(4) = \dfrac{W}{6}$

$\dfrac{W}{t} = \dfrac{Fd}{t}$

$mV = W$

$(t)\dfrac{t}{v}$ $\dfrac{W}{6}$

557

$$Fg = \frac{Gm_1m_2}{r^2}$$

19 The gravitational force of attraction between two objects would be increased by

   1 doubling the mass of both objects, only

   2 doubling the distance between the objects, only

   3 doubling the mass of both objects and doubling the distance between the objects

   4 doubling the mass of one object and doubling the distance between the objects

19 _1_

20 A 5.0×10^2-newton girl takes 10. seconds to run up two flights of stairs to a landing, a total of 5.0 meters vertically above her starting point. What power does the girl develop during her run?

  (1) 25 W   (3) 250 W

  (2) 50. W   (4) 2,500 W

20 _3_

21 The kinetic energy of a 980-kilogram race car traveling at 90. meters per second is approximately

  (1) 4.4×10^4 J   (3) 4.0×10^6 J

  (2) 8.8×10^4 J   (4) 7.9×10^6 J

21 _3_

$$KE = \frac{1}{2}mv^2$$
$$\frac{1}{2}(980)(90)^2$$

$$P = \frac{W}{t}$$

$$\frac{(5 \times 10^2)(5)}{(10)}$$

22 Two aluminum spheres of identical mass and identical charge q hang from strings of equal length. If the spheres are in equilibrium, which diagram best represents the direction of each force acting on the spheres?

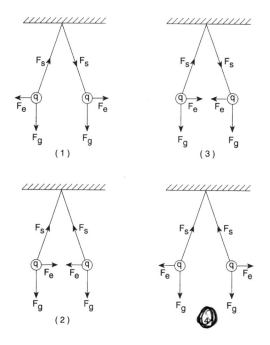

22 ___ 4

23 Moving 2.0 coulombs of charge a distance of 6.0 meters from point A to point B within an electric field requires a 5.0-newton force. What is the electric potential difference between points A and B?

(1) 60. V   (3) 15 V
(2) 30. V   (4) 2.5 V

23 ___ 3

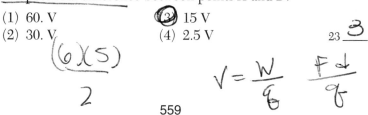

$(6)(5)$

2

$V = \dfrac{W}{q}$ $\dfrac{F \, d}{q}$

24 A metal sphere having an excess of +5 elementary charges has a net electric charge of

(1) 1.6×10^{-19} C
(2) 8.0×10^{-19} C
(3) 5.0×10^0 C
(4) 3.2×10^{19} C

$5 \times (1.6 \times 10^{-19})$

24 __2__

25 The graph below represents the relationship between the force applied to a spring and the compression (displacement) of the spring.

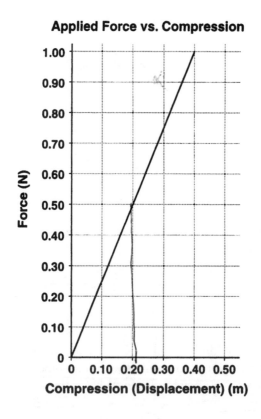

Applied Force vs. Compression

<handwriting>$F_s = kx$

$\cdot 50 = k(\cdot 20)$</handwriting>

What is the spring constant for this spring?

(1) 1.0 N/m   (3) 0.20 N/m

(2) 2.5 N/m   (4) 0.40 N/m

25 __2__

26 A lightning bolt transfers 6.0 coulombs of charge from a cloud to the ground in 2.0×10^{-3} second. What is the average current during this event?

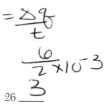

<handwriting>$I = \dfrac{\Delta q}{t}$

$\dfrac{6}{2 \times 10^{-3}} \times 10^{-3}$</handwriting>

(1) 1.2×10^{-2} A   (2) 3.0×10^{3} A

(2) 3.0×10^{2} A   (4) 1.2×10^{4} A

26 __3__

27 Conductivity in metallic solids is due to the presence of free

1 nuclei   3 neutrons

2 protons   (4) electrons

27 __4__

28 The diagram below shows the initial charge and position of three metal spheres, X, Y, and Z, on insulating stands.

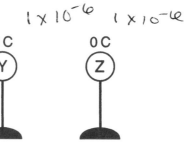

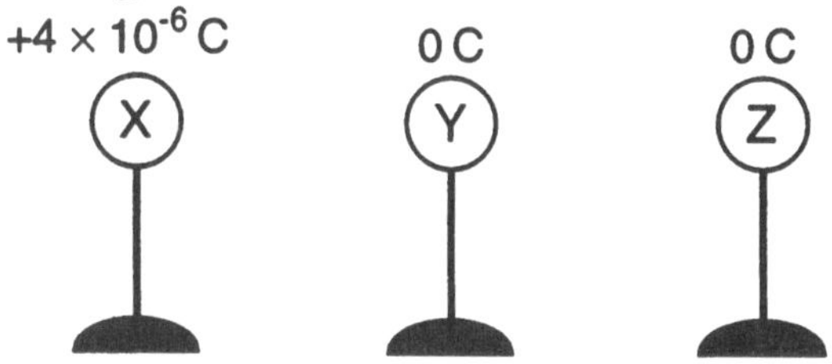

2×10^{-6} 1×10^{-6} 1×10^{-6}

+4 × 10⁻⁶ C   0 C   0 C

X   Y   Z

Sphere X is brought into contact with sphere Y and then removed. Then sphere Y is brought into contact with sphere Z and removed. What is the charge on sphere Z after this procedure is completed?

(1) $+1 \times 10^{-6}$ C   (3) $+3 \times 10^{-6}$ C

(2) $+2 \times 10^{-6}$ C   (4) $+4 \times 10^{-6}$ C

28 __1__

29 In the diagram below, a student compresses the spring in a pop-up toy 0.020 meter.

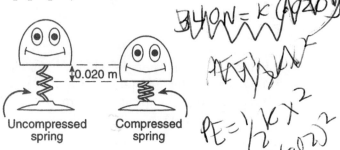

Uncompressed spring   Compressed spring

If the spring has a spring constant of 340 newtons per meter, how much energy is being stored in the spring?

(1) 0.068 J   (3) 3.4 J

(2) 0.14 J   (4) 6.8 J

$W = F d$

29 ___4___

30 Gravitational field strength is to newtons per kilogram as electric field strength is to

1 coulombs per joule
2 coulombs per newton
3 joules per coulomb
4 newtons per coulomb

30 ___4___

$E \neq \dfrac{Fe}{q}$

562

31 Which diagram best represents the magnetic field around a straight wire in which electrons are flowing from left to right?

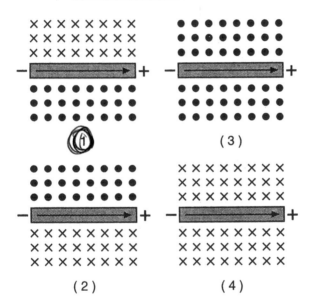

32 The graph below represents the relationship between the potential difference across a metal conductor and the current through the conductor at a constant temperature.

Potential Difference vs. Current

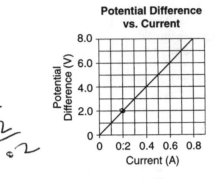

$R = \dfrac{V}{I}$

$R = \dfrac{2}{.2}$

What is the resistance of the conductor?

(1) 1 Ω

(2) 0.01 Ω

(3) 0.1 Ω

(4) 10 Ω

32 ___4___

33 The graph below shows the relationship between the work done on a charged body in an electric field and the net charge on the body.

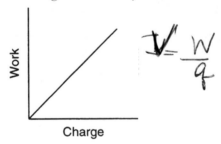

$V = \dfrac{W}{q}$

What does the slope of this graph represent?

1 power

2 potential difference

3 force

4 electric field intensity

33 ___2___

34 In the diagram below, the distance between points
    A and B on a wave is 5.0 meters.

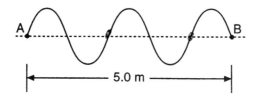

2.5
waves

The wavelength of this wave is

(1) 1.0 m   (3) 5.0 m
(2) 2.0 m   (4) 4.0 m

34 __2__

35 The diagram below shows two resistors connected
    in series to a 20.-volt battery.

I = 1 A

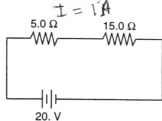

20. V

If the current through the 5.0-ohm resistor is 1.0
ampere, the current through the 15.0-ohm resistor is

(1) 1.0 A   (3) 3.0 A
(2) 0.33 A   (4) 1.3 A

35 __1__

36 Resistors R_1 and R_2 have an equivalent resistance of 6 ohms when connected in the circuit shown below.

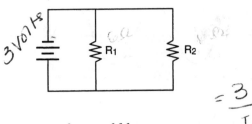

$= \dfrac{3}{I}$

The resistance of R_1 could be

(1) 1 Ω   (3) 8 Ω

(2) 5 Ω   (4) 4 Ω   36 $\underline{3}$

total equivalent always less than any individual resistance

37 The diagram below represents an electric circuit.

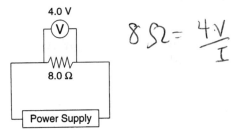

$8\,\Omega = \dfrac{4\,V}{I}$

The total amount of energy delivered to the resistor in 10. seconds is

(1) 3.2 J   (3) 20. J

(2) 5.0 J   (4) 320 J   37 $\underline{3}$

$W = 4\left(\tfrac{1}{2}\right)(10)$

38 The diagram below shows an electron, *e*, located in a magnetic field.

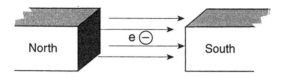

There is *no* magnetic force on the electron when it moves

1 toward the right side of the page
2 toward the top of the page
3 into the page
4 out of the page

electric charge in magnetic field experience no force when more parallel to the magnetic field lines

38 ___1___

39 A copper wire is part of a complete circuit through which current flows. Which graph best represents the relationship between the wire's length and its resistance?

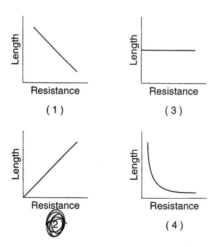

39 ___2___

40 The heating element on an electric stove dissipates 4.0×10^2 watts of power when connected to a 120-volt source. What is the electrical resistance of this heating element?

(1) 0.028 Ω   (3) 3.3 Ω

(2) 0.60 Ω   (4) 36 Ω

40 __4__

$$4 \times 10^2 = \frac{120}{R}$$

41 A monochromatic beam of light has a frequency of 6.5×10^{14} hertz. What color is the light?

1 yellow   3 violet

2 orange   4 blue

41 __4__

42 Two waves having the same amplitude and the same frequency pass simultaneously through a uniform medium. Maximum destructive interference occurs when the phase difference between the two waves is

(1) 0°   (3) 180°

(2) 90°   (4) 360°

42 __3__

by definition maximum destructive interference occurs when 2 waves having 180 phase difference through a uniform medium simultaneously

43 The diagram below shows a tuning fork vibrating in air. The dots represent air molecules as the sound wave moves toward the right.

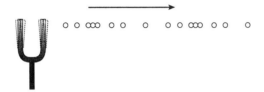

Which diagram best represents the direction of motion of the air molecules?

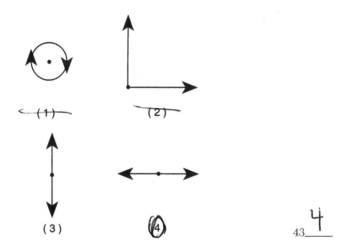

43___

44 The diagram below represents a wave traveling in a uniform medium.

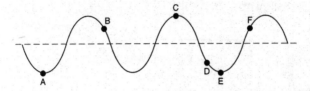

Which two points on the wave are in phase?

(1) A and C   (3) B and D

(2) A and E   (4) B and F   44 _2_

Base your answers to questions 45 through 47 on the diagram below which represents a beam of monochromatic light ($\lambda = 5.9 \times 10^{-7}$ meter) traveling from Lucite into air.

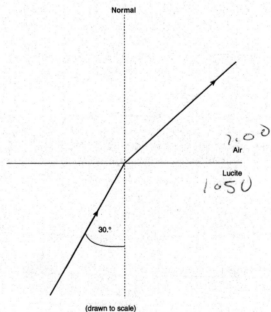

(drawn to scale)

45 What is the measure of the angle of refraction?
[Use a protractor or a mathematical calculation.]

(1) 19°
(2) 30.°

(3) 49°
(4) 60.°

$(1.50)(\sin 30) = (1.00)(\sin \theta)$

45 __3__

46 The speed of the light in Lucite is

(1) 1.5×10^8 m/s
(2) 2.0×10^8 m/s
(3) 3.0×10^8 m/s
(4) 4.5×10^8 m/s

$n = \dfrac{3 \times 10^8}{\text{v}}$

46 __2__

47 The critical angle for the Lucite-air boundary is
approximately

$\sin \theta = \dfrac{1}{1.50}$

(1) 67°
(2) 48°
(3) 42°
(4) 33°

47 __3__

48 A light ray is incident on a plane mirror as shown in
the diagram below.

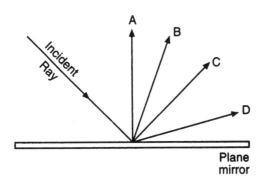

Which ray best represents the reflected ray?

(1) A
(2) B
(3) C
(4) D

48 __3__

49 What occurs as a ray of light passes from air into water?

$1.00 \rightarrow 1.33$

  ① The ray must decrease in speed.
  2 The ray must increase in speed.
  3 The ray must decrease in frequency.
  4 The ray must increase in frequency.

49 ___1___

50 Which waves can be polarized?
  ① light waves from an incandescent bulb
  2 sound waves from a tuba
  3 longitudinal waves
  4 seismic waves (P-waves)

50 ___

Light is a transverse wave - only transverse that can be polarized.

51 In which part of the electromagnetic spectrum does a photon have the greatest energy?

  1 red   3 violet
  2 infrared   ④ ultraviolet

51 ___4___

$E = hf$

52 If all parts of a light beam have a constant phase relationship, with the same wavelength and frequency, the light beam is

  ① monochromatic and coherent
  2 monochromatic and incoherent
  3 polychromatic and coherent
  4 polychromatic and incoherent

52 ___1___

53 A beam of monochromatic light incident on a metal surface causes the emission of photoelectrons. The length of time that the surface is illuminated by this beam is varied, but the intensity of the beam is kept constant. Which graph best represents the relationship between the total number of photoelectrons emitted and the length of time of illumination?

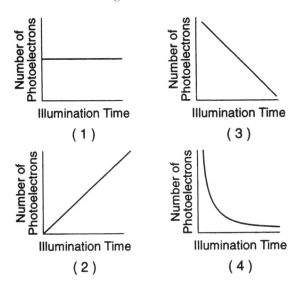

(1)   (3)

(2)   (4)

53 __2__

54 What is the minimum energy required to excite a mercury atom initially in the ground state?

(1) 4.64 eV   (3) 10.20 eV
(2) 5.74 eV   (4) 10.38 eV

54 ____

55 The diagram below represents the hyperbolic path of an alpha particle as it passes very near the nucleus of a gold atom.

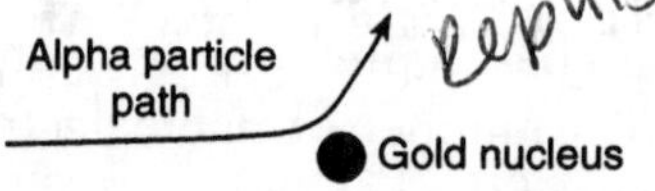

Alpha particle
path

repulsive

● Gold nucleus

The shape of the path is caused by the force between the

1 positively charged alpha particle and the neutral nucleus

② positively charged alpha particle and the positively charged nucleus

3 negatively charged alpha particle and the neutral nucleus

4 negatively charged alpha particle and the positively charged nucleus

55 __2__

PART II

This part consists of six groups, each containing ten questions. Each group tests an optional area of the course. Choose two of these six groups. Be sure that you answer all ten questions in each group chosen. Record the answers to these questions in the spaces provided. [20]

GROUP 1—Motion in a Plane

*If you choose this group, be sure to answer questions **56–65**.*

Base your answers to questions 56 through 58 on the diagram and information below.

A machine launches a tennis ball at an angle of 45° with the horizontal, as shown. The ball has an initial vertical velocity of 9.0 meters per second and an initial horizontal velocity of 9.0 meters per second. The ball reaches its maximum height 0.92 second after its launch. [Neglect air resistance and assume the ball lands at the same height above the ground from which it was launched.]

$V_{iv} = 9 m/s$

$V_{iH} = 9 m/s$

$t = 0.92 s$

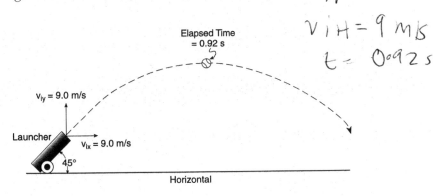

Elapsed Time = 0.92 s

$v_{iy} = 9.0$ m/s

Launcher

$v_{ix} = 9.0$ m/s

45°

Horizontal

$V_{1y} = V_1 \sin \theta$

$V_1 = \dfrac{90 m/s}{\sin 45}$

56 The speed of the tennis ball as it leaves the launcher is approximately

(1) 4.5 m/s   (3) 13 m/s
(2) 8.3 m/s   (4) 18 m/s

56 __3__

$V_1 = (13)(\sin 45)$

57 The total horizontal distance traveled by the tennis ball during the entire time it is in the air is approximately

(1) 23 m   (3) 8.3 m
(2) 17 m   (4) 4.1 m

2

57 __X__

$d = (13)$

Note that question 58 has only three choices.

58 The speed at which the launcher fires tennis balls is constant, but the angle between the launcher and the horizontal can be varied. As the angle is decreased from 45° to 30.°, the range of the tennis balls

1 decreases
2 increases
3 remains the same

58 __1__

59 A 2-kilogram block is dropped from the roof of a tall building at the same time a 6-kilogram ball is thrown horizontally from the same height. Which statement best describes the motion of the block and the motion of the ball? [Neglect air resistance.]

1 The 2-kg block hits the ground first because it has no horizontal velocity.
2 The 6-kg ball hits the ground first because it has more mass.
3 The 6-kg ball hits the ground first because it is round.
④ The block and the ball hit the ground at the same time because they have the same vertical acceleration.

59 __4__

60 As a cart travels around a horizontal circular track, the cart *must* undergo a change in

1 velocity   3 speed
2 inertia   4 weight

60 __1__

Base your answers to questions 61 and 62 on the diagram below which represents the orbit of a comet about the Sun.

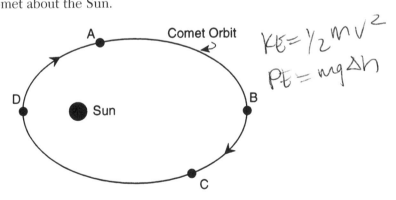

$$KE = \tfrac{1}{2}mv^2$$
$$PE = mg\Delta h$$

61 At which position in its orbit is the comet's speed greatest?

(1) A
(2) B
(3) C
(4) D

61 ___

Note that question 62 has only three choices.

62 As the comet moves from point A to point B, its potential energy

1 decreases
2 increases
3 remains the same

62 ___

63 The diagram below shows a satellite of mass m orbiting Earth in a circular path of radius R.

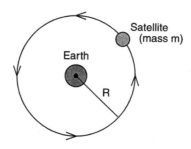

Earth

Satellite
(mass m)

R

If centripetal force F_c is acting on the satellite, its speed is equal to

(1) $\sqrt{\dfrac{F_c R}{m}}$

(2) $\dfrac{F_c R}{m}$

(3) $\sqrt{\dfrac{F_c m}{R}}$

(4) $F_c mR$

63 ___

$F_c = ma_c$

$F_c = \dfrac{mv^2}{r}$

$\sqrt{\dfrac{F_c r}{m}}$

578

64 A ball attached to a string is whirled at a constant speed of 2.0 meters per second in a horizontal circle of radius 0.50 meter. What is the magnitude of the ball's centripetal acceleration?

(1) 1.0 m/s²   (3) 8.0 m/s²

(2) 2.0 m/s²   (4) 4.0 m/s²

64 ___3___

$$ac = \frac{2^2}{.50}$$

Note that question 65 has only three choices.

65 The symbols below are terms for Earth orbiting the Sun and a comet orbiting the Sun.

R_e = the mean radius of Earth's orbit

T_e = the period of Earth's orbit

R_c = the mean radius of the comet's orbit

T_c = the period of the comet's orbit

Compared to the value of $\frac{R_e^3}{T_e^2}$, the value of $\frac{R_c^3}{T_c^2}$ is

1 smaller

2 larger

3 the same

65 ___3___

Kepler's 3rd Law

cube of a planet's average distance from the sun divided by the square of its period (T^2) is constant for all planetary bodies

GROUP 2—Internal Energy

If you choose this group, be sure to answer questions 66–75.

66 Which substance remains a liquid over the *smallest* temperature range?

1 copper   3 lead

2 silver   4 iron

66 _____

579

67 While orbiting Earth, the space shuttle has recorded temperatures ranging from 398 K to 118 K. These temperatures correspond to Celsius temperatures ranging from

(1) 125°C to −391°C   (3) 671°C to 391°C

(2) 125°C to −155°C   (4) 671°C to 155°C   67 ____

68 What is the total amount of energy needed to change the temperature of 0.20 kilogram of lead from 20.°C to 30.°C?

(1) 0.26 kJ   (3) 0.84 kJ

(2) 0.65 kJ   (4) 1.3 kJ   68 ____

69 When a solid sample was heated, its temperature increased but it did not melt. Which statement best describes the changes in the average kinetic and potential energies of the molecules of the sample?

1 Potential energy decreased and kinetic energy remained the same.

2 Potential energy increased and kinetic energy remained the same.

3 Kinetic energy decreased and potential energy remained the same.

4 Kinetic energy increased and potential energy remained the same.   69 ____

70 How are the boiling point of water and the melting point of ice affected by a decrease in pressure?

  1 The boiling point of water increases, and the melting point of ice increases.
  2 The boiling point of water increases, and the melting point of ice decreases.
  3 The boiling point of water decreases, and the melting point of ice increases.
  4 The boiling point of water decreases, and the melting point of ice decreases.   70 _____

71 A 1.0-kilogram sample of water is boiling at 100.°C in an open container. If a 0.50-kilogram piece of lead at 300.°C is placed in the boiling water, how will the temperature of the two substances be affected?

  1 The temperature of the water will decrease, and the temperature of the lead will remain the same.
  2 The temperature of the water will increase, and the temperature of the lead will remain the same.
  3 The temperature of the water will remain the same, and the temperature of the lead will decrease.
  4 The temperature of the water will remain the same, and the temperature of the lead will increase.   71 _____

72 Which graph best represents the relationship between pressure (P) and volume (V) for a fixed mass of an ideal gas at constant temperature?

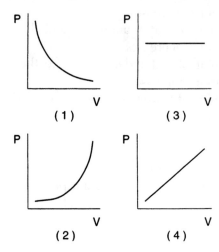

(1)

(3)

(2)

(4)

72 ____

73 A given mass of an ideal gas is enclosed in a rigid-walled container. If the Kelvin temperature of the gas is doubled, its pressure will be

1 halved   3 quartered
2 doubled   4 quadrupled

73 ____

74 In a diesel engine, the piston compresses gases in a cylinder. Why does the temperature of the gases rise during this process?

1 Heat enters the cylinder from the surroundings.
2 Heat is expelled through the exhaust system.
3 Work is done on the surroundings by the gases.
4 Work is done by the piston on the gases.

74 ____

75 A commercial freezer vaporizes ammonia in its cooling coils to remove heat from an ice machine. How much ammonia at $-33°C$ must be vaporized to remove 6,850 kilojoules of heat from the ice machine?

(1) 0.200 kg   (3) 20.6 kg

(2) 5.00 kg   (4) 1370 kg   75 ____

GROUP 3—**Electromagnetic Applications**

If you choose this group, be sure to answer questions **76–85.**

76 A student uses identical field magnets and coils of wire, as well as additional components, to make the electric motors shown in the diagrams below. Which combination of core and current through the coil of wire will produce the greatest torque on the motor's armature?

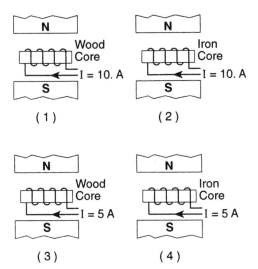

76 ____

77 A simple electrical circuit contains a battery, a light bulb, and a properly connected ammeter. The ammeter has a very low internal resistance because it is connected in

1 parallel with the bulb to have little effect on the current through the bulb
2 parallel with the bulb to prevent current flow through the bulb
3 series with the bulb to have little effect on the current through the bulb
4 series with the bulb to prevent current flow through the bulb   77 _____

78 Which expression is a unit of potential difference equivalent to a volt?

(1) $\dfrac{\text{tesla} \times \text{meter}}{\text{second}}$ (3) $\dfrac{\text{tesla} \times \text{meter}^2}{\text{second}}$

(2) $\dfrac{\text{tesla} \times \text{second}}{\text{meter}}$ (4) $\dfrac{\text{tesla} \times \text{second}}{\text{meter}^2}$ 78 _____

79 An operating electric motor has a back electromotive force because, in addition to acting as a motor, it acts as

1 a split-ring commutator
2 a transformer
3 an induction coil
4 a generator   79 _____

80 In a mass spectrometer, the strength of the magnetic field is 1.0×10^{-1} tesla. Upon entering the chamber of the spectrometer, a positive ion traveling at 2.0×10^{6} meters per second perpendicular to the magnetic field experiences a magnetic force having a magnitude of 3.2×10^{-14} newton. The charge on this positive ion is

(1) 6.4×10^{-21} C   (3) 6.4×10^{-9} C

(2) 1.6×10^{-19} C   (4) 1.6×10^{-9} C   80 ____

81 The Millikan oil drop experiment was designed to determine the

1 sign of the charge of an electron
2 mass of a proton
3 ratio of charge to mass of an electron
4 magnitude of the charge of an electron   81 ____

82 A transformer plugged into a 120-volt household electrical outlet is used to operate a doorbell at a potential difference of 12 volts. What is the ratio of the number of turns in the primary coil to the number of turns in the secondary coil of the transformer?

(1) 10:1   (3) 120:1

(2) 12:1   (4) 1440:1   82 ____

83 The diagram below shows an evacuated cathode ray tube consisting of a source of electrons at one end, a fluorescent screen at the other end, and a pair of oppositely charged parallel plates in between.

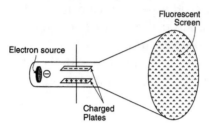

Which diagram best represents the motion of the electron beam in the tube as it passes between the oppositely charged plates?

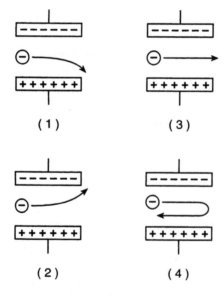

83 _____

84 In the diagram below, a potential difference is induced in a rectangular wire loop as it is rotated at constant speed between two magnetic poles.

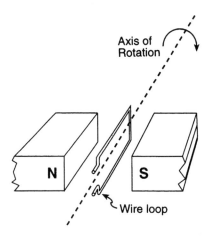

Axis of Rotation

N   S

Wire loop

If the direction of the field is reversed and the speed of rotation is doubled, the magnitude of the maximum induced potential difference will be

1 one-half as great   3 the same
2 twice as great   4 four times as great   84 ____

85 A 100% efficient transformer has 40. turns of wire in the primary coil and 80. turns of wire in the secondary coil. If 20. watts of power is supplied to the primary coil, the power developed in the secondary coil will be

(1) 10. W   (3) 80. W
(2) 20. W   (4) 160 W   85 ____

GROUP 4—Geometric Optics

*If you choose this group, be sure to answer questions **86–95**.*

86 A candle is located beyond the principal focus, *F*, of a concave spherical mirror. Two light rays originating from the same point on the candle are incident on the mirror, as shown in the diagram below.

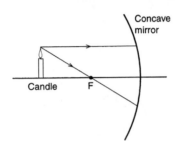

After reflecting from the mirror, the light rays will

1 diverge to form a virtual image
2 diverge to form a real image
3 converge to form a virtual image
4 converge to form a real image   86 _____

87 The radius of curvature of a spherical mirror is *R*. The focal length of this mirror is equal to

(1) $\dfrac{R}{2}$ (3) $\dfrac{R}{4}$

(2) 2*R*   (4) 4*R*   87 _____

88 A candle is placed 0.24 meter in front of a converging mirror that has a focal length of 0.12 meter. How far from the mirror is the image of the candle located?

(1) 0.08 m   (3) 0.24 m

(2) 0.12 m   (4) 0.36 m   88 _____

89 A converging lens forms a real image that is four times larger than the object. If the image distance is 0.16 meter, what is the object distance?

(1) 0.040 m   (3) 0.16 m

(2) 0.080 m   (4) 0.64 m   89 _____

90 In the diagram below, a person is standing 5 meters from a plane mirror. The chair in front of the person is located 2 meters from the mirror.

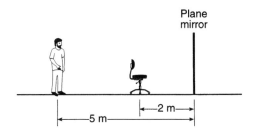

What is the distance between the person and the image he observes of the chair?

(1) 7 m   (3) 3 m

(2) 2 m   (4) 10. m   90 _____

91 The diagram below shows parallel monochromatic light rays being reflected from a concave mirror.

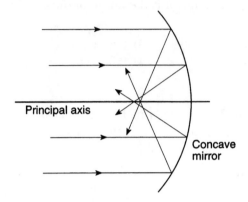

The mirror fails to produce a sharp focal point as a result of

1 dispersion
2 diffuse reflection
3 spherical aberration
4 chromatic aberration

91 ____

92 Which glass lens in air can produce an enlarged real image of an object?

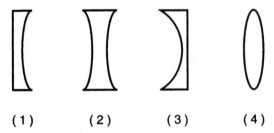

(1)   (2)   (3)   (4)

92 ____

93 In the diagram below, parallel light rays in air diverge as a result of interacting with an optical device.

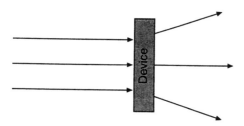

The device could be a

1 convex glass lens
2 rectangular glass block
3 plane mirror
4 concave glass lens   93 ____

94 A person is standing in front of a diverging (convex) mirror. What type of image does the mirror form of the person?

1 erect, virtual, and smaller than the person
2 erect, virtual, and the same size as the person
3 erect, real, and smaller than the person
4 erect, real, and the same size as the person   94 ____

95 Which graph best represents the relationship between image size (S_i) and image distance (d_i) for real images formed by a converging lens?

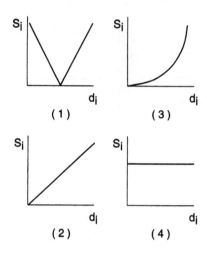

(1)

(3)

(2)

(4)

95 ____

GROUP 5—**Solid State**

If you choose this group, be sure to answer questions **96–105.**

96 Which statement best explains how the resistivity of glass compares to the resistivity of copper?

1 Glass has a lower resistivity and is a poor conductor.

2 Glass has a lower resistivity and is a good conductor.

3 Glass has a higher resistivity and is a poor conductor.

4 Glass has a higher resistivity and is a good conductor.

96 ____

97 Metals that are excellent conductors have valence electrons that are

   1 difficult to dislodge and difficult to move through the crystal

   2 difficult to dislodge but easy to move through the crystal

   3 easy to dislodge but difficult to move through the crystal

   4 easy to dislodge and easy to move through the crystal   97 ____

98 Which energy band diagram best represents a semiconductor?

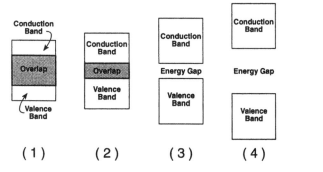

99 Alternating current from a wall outlet can be converted to direct current by

   1 an N-type semiconductor

   2 a P-type semiconductor

   3 an emitter

   4 a diode   99 ____

100 The table below lists the number of valence electrons for some elements.

Element	Number of Valence Electrons
antimony	5
silicon	4
germanium	4
aluminum	3

Which combination of elements could be used to make an *N*-type semiconductor?

1 antimony and silicon
2 silicon and germanium
3 germanium and aluminum
4 aluminum and antimony

100 _____

101 Which diagram below shows a correctly labeled *P–N–P* transistor?

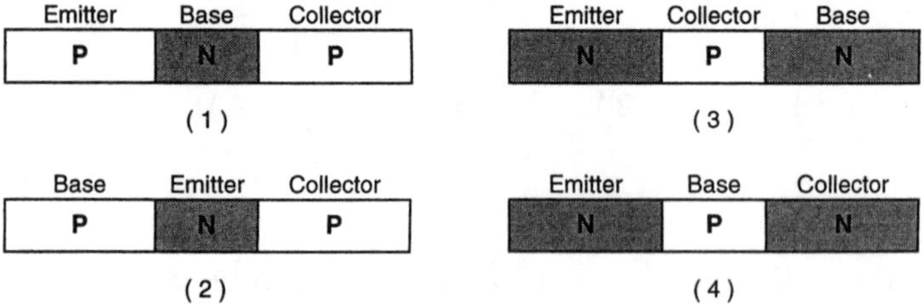

101 _____

102 The diagram below represents an *N*-type semiconductor connected to a battery.

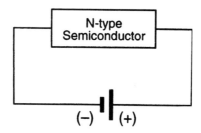

Which phrase best describes the majority charge carriers within the semiconductor?

1 electrons moving to the left
2 electrons moving to the right
3 holes moving to the left
4 holes moving to the right

102 ____

103 In the circuit diagram below, a diode and an incandescent light bulb are connected in series with a source of alternating current having a frequency of 60 hertz.

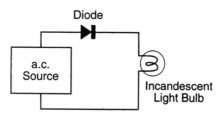

How many times per second does a maximum current exist in the light bulb?

(1) 30   (3) 120
(2) 60   (4) 240

103 ____

104 Which part of an *N–P–N* transistor is forward biased?

   1 an integrated circuit
   2 a parallel circuit
   3 an emitter-base combination
   4 a collector-base combination   104 _____

Note that question 105 has only three choices.

105 The transistor shown in the circuit diagram below is being used as an amplifier.

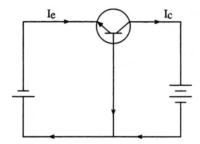

When the emitter current (I_e) increases, the collector current (I_c)

   1 decreases
   2 increases
   3 remains the same   105 _____

GROUP 6—**Nuclear Energy**

If you choose this group, be sure to answer questions **106–115.**

106 How much energy would be generated if a 1.0×10^{-3}-kilogram mass were completely converted to energy?

(1) 9.3×10^{-1} MeV   (3) 9.0×10^{13} J

(2) 9.3×10^{2} MeV   (4) 9.0×10^{16} J   106 ____

107 One isotope of uranium is $^{238}_{92}$U. Any other isotope of uranium must have

(1) 92 protons   (3) 92 neutrons

(2) 146 protons   (4) 146 neutrons   107 ____

108 A cyclotron is used in medical research to make radioisotopes. The primary function of a cyclotron is to

1 determine the mass of an atom
2 determine the half-life of a nuclide
3 accelerate neutrons
4 accelerate charged particles   108 ____

109 As the nucleus of an unstable atom emits only gamma radiation, the nucleus must

1 gain energy   3 lose protons
2 lose energy   4 gain protons   109 ____

110 In the reaction $^{24}_{11}$Na $\rightarrow$ $^{24}_{12}$Mg $+ X$, particle X is a

1 positive electron   3 proton
2 negative electron   4 neutron   110 ____

111 A 24-gram sample of a radioactive nuclide decayed to 3.0 grams of the nuclide in 36 minutes. How much of the original nuclide sample remained after the first 12 minutes?

(1) 12 g   (3) 6.0 g

(2) 2.0 g   (4) 8.0 g   111 _____

112 A fusion reactor for commercial production of energy has *not* yet been developed. The best explanation for this situation is that fusion reactions

1 occur at extremely low temperatures
2 form highly radioactive products
3 require very high energies
4 need fuels unavailable on Earth   112 _____

113 According to the Uranium Disintegration Series, how many beta particles are emitted when an atom of $^{218}_{84}$Po decays to $^{206}_{82}$Pb?

(1) 7   (3) 3

(2) 6   (4) 4   113 _____

114 In which nuclear equation does X represent a neutron?

(1) $^{3}_{1}\text{H} + ^{1}_{1}\text{H} \rightarrow ^{4}_{2}\text{H} + X$

(2) $^{2}_{1}\text{H} + ^{2}_{1}\text{H} \rightarrow ^{3}_{1}\text{H} + ^{1}_{1}\text{H} + X$

(3) $^{2}_{1}\text{H} + ^{3}_{1}\text{H} \rightarrow ^{4}_{2}\text{He} + X$

(4) $^{12}_{6}\text{C} + ^{1}_{0}\text{n} \rightarrow ^{13}_{7}\text{N} + X$ 114 _____

115 Which statement best describes what occurs when the control rods are inserted into a nuclear reactor?

   1 The number of fission reactions decreases because the control rods absorb neutrons.

   2 The number of fission reactions decreases because the control rods absorb electrons.

   3 The number of fission reactions increases because the control rods release neutrons.

   4 The number of fission reactions increases because the control rods release electrons.   115 _____

PART III

You must answer *all* questions in this part.   [15]

Base your answers to questions 116 through 118 on the information and diagram below, which is drawn to a scale of 1.0 centimeter = 3.0 meters.

A 650-kilogram roller coaster car starts from rest at the top of the first hill of its track and glides freely. [Neglect friction.]

$$m = 650 \, kg \quad V_i = 0$$

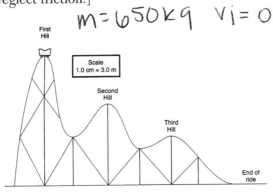

116 Using a metric ruler and the scale of 1.0 cm = 3.0 m, determine the height of the first hill.   [1]

117 Determine the gravitational potential energy of the car at the top of the first hill. [Show all calculations, including the equation and substitution with units.]   [2]

$$\Delta PE = mg \Delta h$$
$$(650 \, kg)(9.81 \, m/s^2)(\Delta h)$$

118 Using one or more complete sentences, compare
the kinetic energy of the car at the top of the hill to
its kinetic energy at the top of the third hill. [1]

KE energy is greater @ the
top of the 3rd hill -b/c good
potential energy must be proved le
KE is less

Base your answers to questions 119 through 121
on the information and diagram below.

Two small charged spheres, A and B, are separated
by a distance of 0.50 meter. The charge on sphere
A is $+2.4 \times 10^{-6}$ coulomb and the charge on
sphere B is -2.4×10^{-6} coulomb.

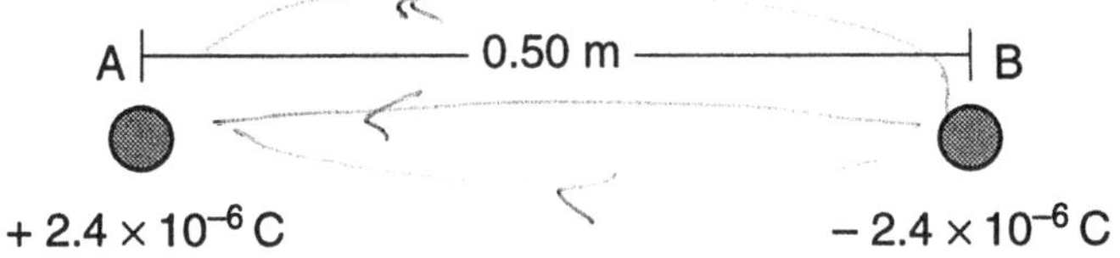

A ├──────── 0.50 m ────────┤ B

$+ 2.4 \times 10^{-6}$ C $- 2.4 \times 10^{-6}$ C

119 On the diagram, sketch *three* electric field lines to
represent the electric field in the region between
sphere A and sphere B. [Draw an arrowhead on
each field line to show the proper direction.] [2]

120 Calculate the magnitude of the electrostatic force
that sphere A exerts on sphere B. [Show all calculations, including the equation and substitution
with units.] [2]

$$Fe = \frac{kq_1q_2}{r^2}$$

$$\frac{(8.99 \times 10^9)(2.4 \times 10^{-6})(-2.4 \times 10^{-6})}{0.50^2}$$

$-.207$

121 Using the axes, sketch the general shape of the graph that shows the relationship between the magnitude of the electrostatic force between the two charged spheres and the distance separating them. The charge on each sphere remains constant as the distance separating them is varied. [1]

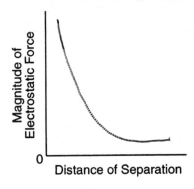

Base your answers to questions 122 through 126 on the information and table below.

A photoemissive metal was illuminated successively by photons of various frequencies. The maximum kinetic energies of the emitted photoelectrons were measured and recorded in the table below.

FREQUENCY ($\times 10^{14}$ Hz)	MAXIMUM KINETIC ENERGY ($\times 10^{-9}$ J)
5.3	0.58
6.0	1.08
6.9	1.73
7.6	2.07

Using the information in the data table, construct a graph on the grid provided.

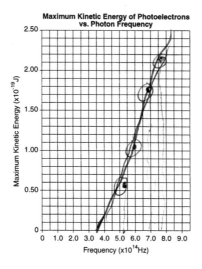

Maximum Kinetic Energy of Photoelectrons vs. Photon Frequency

Maximum Kinetic Energy (x10⁻¹⁹ J) vs. Frequency (x10¹⁴ Hz)

122 Plot the data points for maximum kinetic energy versus frequency. [1]

123 Draw the best-fit line. [1]

124 Using the graph, determine the threshold frequency of the metal. [1]

125 Determine the slope of the graph. [Show all calculations, including the equation and substitution with units.] [2]

$$\frac{2.07 - 0.58}{7.06 - 5.03}$$

_____

_____

_____

$$\frac{\Delta KE_{max}}{\Delta F}$$

$$1.49 = 0.647$$

$$\frac{2 \times 10^{-19} J - 0.4 \times 10^{-19}}{8 \times 10^{14} - 5 \times 10^{14}} \quad \frac{2.03}{603}$$

126 Name the physical constant represented by the slope of a graph of maximum kinetic energy of photoelectrons versus photon frequency. [1]

plancks constant.

Examination
January 2001
Physics

PART I

Answer all 55 questions in this part. [65]

Directions (1–55): For *each* statement or question, select the word or expression that, of those given, best completes the statement or answers the question. Record the answers to those questions in the spaces provided.

1 The maximum time allowed for the completion of this examination is approximately

  (1) 10^2 s   (3) 10^4 s

  (2) 10^3 s   (4) 10^5 s   1 ____

2 The diagram below represents the relationship between velocity and time of travel for four cars, *A, B, C,* and *D,* in straight-line motion.

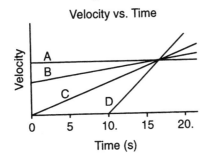

605

Which car has the greatest acceleration during the time interval 10. seconds to 15 seconds?

(1) *A*   (3) *C*

(2) *B*   (4) *D*   2 _____

3 The diagram below shows a block on a horizontal frictionless surface. A 100.-newton force acts on the block at an angle of 30.° above the horizontal.

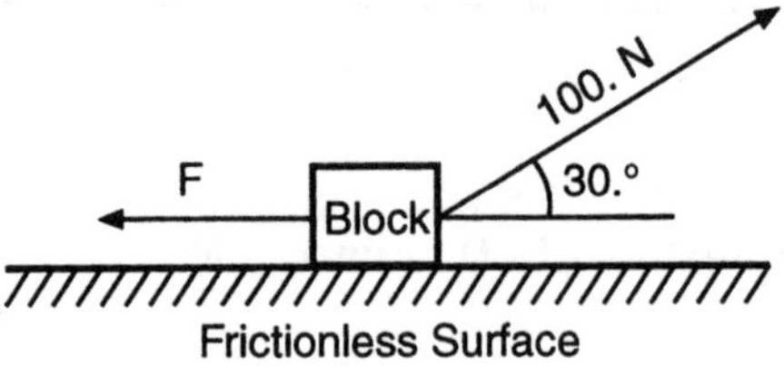

What is the magnitude of force *F* if it establishes equilibrium?

(1) 50.0 N   (3) 100. N

(2) 86.6 N   (4) 187 N   3 _____

4 Approximately how far will an object near Earth's surface fall in 3.0 seconds?

(1) 88 m   (3) 29 m

(2) 44 m   (4) 9.8 m   4 _____

5 A 5-newton force directed east and a 5-newton force directed north act concurrently on a point. The resultant of the two forces is

(1) 5 N northeast   (3) 7 N northeast

(2) 10. N southwest   (4) 7 N southwest   5 _____

6 Into how many possible components can a single force be resolved?

  1 an unlimited number
  2 two components
  3 three components
  4 four components at right angles to each other   6 ____

7 The graph below represents the relationship between the forces applied to an object and the corresponding accelerations produced.

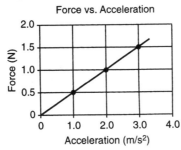

What is the inertial mass of the object?
  (1) 1.0 kg   (3) 0.50 kg
  (2) 2.0 kg   (4) 1.5 kg   7 ____

8 Which is a derived unit?

  1 meter   3 kilogram
  2 second   4 newton   8 ____

9 What is the magnitude of the gravitational force between two 5.0-kilogram masses separated by a distance of 5.0 meters?

  (1) 5.0×10^0 N   (3) 6.7×10^{-11} N
  (2) 3.3×10^{-10} N   (4) 1.3×10^{-11} N   9 ____

10 What is the displacement of the mass hanger (H) shown in the diagram after a 0.20-kilogram mass is loaded on it? [Assume the hanger is at rest in both positions.]

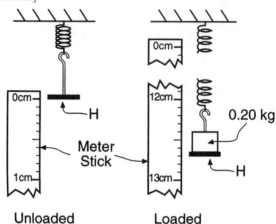

Unloaded	Loaded

(1) 12.30 cm   (3) 12.70 cm
(2) 12.50 cm   (4) 13.30 cm   10 _____

11 Two cars having different weights are traveling on a level surface at different constant velocities. Within the same time interval, greater force will always be required to stop the car that has the greater

1 weight   3 velocity
2 kinetic energy   4 momentum   11 _____

12 A 0.050-kilogram bullet is fired from a 4.0-kilogram rifle that is initially at rest. If the bullet leaves the rifle with momentum having a magnitude of 20. kilogram•meters per second, the rifle will recoil with a momentum having a magnitude of

(1) 1,600 kg•m/s   (3) 20. kg•m/s
(2) 80. kg•m/s   (4) 0.25 kg•m/s   12 _____

13 A 2.0-kilogram mass weighs 10. newtons on planet X. The acceleration due to gravity on planet X is approximately

(1) 0.20 m/s^2 (3) 9.8 m/s^2

(2) 5.0 m/s^2 (4) 20. m/s^2 13 ____

Note that questions 14 and 15 have only three choices.

14 The graph below represents the motion of an object.

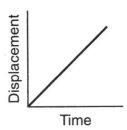

According to the graph, as time increases, the velocity of the object

1 decreases

2 increases

3 remains the same   14 ____

15 A wooden block is at rest on a horizontal steel surface. If a 10.-newton force applied parallel to the surface is required to set the block in motion, how much force is required to keep the block moving at constant velocity?

(1) less than 10. N

(2) greater than 10. N

(3) 10. N   15 ____

16 In the diagram below, a 20.0-newton force is used to push a 2.00-kilogram cart a distance of 5.00 meters.

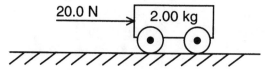

The work done on the cart is

(1) 100. J   (3) 150. J

(2) 200. J   (4) 40.0 J   16 _____

Note that questions 17 and 18 have only three choices.

17 The diagram below shows two identical wooden planks, *A* and *B*, at different incline angles, used to slide concrete blocks from a truck.

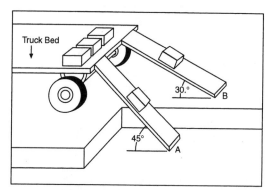

Compared to the amount of work done against friction by a block sliding down plank *A*, the work done against friction by a block sliding down plank *B* is

1 less

2 more

3 the same   17 _____

18 Two vacationers walk out on a horizontal pier as shown in the diagram below.

As they approach the end of the pier, their gravitational potential energy will

1 decrease
2 increase
3 remain the same

18 ____

19 A girl weighing 500. newtons takes 50. seconds to climb a flight of stairs 18 meters high. Her power output vertically is

(1) 9,000 W   (3) 1,400 W
(2) 4,000 W   (4) 180 W

19 ____

20 The graph below represents the elongation of a spring as a function of the applied force.

Force vs. Elongation

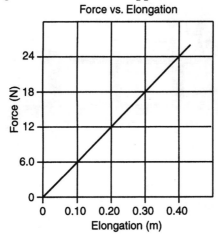

How much work must be done to stretch the spring 0.40 meter?

(1) 4.8 J   (3) 9.8 J

(2) 6.0 J   (4) 24 J   20 _____

21 Metal sphere A has a charge of +12 elementary charges and identical sphere B has a charge of +16 elementary charges. After the two spheres are brought into contact, the charge on sphere A is

(1) –2 elementary charges

(2) +2 elementary charges

(3) +14 elementary charges

(4) +28 elementary charges   21 _____

22 Electrostatic force F exists between two point charges. If the distance between the charges is tripled, the force between the charges will be

(1) $\dfrac{F}{9}$ (3) $3F$

(2) $\dfrac{F}{3}$ (4) $9F$ 22 _____

23 Which net charge could be found on an object?

(1) $+3.2 \times 10^{-18}$ C   (3) -1.8×10^{-18} C

(2) $+2.4 \times 10^{-19}$ C   (4) -0.80×10^{-19} C   23 _____

24 In the diagram below, two identical spheres, A and B, have equal net positive charges.

P

A   B

Which arrow best represents the direction of their resultant electric field at point P?

(1) ⟶   (3)

(2)   (4)

                                         24 _____

25 How much work is done in moving 5.0 coulombs of
    charge against a potential difference of 12 volts?

    (1) 2.4 J   (3) 30. J
    (2) 12 J   (4) 60. J   25 _____

26 Compared to insulators, metals are better conduc-
    tors of electricity because metals contain more free

    1 protons   3 positive ions
    2 electrons   4 negative ions   26 _____

27 If a 15-ohm resistor is connected in parallel with a
    30.-ohm resistor, the equivalent resistance is

    (1) 15 Ω   (3) 10. Ω
    (2) 2.0 Ω   (4) 45 Ω   27 _____

28 Identical charges A, B, and C are located between
    two oppositely charged parallel plates, as shown in
    the diagram below.

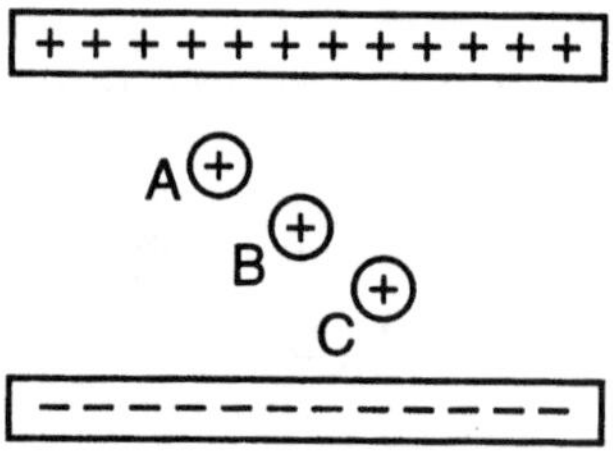

    The magnitude of the force exerted on the charges
    by the electric field between the plates is

    1 least on A and greatest on C
    2 greatest on A and least on C
    3 the same on A and C, but less on B
    4 the same for A, B, and C   28 _____

29 A metal wire has length L and cross-sectional area A. The resistance of the wire is directly proportional to

(1) $\dfrac{L}{A}$

(3) $\dfrac{A}{L}$

(2) $L \times A$

(4) $L + A$

29 ____

30 The diagram below shows electric currents in conductors that meet at junction P.

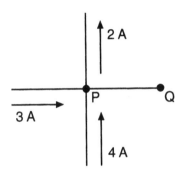

What are the magnitude and direction of the current in conductor PQ?

(1) 9 A toward P

(3) 5 A toward P

(2) 9 A toward Q

(4) 5 A toward Q

30 ____

31 A glass rod becomes positively charged when it is rubbed with silk. This net positive charge accumulates because the glass rod

1 gains electrons

3 loses electrons

2 gains protons

4 loses protons

31 ____

Base your answers to questions 32 and 33 on the diagram below, which shows two resistors and three ammeters connected to a voltage source.

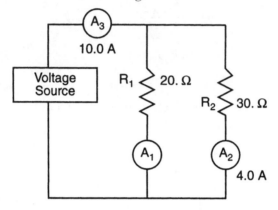

32 What is the potential difference across the source?

(1) 440 V   (3) 120 V
(2) 220 V   (4) 60. V   32 _____

33 What is the current reading of ammeter A_1?

(1) 10.0 A   (3) 3.0 A
(2) 6.0 A   (4) 4.0 A   33 _____

34 In the diagram below, a wire carrying an electron current into the page, as denoted by X, is placed in a magnetic field.

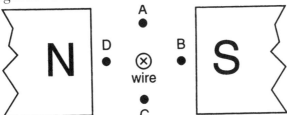

The magnetic field exerts a force on the wire toward point

(1) A   (3) C
(2) B   (4) D   34 _____

35 A wire carries a current of 2.0 amperes. How many electrons pass a given point in this wire in 1.0 second?

(1) 1.3×10^{18} (3) 1.3×10^{19}
(2) 2.0×10^{18} (4) 2.0×10^{19} 35 _____

36 How much time is required for an operating 100-watt light bulb to dissipate 10 joules of electrical energy?

(1) 1 s   (3) 10 s
(2) 0.1 s   (4) 1000 s   36 _____

Note that question 37 has only three choices.

37 A wire conductor is moved at constant speed perpendicularly to a uniform magnetic field. If the strength of the magnetic field is increased, the induced potential across the ends of the conductor

1 decreases
2 increases
3 remains the same   37 _____

38 A light spring is attached to a heavier spring at one end. A pulse traveling along the light spring is incident on the boundary with the heavier spring. At this boundary, the pulse will be

1 totally reflected
2 totally absorbed
3 totally transmitted into the heavier spring
4 partially reflected and partially transmitted into the heavier spring   38 _____

39 As a wave travels through a medium, the particles of the medium vibrate in the direction of the wave's travel. What type of wave is traveling through the medium?

1 longitudinal   3 transverse
2 torsional   4 hyperbolic   39 _____

40 A wave completes one vibration as it moves a distance of 2 meters at a speed of 20 meters per second. What is the frequency of the wave?

(1) 10 Hz   (3) 20 Hz
(2) 2 Hz   (4) 40 Hz   40 _____

41 What is the period of a wave if 20 crests pass an observer in 4 seconds?

(1) 80 s   (3) 5 s

(2) 0.2 s   (4) 4 s   41 _____

42 A beam of green light may have a frequency of

(1) 5.0×10^{-7} Hz   (3) 3.0×10^{8} Hz

(2) 1.5×10^{2} Hz   (4) 6.0×10^{14} Hz   42 _____

43 The diagram below shows a pulse moving to the right in a rope. A is a point on the rope.

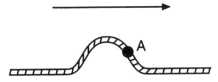

Which arrow best shows the direction of movement of point A at this instant?

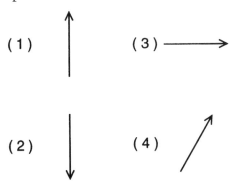

43 _____

44 The diagram below represents a periodic wave.

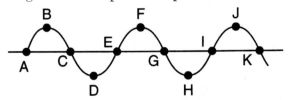

Which two points on the wave are in phase?

(1) A and D (3) C and K

(2) A and G (4) D and I 44 ____

45 Two waves traveling in the same medium interfere to produce a standing wave. What is the phase difference between the two waves at a node?

(1) $0°$ (3) $180°$

(2) $90°$ (4) $360°$ 45 ____

46 When yellow light having a wavelength of 5.8×10^{-7} meter shines through two slits 2.0×10^{-4} meter apart, an interference pattern is formed on a screen 2.0 meters from the slits. What distance separates the first-order maximum and the central maximum?

(1) 5.8×10^{-11} m   (3) 5.8×10^{-3} m

(2) 1.5×10^{-3} m   (4) 6.9×10^{2} m   46 ____

47 The diagram below shows parallel rays of light incident on an irregular surface.

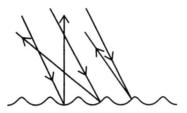

Which phenomenon of light is illustrated by the diagram?

1 diffraction   3 regular reflection
2 refraction   4 diffuse reflection   47 ____

48 What is the speed of light in a medium having an absolute index of refraction of 2.3?

(1) 0.77×10^8 m/s   (3) 1.5×10^8 m/s
(2) 1.3×10^8 m/s   (4) 2.3×10^8 m/s   48 ____

Note that questions 49 through 52 have only three choices.

49 The driver of a car blows the horn as the car approaches a crosswalk. Compared to the actual pitch of the horn, the pitch observed by a pedestrian in the crosswalk is

1 lower
2 higher
3 the same   49 ____

50 Compared to wavelengths of visible light, the wavelengths of ultraviolet light are

   1 shorter
   2 longer
   3 the same   50 _____

51 As a pulse travels along a rope, the pulse loses energy and its amplitude

   1 decreases
   2 increases
   3 remains the same   51 _____

52 The diagram below shows a ray of light passing through two media.

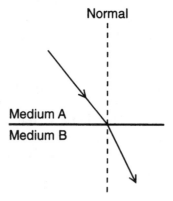

When the wave travels from medium A into medium B, its speed

   1 decreases
   2 increases
   3 remains the same   52 _____

53 Experiments performed with light indicate that light exhibits

1 particle properties, only
2 wave properties, only
3 both particle and wave properties
4 neither particle nor wave properties   53 _____

54 What is the energy of a quantum of light having a frequency of 6.0×10^{14} hertz?

(1) 1.6×10^{-19} J   (3) 3.0×10^{8} J
(2) 4.0×10^{-19} J   (4) 5.0×10^{-7} J   54 _____

55 Photons with an energy of 7.9 electronvolts strike a zinc plate, causing the emission of photoelectrons with a maximum kinetic energy of 4.0 electronvolts. The work function of the zinc plate is

(1) 11.9 eV   (3) 3.9 eV
(2) 7.9 eV   (4) 4.0 eV   55 _____

PART II

This part consists of six groups, each containing ten questions. Each group tests an optional area of the course. Choose two of these six groups. Be sure that you answer all ten questions in each group chosen. Record the answers to these questions in the spaces provided. [20]

GROUP 1—**Motion in a Plane**

*If you choose this group, be sure to answer questions **56–65**.*

Base your answers to questions 56 through 59 on the information and diagram below.

A 1.00×10^3-kilogram car is driven clockwise around a flat circular track of radius 25.0 meters. The speed of the car is a constant 5.00 meters per second.

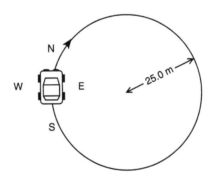

56 What minimum friction force must exist between the tires and the road to prevent the car from skidding as it rounds the curve?

(1) 1.25×10^5 N   (3) 5.00×10^3 N
(2) 9.80×10^4 N   (4) 1.00×10^3 N   56 _____

57 If the circular track were to suddenly become frictionless at the instant shown in the diagram, the car's direction of travel would be

1 toward *E*   3 toward *W*
2 toward *N*   4 a clockwise spiral   57 ____

58 At the instant shown in the diagram, the car's centripetal acceleration is directed

1 toward *E*   3 toward *W*
2 toward *N*   4 clockwise   58 ____

59 Which factor, when doubled, would produce the greatest change in the centripetal force acting on the car?

1 mass of the car   3 velocity of the car
2 radius of the track   4 weight of the car   59 ____

_____

60 The diagram below shows four different locations of a satellite in its elliptical orbit about Earth.

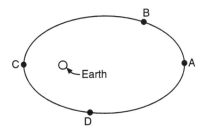

At which location is the magnitude of the satellite's velocity greatest?

(1) *A*   (3) *C*
(2) *B*   (4) *D*   60 ____

625

61 Which statement is consistent with Kepler's laws of planetary motion?

1 The planets move at a constant speed around the Sun.
2 The speed of a planet is directly proportional to the radius of the path of motion.
3 The more massive the planet, the slower the planet moves around the Sun.
4 An imaginary line from a planet to the Sun sweeps out equal areas in equal time intervals.   61 ____

62 The path of a projectile fired at a 30° angle to the horizontal is best described as

1 parabolic   3 circular
2 linear   4 hyperbolic   62 ____

63 A projectile is launched with an initial velocity of 200 meters per second at an angle of 30° above the horizontal. What is the magnitude of the vertical component of the projectile's initial velocity?

(1) 200 m/s × cos 30°   (3) $\dfrac{200 \text{ m/s}}{\sin 30°}$

(2) 200 m/s × sin 30°   (4) $\dfrac{200 \text{ m/s}}{\cos 30°}$ 63 ____

Note that questions 64 and 65 have only three choices.

64 The table below gives the mean radius of orbit and orbital period for two moons of a planet.

Moon	Mean Radius of Orbit	Orbital Period
A	R	T
B	4R	8T

Compared to the ratio of the orbital radius cubed to the period squared for Moon A, the ratio of the orbital radius cubed to the period squared for Moon B is

1 less
2 greater
3 the same   64 _____

65 A satellite is in geosynchronous orbit. Compared to Earth's period of rotation, the satellite's period of revolution is

1 less
2 greater
3 the same   65 _____

GROUP 2—**Internal Energy**

If you choose this group, be sure to answer questions **66–75.**

Base your answers to questions 66 through 68 on the graph below, which represents changes in a 5.00-kilogram sample of a substance as it absorbs heat at a constant rate of 41.9 kilojoules per minute.

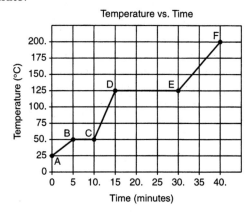

Temperature vs. Time

66 What is the minimum amount of heat absorbed by 1.00 kilogram of the substance during interval *BE*?

(1) 105 kJ   (3) 419 kJ
(2) 210. kJ   (4) 1050 kJ   66 _____

67 The heat of vaporization of the substance is

(1) 210. kJ/kg   (3) 12.6 kJ/kg
(2) 126 kJ/kg   (4) 629 kJ/kg   67 _____

68 The boiling point of the substance is

(1) 25°C   (3) 125°C

(2) 50.°C   (4) 150.°C   68 ____

69 Gas molecules at the same temperature are always assumed to have

1 uniform velocity
2 uniform acceleration
3 straight-line motion
4 random motion   69 ____

70 A 1.0×10^3-kilogram block absorbs 2.4×10^3 kilojoules of heat as its temperature rises from 710.°C to 720.°C. What is the specific heat of the block?

(1) 2.4×10^5 kJ/kg•C°   (3) 4.2×10^{-7} kJ/kg•C°

(2) 0.24 kJ/kg•C°   (4) 4.2×10^{-8} kJ/kg•C°   70 ____

71 Which property determines the direction of the exchange of internal energy between two objects?

1 temperature   3 mass
2 specific heat   4 density   71 ____

72 Which two quantities are measured in the same units?

1 mechanical energy and heat
2 energy and power
3 momentum and work
4 work and power   72 ____

73 Which graph best represents the relationship between the average molecular kinetic energy ($\overline{KE}$) of an ideal gas and its absolute temperature (T)?

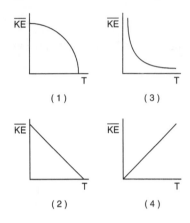

(1)

(3)

(2)

(4)

73 _____

Note that questions 74 and 75 have only three choices.

74 Equal masses of zinc and copper at room temperature are placed in an oven that supplies heat energy at a rate of 1 kilojoule per minute. Compared to the time needed for the zinc sample to reach its melting point, the time needed for the copper sample to reach its melting point will be

1 less
2 greater
3 the same

74 _____

75 As the volume of a fixed mass of an ideal gas increases at constant temperature, the product of the pressure and the volume of the gas

1 decreases
2 increases
3 remains the same

75 _____

GROUP 3—**Electromagnetic Applications**

If you choose this group, be sure to answer questions **76–85.**

Base your answers to questions 76 through 78 on the information and diagram below.

Two coils of wire are wound around the same permeable core. Coil I has 200. turns and coil II has 3000. turns. Coil I is supplied with 30.0 amperes of alternating current at 90.0 volts.

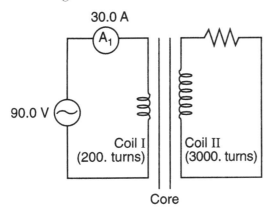

76 What does the diagram above represent?
   1 a step-up transformer
   2 a step-down transformer
   3 a transistor
   4 an ammeter   76 ____

77 What potential difference is induced in coil II?
   (1) 6.00 V   (3) 1350 V
   (2) 90.0 V   (4) 2700 V   77 ____

78 What is the power developed in coil I?
   (1) 1.80×10^2 W   (3) 4.05×10^3 W
   (2) 2.70×10^3 W   (4) 4.00×10^2 W   78 ____

79 A torque exists on the armature of an operating electric motor. The magnitude of this torque would decrease if there were an increase in the

1 current in the armature coil
2 magnetic field strength of the field magnet
3 potential difference applied to the armature coil
4 resistance of the wire in the armature   79 _____

80 In order to measure the current through an electrical device, an ammeter is placed in series with the device. Compared to the electrical device, the ammeter should have a much

1 lower permeability   3 lower resistance
2 higher permeability   4 higher resistance   80 _____

81 As the armature of an operating electric motor turns, a voltage is induced. This voltage is opposite in direction to the applied voltage and referred to as

1 conduction   3 magnetic levitation
2 reverse current   4 back emf   81 _____

82 A voltmeter is made by connecting the current-carrying wire loop of a galvanometer in

1 series with a high resistance
2 series with a low resistance
3 parallel with a high resistance
4 parallel with a low resistance   82 _____

83 The phenomenon by which an incandescent object gives off electrons is known as

1 thermionic emission   3 induction

2 laser emission   4 spectroscopy   83 _____

84 A single loop of wire is placed between the poles of permanent magnets, as shown in the diagram below.

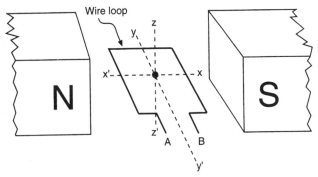

If a potential difference is applied to the ends of loop *AB*, in which direction will the loop move?

1 up toward z 3 around the y–$y´$-axis

2 down toward $z´$ 4 around the x–$x´$-axis   84 _____

85 The diagram below shows electron *e* about to enter
    the region between the poles of two magnets.

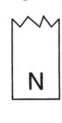

N

S

Upon entering the region between the poles, the
moving electron will experience a magnetic force
directed

1 toward the north pole   3 into the page
2 toward the south pole   4 out of the page   85 _____

GROUP 4—**Geometric Optics**

If you choose this group, be sure to answer questions **86–95.**

Base your answers to questions 86 through 90 on the information and diagram below.

A converging lens has a focal length of 0.080 meter. A light ray travels from the object to the lens parallel to the principal axis.

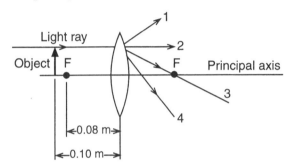

86 Which line best represents the path of the ray after it leaves the lens?

(1) 1   (3) 3

(2) 2   (4) 4   86 _____

87 How far from the lens is the image formed?

(1) 0.020 m   (3) 0.40 m

(2) 0.18 m   (4) 0.80 m   87 _____

88 If the lens is made of crown glass, the speed of light in the lens is closest to

(1) 1.5×10^8 m/s   (3) 3.0×10^8 m/s

(2) 2.0×10^8 m/s   (4) 4.0×10^8 m/s   88 _____

89 Which phenomenon best explains the path of the light ray through the lens?

1 diffraction   3 reflection

2 dispersion   4 refraction   89 ____

Note that question 90 has only three choices.

90 If the lens were placed in water, its focal length would

1 decrease

2 increase

3 remain the same   90 ____

91 Which characteristics best describe the image produced by a plane mirror?

1 real and inverted   3 virtual and inverted

2 real and erect   4 virtual and erect   91 ____

92 When an object is placed at the focal point of a concave mirror, the mirror produces

1 an image that is smaller than the object

2 an image that is larger than the object

3 an image that is the same size as the object

4 no image of the object   92 ____

93 Which optical device causes parallel light rays to diverge?

1 convex mirror   3 plane mirror
2 concave mirror   4 convex lens   93 _____

94 When a boy who is 1.00 meter tall stands in front of a vertical plane mirror, he is able to see the image of his entire body. What is the *minimum* height, from top to bottom, of the mirror?

(1) 1.00 m   (3) 0.50 m
(2) 2.00 m   (4) 0.25 m   94 _____

95 An object arrow is placed in front of a concave mirror having center of curvature C and principal focus F. Which diagram best shows the location of point I, the image of the tip of the object arrow?

(1)

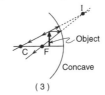

(3)

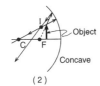

(2)

(4)   95 _____

637

GROUP 5—**Solid State**

If you choose this group, be sure to answer questions **96–105.**

Base your answers to questions 96 through 98 on the diagram below of a semiconductor.

96 The semiconductor represented in the diagram is a

   1 transistor   3 emitter

   2 resistor   4 diode   96 ____

97 This device is called a semiconductor because

   1 positive holes move from X to Z

   2 positive holes move from Z to X

   3 negative holes flow from Z to X

   4 negative electrons flow from X to Z 97 ____

98 The part of the semiconductor labeled X is called the

   1 cathode   3 anode

   2 emitter   4 collector   98 ____

99 A diode can be used in a circuit to

   1 convert direct current to alternating current

   2 convert alternating current to direct current

   3 amplify voltage

   4 amplify current   99 ____

100 Which diagram represents an *N–P–N* transistor?

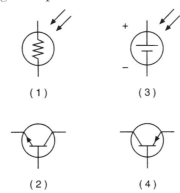

(1)   (3)

(2)   (4)   100 ____

101 An impurity that is added to a semiconductor in order to provide holes is classified as

1 a donor   3 an acceptor
2 a receptor   4 a bias   101 ____

102 Current in a semiconductor is caused by the movement of

1 electrons, only
2 holes, only
3 isotopes
4 both electrons and holes   102 ____

103 In a *P–N–P* transistor, the section that has the thinnest segment is the

1 emitter   3 collector
2 acceptor   4 base   103 ____

Note that questions 104 and 105 have only three choices.

104 Compared to the current flow when a forward bias is applied to a *P–N* junction, the current flow when a reverse bias is applied to a *P–N* junction is

1 less
2 greater
3 the same

104 _____

105 Donor materials are added to semiconductors so that the number of available electrons will

1 decrease
2 increase
3 remain the same

105 _____

GROUP 6—**Nuclear Energy**

If you choose this group, be sure to answer questions **106–115.**

Base your answers to questions 106 through 109 on the diagram below, which shows a nuclear reactor designed to obtain energy in the form of heat from a nuclear fission reaction.

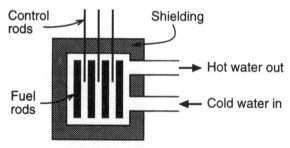

106 In the fission reaction

$$^{235}_{92}U + ^{1}_{0}n \rightarrow F_1 + F_2 + 3^{1}_{0}n + \text{heat},$$

the fission fragments F_1 and F_2 might be

(1) $^{141}_{56}$Ba and $^{92}_{36}$Kr   (3) $^{131}_{51}$Sb and $^{99}_{41}$Nb

(2) $^{141}_{56}$Ba and $^{92}_{36}$Kr   (4) $^{132}_{51}$Sb and $^{99}_{41}$Nb

106 _____

107 What is the total energy produced by converting
1.0 kilogram of $^{235}_{92}$U to energy in the reactor?

(1) 9.0×10^{16} J   (3) 9.0×10^{8} J

(2) 2.4×10^{16} J   (4) 3.0×10^{8} J   107 _____

108 One of the radioactive waste products of the reactor has a half-life of 250 years. What fraction of a given sample of this product will remain after 1,000 years?

(1) $\frac{1}{2}$ (3) $\frac{1}{8}$

(2) $\frac{1}{4}$ (4) $\frac{1}{16}$ 108 _____

109 The water in the reactor acts both as a heat transfer agent and a moderator. In its capacity as a moderator, the water

1 accelerates the neutrons to higher speeds so that they can interact with nuclei more energetically
2 slows the neutrons to increase the probability of nuclear interaction
3 prevents a chain reaction from occurring
4 absorbs neutrons and slows the nuclear reaction   109 _____

110 The number of nucleons in a $^{206}_{82}$Pb nucleus is

(1) 0   (3) 124

(2) 82   (4) 206   110 _____

111 An atomic nucleus emits energy as it decays from an excited state to a more stable state without a change in its atomic number. This energy is emitted in the form of

1 an alpha particle   3 a gamma ray
2 an electron   4 a beta particle   111 _____

112 Which process occurs as nitrogen nuclei are bombarded by alpha particles to produce an isotope of oxygen?

    1 photoelectric emission
    2 thermionic emission
    3 fission
    4 transmutation   112 ____

113 The particle $_{+1}^{0}e$ is called a

    1 positron   3 neutron
    2 proton   4 photon   113 ____

114 High temperatures are required for controlled nuclear fusion because nuclei must overcome the forces of

    1 electrostatic attraction
    2 electrostatic repulsion
    3 magnetic attraction
    4 magnetic repulsion   114 ____

115 The nuclear force that holds nucleons together is

    1 weak and short range
    2 weak and long range
    3 strong and short range
    4 strong and long range   115 ____

PART III

You must answer *all* questions in this part. [15]

Base your answers to questions 116 through 118 on the information and diagram below.

A 20.-kilogram block is placed at the top of a 10.-meter-long inclined plane. The block starts from rest and slides without friction down the length of the incline.

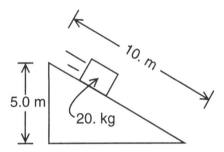

116 Determine the gravitational potential energy of the block at the top of the incline. [Show all calculations, including the equation and substitution with units.] [2]

_____

_____

_____

117 Determine the kinetic energy of the block just as it reaches the bottom of the incline [1]

_____

118 Using the axes provided, sketch a graph of the gravitational potential energy of the block as a function of its kinetic energy for the complete slide. *Label your graph with appropriate values and units.* [2]

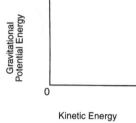

Kinetic Energy

Base your answers to questions 119 through 121 on the information and wave diagrams below.

Two waves, *A* and *B*, pass through the same medium at the same time.

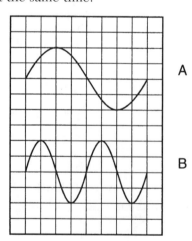

119 On the grid provided, sketch the wave pattern produced when the two waves interfere. [3]

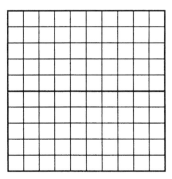

120 Name a wave characteristic that is the same for
both wave *A* and wave *B*. [1]

121 Name a wave characteristic that is different for
wave *A* and wave *B*. [1]

Base your answers to questions 122 through 125 on the information below.

A mercury atom makes a direct transition from energy level *e* to energy level *b*.

122 Determine the energy, in electronvolts, given off in the transition.   [1]

_____

123 What is the energy of the emitted photon in joules?
[1]

_____

124 Determine the frequency of the radiation corresponding to the emitted photon. [Show all calculations, including the equation and substitution with units.]   [2]

_____

_____

_____

125 Explain what would happen if a 4.50-electronvolt photon were incident on a mercury atom in the ground state.   [1]

_____

Answers to the Regents Examinations

JUNE 2000

Part I

1. 3	11. 1	20. 3	29. 1	38. 1	47. 3
2. 1	12. 2	21. 3	30. 4	39. 2	48. 3
3. 2	13. 4	22. 4	31. 1	40. 4	49. 1
4. 4	14. 2	23. 3	32. 4	41. 4	50. 1
5. 3	15. 4	24. 2	33. 2	42. 3	51. 4
6. 3	16. 3	25. 2	34. 2	43. 4	52. 1
7. 2	17. 2	26. 3	35. 1	44. 2	53. 2
8. 4	18. 4	27. 4	36. 3	45. 3	54. 1
9. 3	19. 1	28. 1	37. 3	46. 2	55. 2
10. 4					

Part II

Group 1	Group 2	Group 3	Group 4	Group 5	Group 6
56. 3	66. 4	76. 2	86. 4	96. 3	106. 3
57. 2	67. 2	77. 3	87. 1	97. 4	107. 1
58. 1	68. 1	78. 3	88. 3	98. 3	108. 4
59. 4	69. 4	79. 4	89. 1	99. 4	109. 2
60. 1	70. 3	80. 2	90. 1	100. 1	110. 2
61. 4	71. 3	81. 4	91. 3	101. 1	111. 1
62. 2	72. 1	82. 1	92. 4	102. 2	112. 3
63. 1	73. 2	83. 1	93. 4	103. 2	113. 4
64. 3	74. 4	84. 2	94. 1	104. 3	114. 3
65. 3	75. 2	85. 2	95. 2	105. 2	115. 1

JANUARY 2001

Part I

1. 3	11. 4	20. 1	29. 1	38. 4	47. 4
2. 4	12. 3	21. 3	30. 4	39. 1	48. 2
3. 2	13. 2	22. 1	31. 3	40. 1	49. 2
4. 2	14. 3	23. 1	32. 3	41. 2	50. 1
5. 3	15. 1	24. 4	33. 2	42. 4	51. 1
6. 1	16. 1	25. 4	34. 1	43. 1	52. 1
7. 3	17. 2	26. 2	35. 3	44. 3	53. 3
8. 4	18. 3	27. 3	36. 2	45. 3	54. 2
9. 3	19. 4	28. 4	37. 2	46. 3	55. 3
10. 3					

Part II

Group 1	Group 2	Group 3	Group 4	Group 5	Group 6
56. 4	**66.** 2	**76.** 1	**86.** 3	**96.** 4	**106.** 1
57. 2	**67.** 2	**77.** 3	**87.** 3	**97.** 1	**107.** 1
58. 1	**68.** 3	**78.** 2	**88.** 2	**98.** 3	**108.** 4
59. 3	**69.** 4	**79.** 4	**89.** 4	**99.** 2	**109.** 2
60. 3	**70.** 2	**80.** 3	**90.** 2	**100.** 2	**110.** 4
61. 4	**71.** 1	**81.** 4	**91.** 4	**101.** 3	**111.** 3
62. 1	**72.** 1	**82.** 1	**92.** 4	**102.** 4	**112.** 4
63. 2	**73.** 4	**83.** 1	**93.** 1	**103.** 4	**113.** 1
64. 3	**74.** 2	**84.** 3	**94.** 3	**104.** 1	**114.** 2
65. 3	**75.** 3	**85.** 4	**95.** 3	**105.** 2	**115.** 3

Index

NOTES

NOTES

NOTES

NOTES

NOTES

NOTES

NOTES

NOTES

NOTES